Traditional Telecommunications Networks

Traditional Telecommunications Networks

The International Handbook of
Telecommunications Economics, Volume I

Edited by

Gary Madden

*Professor of Economics, Curtin University of Technology and
Director, Communications Economics and Electronic Markets
Research Centre, Australia*

Edward Elgar
Cheltenham, UK • Northampton, MA, USA

384
T 763

Published by
Edward Elgar Publishing Limited
Glensanda House
Montpellier Parade
Cheltenham
Glos GL50 1UA
UK

Edward Elgar Publishing, Inc.
136 West Street
Suite 202
Northampton
Massachusetts 01060
USA

A catalogue record for this book
is available from the British Library

Library of Congress Cataloguing in Publication Data

Traditional telecommunications networks/edited by Gary Madden.
     p. cm.
    (The international handbook of telecommunications economics, vol. I)
    Includes bibliographical references and index.
    1. Telecommunication. 2. Telecommunication systems.
    I. Madden, Gary. II. Series.
    HE7631.T73   2003
    384—dc21   2002032056

ISBN 1 84064 272 6

Printed and bound in Great Britain by MPG Books Ltd, Bodmin, Cornwall

Contents

v

Contents of the Handbook

VOLUME II   EMERGING TELECOMMUNICATIONS NETWORKS

VOLUME III   WORLD TELECOMMUNICATIONS MARKETS

Part 1   From Telecommunications Reform to Information Social Policy

Part 4   Regional Developments

Figures

Tables

Contributors

James H. Alleman is a Visiting Associate Professor in the Media, Communications and Entertainment Program at Columbia Business School, and Director of Research, Columbia Institute for Tele-Information, New York and Associate Professor for the Interdisciplinary Telecommunications Program at the University of Colorado, USA. Dr Alleman founded Paragon Service International – a telecommunications call-back firm. He has been granted patents on the call-back process widely used by the industry.

T. Randolph Beard was educated at Tulane University and Vanderbilt University, where he received a PhD in Economics in 1988. He is Associate Professor of Economics at Auburn University, where he specializes in microeconomics, regulation and industrial economics. He is author of two books, *Economics, Entropy and the Environment* (Edward Elgar, 2000) and *Initial Public Offerings: Findings and Theories* (Kluwer, 1995). His work has appeared in *RAND Journal of Economics*, *Review of Economics and Statistics*, *Journal of Business*, *Management Science*, *Journal of Industrial Economics*, and many other academic outlets. His current research focuses on telecommunications deregulation and imperfect competition.

Yale M. Braunstein is a Professor at the School of Information Management and Systems of the University of California, Berkeley. Dr Braunstein received a BSc degree from Rensselaer Polytechnic Institute and a PhD in Economics from Stanford University. Prior to his appointment at Berkeley, he was a member of the economics faculties at New York and Brandeis Universities. Dr Braunstein has served as a consultant for several corporations and government agencies, including the US Department of Commerce, the FCC, the Ministry of Communications of the State of Israel, and the national telecommunications regulator in Sweden. He is currently a member of the editorial board of *Information Economics and Policy*.

Laurits R. Christensen is Chairman of Laurits R. Christensen Associates in Madison, Wisconsin. In his 20 years as an economics professor Dr Christensen published in leading economic journals, and among his areas of expertise are productivity measurement, applied microeconomics and econometric modelling. In the telecommunications industry, he is a noted authority on productivity studies and incentive regulation. Dr Christensen testified in the landmark DOJ v. AT&T, and MCI v. AT&T cases, which led to the divestiture of AT&T.

Grant Coble-Neal is a Research Fellow at the Communication Economics and Electronic Markets Research Centre located at Curtin University of Technology. Grant recently received the Best Student Paper award from International Telecommunications Society Asia-Indian Ocean Regional Conference. He has co-authored with Gary Madden several articles analysing aspects of the telecommunications industry. His work has appeared in *Applied Economics*, *The Economic Record*, *Information Economics and Policy*, *International Journal of Forecasting*, *Prometheus* and *Telecommunications Policy*.

Russel J. Cooper is Foundation Professor of Economics at the University of Western Sydney, Nepean. His research interests include inter-temporal optimization models, duality theory and applications, inter-industry modelling, applied consumer demand studies and spatial economics. He has published in *Econometrica*, *International Economic Review*, *Review of Economics and Statistics*, *Macroeconomic Dynamics*, *Canadian Journal of Economics*, *Economics Letters* and *Economic Record*. His current research concerns new growth economics, the economic modelling of optimization decisions of firms and households under uncertainty, applications of duality theory for estimation of cost functions, investment and pricing decisions of high-technology firms and the rational economic modelling of technology transfer across the digital divide.

W. Erwin Diewert is a Professor of Economics at the University of British Columbia. He has published widely in journals and books. His main areas of research include duality theory, flexible functional forms, index number theory (including the concept of a superlative index number formula), the measurement of productivity and the calculation of excess burdens of taxation. He has acted as a consultant on measurement and regulatory issues for the IMF, the World Bank, the Bureau of Labor Statistics, the Bureau of Economic Analysis, the OECD, the New Zealand Treasury, the Business Roundtable in New Zealand, Bell Canada, B.C. Telephone, the American Association of Railways, the Victorian Treasury and Industry Canada.

Michael A. Einhorn is Principal, LECG, LLC, New York and an Adjunct Professor at the Fordham Graduate School of Business, New York. Previous employment includes Bell Laboratories, Rutgers University and the Antitrust Division of the US Department of Justice. Dr Einhorn's general interests include applied microeconomics, intellectual property, media economics and public utility regulation. He has published articles on incentive regulation, regulated pricing, antitrust, copyright, market compatibility and international settlements. The views expressed here are those of the author and do not necessarily represent the positions of other experts at LECG.

George S. Ford is Chief Economist of Strategic Policy and Planning at Z-Tel Communications, and is responsible for performing and evaluating economic analyses pertaining to Z-Tel's strategic plans and public policy positions. Dr Ford began his career as a professional economist in the Competition Division of the FCC, where he became involved with competition and regulation issues across all industries under the FCC's jurisdiction. After serving with the FCC, Dr Ford became a Senior Economist at MCI-Worldcom. Dr Ford received his PhD in Economics from Auburn University.

Douglas A. Galbi is a Senior Economist at the FCC. After earning a PhD in Economics from MIT, he did economic research and policy analysis at King's College Cambridge, at the Macroeconomic and Finance Unit, Moscow, and at the World Bank, Washington. He has served as the Chief Economist for the FCC's International Bureau. His research interests include network interconnection, media development, and the social and political effects of communications technology.

Benjamin E. Hermalin is the Willis H. Booth Professor of Banking and Finance at the Walter A. Haas School of Business, University of California, Berkeley, where he is also currently serving as Associate Dean for Academic Affairs. In addition, he holds an appointment at the Department of Economics, University of California, Berkeley. Professor Hermalin has published on a wide range of topics within economics. His primary research area, however, is the economics of organization, in which he has studied a variety of questions including issues of contract renegotiation, boards of directors, and, most recently, corporate culture and emotional responses.

Michael L. Katz is the Edward J. and Mollie Arnold Professor of Business Administration at the University of California, Berkeley. He has published articles on the economics of network industries, intellectual property licensing, telecommunications policy, and cooperative research and development. He is a co-editor of *The California Management Review* and serves on the editorial board of *The Journal of Economics and Management Strategy*. Dr Katz is a member of the Computer Science and Telecommunications Board of the National Academy of Sciences. Dr Katz was Chief Economist of the FCC from January 1994 through January 1996.

Mark E. Meitzen is a Vice President at Laurits R. Christensen Associates in Madison, Wisconsin. He has expertise in telecommunications costing methods, productivity measurement and incentive regulation. Among his recent assignments Dr Meitzen directed a project that helped a Latin American regulator implement price-cap regulation in that country's telecommunications industry. He has also directed projects that measure the costs of universal service.

M. Ishaq Nadiri is the Jay Gould Professor of Economics at New York University and Faculty Research Associate of the NBER. He has taught at University of California, Berkeley, Northwestern University, University of Chicago and Columbia University. He was the chairman of the Economics Department at New York University and the Director of C.V. Starr Center. His main research interests and fields of specialization include productivity analysis, telecommunications economics, investment theory and modelling, monetary economics, quantitative analysis, applied economics, economics of technical change and investment in R&D. He has served as a consultant to a number of corporations and organizations, including the Ford Foundation, the UNCTAD, several governmental agencies and foreign governments, the United Nations Agencies, World Bank and IMF.

Banani Nandi is a Principal Technical Staff Member with AT&T Laboratories (formerly AT&T Bell Laboratories). She received her PhD degree from New York University and her PhD dissertation was awarded the Otto Ehrlich Memorial Award. Before joining AT&T, she taught at Calcutta University, Columbia University, New York University and Rutgers University. Dr Nandi's fields of specialization include international trade and econometrics. At present, her research focuses on the impact of telecommunications and information technologies on productivity and economic growth.

Paul N. Rappoport is an Associate Professor of Economics in the Department of Economics at Temple University, Philadelphia, USA. He is the founder and former Chairman of PNR and Associates. He is also a Senior Academic Fellow in Temple's E-Commerce Institute. His current research interests include modelling business demand for bandwidth, measuring Internet transactions, developing Internet metrics and e-commerce benchmarks.

Philip E. Schoech is a Vice President at Laurits R. Christensen Associates in Madison, Wisconsin. His research interests are in the areas of productivity measurement, econometric cost analysis and incentive regulation. Dr Schoech has conducted a variety of productivity and econometric cost studies of the telecommunications, postal, electric and gas utility, transportation and manufacturing industries.

Lester D. Taylor is Professor of Economics and Professor of Agricultural and Natural Resource Economics at the University of Arizona, Tucson, USA. He has a PhD in Economics from Harvard, and taught at Harvard and the University of Michigan before moving to Arizona in 1972. For the past 25 years, his published research has focused heavily on the telecommunications industry. His most recent book in telecommunications, *Telecommunications Demand in Theory and Practice*, was published in 1994.

Terence J. Wales is Professor Emeritus of Economics at the University of British Columbia. He has published papers in journals and books. His research interests are primarily in the areas of applied consumer and producer demand analysis. In particular they focus on the myriad of econometric issues that arise in applying the appropriate standard economic theory to the data including the introduction of demographics, use of flexible functional forms and the imposition of curvature conditions required by the theory.

Dennis L. Weisman is a Professor of Economics at Kansas State University and a former Director of Strategic Marketing for SBC Communications. His primary research interests are in the areas of government regulation and strategic firm behaviour. In addition to publishing journal articles and book chapters, he is the co-author of *Designing Incentive Regulation for the Telecommunications Industry* (MIT Press and AEI Press, 1996) and *The Telecommunications Act of 1996: The 'Costs' of Managed Competition* (Kluwer Academic Press, 2000). He also serves on the editorial boards of the *Journal of Regulatory Economics* and *Information Economics and Policy*.

BIS Title: NIA

Preface   ed.

PURPOSE AND MOTIVATION

The *International Handbook of Telecommunications Economics* aims to serve as a source of reference and a technical supplement for the field of telecommunications economics. The intention is to provide both comprehensive and up-to-date surveys of recent developments and the state of various aspects of telecommunications markets. The Handbook is meant for professional use by economists, advanced undergraduate and graduate students, and also to prove useful to policy analysts, engineers and managers within the industry. With the last two decades of the twentieth century witnessing revolutionary technology and institutional change in telecommunications markets, publication of the Handbook is timely. In terms of organization it has proved convenient to categorize material into volumes on the basis of the economics pertaining to traditional 'narrowband technology' markets, markets associated with emerging 'broadband technologies' and the political economy and institutions that structure markets within which telecommunications companies operate and evolve.

The industrial organization of traditional or circuit-switched markets was primarily that of mandated monopolies offering a small array of narrowband services such as local, long-distance and international calling. This structure was supported by natural monopoly arguments. Not surprisingly, research in this era often concerned the analysis and review of production, costs and the productivity of operators. Further, the increasing privatization of incumbent operators required them to produce closer to minimum cost. Accordingly, studies examining the economies of scale and scope were prominent. Market deregulation and liberalization led to the rebalancing of rates so as to reduce cross-subsidization. In this changing environment, the analysis of competition, demand, pricing and aspects of industrial organization were important ongoing research themes. With the growth of global telecommunications, the organization of international markets and their pricing and other arrangements came under increasing scrutiny. The role of telecommunications in economic development was also considered an important discourse. During this regime welfare aspects of telecommunications markets were concerned with identifying and fulfilling universal service obligations.

In the face of developments in digital technology (mobile, satellite and terrestrial), telecommunications, cable TV, broadcasting and computers are fast converging into a single industry. The advent of these high-capacity and intelligent networks has resulted in the development of new services, such as SMS, home banking and Internet services. The proliferation of networks and their need for, more or less, seamless interconnection has led to industrial restructuring through mergers and alliances. Such reorganization is also intended to allow operators to offer a full range of services. Growth in Internet network infrastructure and subscription has provided a base for the development of e-commerce markets. Accordingly much recent research on broadband networks is forward-looking and concerned with innovation and the future of these networks, for example, the forecasting of Internet telephony adoption, the impact of e-commerce on industrial organization and the structure of future retail markets. Fixed, mobile and satellite market segments, their interaction, regulation and pricing are also matters of importance to network planners and other market participants. When faced with competing technology platforms, practically for the first time, investment decisions are more complex. Appropriate investment decisions, inherently difficult, are made more so because of the price volatility of extant technology due to ongoing innovation to enhance its productivity. In this environment new treatments of uncertainty in telecommunications are being developed. This broadband regime brings with it concerns of identifying appropriate standards and delivery for universal service.

Finally, the structure within which modern communications companies operate and evolve is complex, Accordingly, corporate strategy has to evolve beyond that relevant to a national monopoly, which may provide international service, to deal optimally with new technology and its associated service innovations, and changing composition and patterns of demand. Corporations must also account for multiple objectives associated with both national economic and social policy. Further, the subtle transition from communication to information markets places additional demands on corporate trajectories. In this light the recent corporate experience of major international telecommunications companies can provide useful lessons. Exploring the interaction of diversity in national approaches with the continuing need for international cooperation and coordination continues to be an important area of debate.

ORGANIZATION

The organization of the *International Handbook of Telecommunications Economics* has sought to reflect principal topics in the field that have either received intensive research attention or are in need of an integrative survey.

The chapters of the Handbook can be read independently, though many are complementary. Volume I begins the Handbook with three chapters on production, cost and productivity. Russel Cooper, Erwin Diewert and Terry Wales (Chapter 1) focus on the subadditivity of cost functions. To determine whether a multi-product regulated industry is a natural monopoly, it is helpful to estimate a multiple output cost function for the incumbent firm and then determine whether this estimated cost function is subadditive. This chapter considers issues in the development of functional forms capable of addressing the subadditivity question in a theoretically valid context. Yale Braunstein and Grant Coble-Neal (Chapter 2) consider issues related to cost function estimation in the telecommunications industry. Their coverage encompasses concerns associated with specification of functional forms, problems with data and approaches to addressing them, and the treatment of technical change. Estimation and testing of economic theory are also discussed. Laurits Christensen, Philip Schoech and Mark Meitzen (Chapter 3) present a survey of TFP methods and studies for the telecommunications industry. TFP is widely recognized as a comprehensive measure of productive efficiency. Further, methodological issues in measuring TFP are considered. The relationship between TFP and output growth is also examined. Finally, a discussion of TFP and its application to price cap regulation concludes the chapter.

The next four chapters of Volume I analyse the traditional areas of concern in telecommunications markets, that is, competition and market structure, demand, pricing of network services and access, and vertical integration. T. Randolph Beard and George Ford (Chapter 4) focus on competition and structure in local and toll markets. They provide a review of conceptual and empirical analyses of competition relevant to the telecommunications industry. A peculiar result they provide is that competition may increase market cross-subsidy. Competition in parts of a geographic market, or fragmented duopoly, is also formally considered. Finally, merger effects are analysed using an empirical simulation model to evaluate competition in the long-distance industry. Lester Taylor (Chapter 5) notes that demand analysts must contend with shifting industry boundaries, new services, mobile telephony growth, disappearing data sources and emergence of the Internet. He describes in detail several studies that illustrate fundamental features of telecommunications demand, with regard to both theory and technique, and transcend their particular applications. Benjamin Hermalin and Michael Katz (Chapter 6) summarize and integrate the literature on retail pricing of telecommunications network services. Their central concerns are how to price network access and the exchange of information based on use. They focus on four characteristics and their implications for socially and privately optimal pricing to end-users: consumption decisions have external effects, parties exchanging messages often have an ongoing relationship, service providers can identify customers, and incremental costs are often below average costs. Dennis Weisman (Chapter 7) considers that the dominant trends in the telecommunications industry with respect to market structure are consolidation and vertical integration. Such developments contain complex public policy and regulatory issues that both renew old debates and generate new ones. This chapter focuses on the strategic motivation for vertical integration in telecommunications markets and the corresponding regulatory and public policy issues.

The following three complementary chapters are concerned with international telecommunications markets. Douglas Galbi (Chapter 8) outlines key aspects of industry economics, firm organization and public policy relevant to understanding the globalization of telecommunications competition. Michael Einhorn (Chapter 9) reviews contemporary US policy concerning international settlements of voice and data telephone traffic. Accordingly, the FCC's previous settlements policies and the problems they posed to US carriers and callers are examined and FCC's strategy for reform is then detailed. The market under more competitive regimes is also considered. M. Ishaq Nadiri and Banani Nandi (Chapter 10) examine the contribution of telecommunications infrastructure to economic growth. Studies based on an I–O framework, and econometric cost models that describe telecommunications capital effects on industry cost structures and national productivity growth are considered. The social rate of return to telecommunications investment is discussed therein.

James Alleman and Paul Rappoport (Chapter 11) conclude Volume I with a review of the universal service concept. They argue that policy is inefficient as it is not directed to marginal subscribers, and it is costly to support because it is not targeted. Further, they consider that cross-subsidy reduces the demand for subscriber access from the group that it is intended to aid. Finally, subsidy inhibits effective competition by preventing the market from testing the efficiency of providers.

Volume II is devoted to the economics of high capacity and intelligent networks, and commences with three related but diverse chapters that cover information market innovation, Internet network history and evolution, and forecast its future. Cristiano Antonelli (Chapter 1) provides a framework to consider the accumulation of localized knowledge and competence dynamics. Such dynamics are from innovative agent activity. Agents are bounded by their competence and exposed to changing price and non-price competition. Michael Pelcovits and Vinton Cerf (Chapter 2) describe the Internet and outline its history to enable discussion of the economic forces that shape the network, and how the Internet's economics are different from those of traditional telecommunications networks. The chapter concludes with a gaze into the future of the Internet. Paul Rappoport, Donald Kridel, Lester Taylor, James Alleman and Kevin Duffy-Deno (Chapter 3) use data from an omnibus survey to empirically estimate models of consumer Internet access. They consider dial-up access only, dial-up or broadband access and broadband but cable modem or ADSL choice situations. Results confirm that availability matters. When all forms of access are available, ADSL and cable are strong substitutes to dial-up access. ADSL and cable modems are substitutes. Other factors affecting broadband adoption are the type, duration and reach of use.

The next four chapters in Volume II concern e-commerce markets and, more generally, online activity and its regulation. Steven Globerman (Chapter 4) provides a description of e-commerce, its current state and prospects. Identification and assessment of hypotheses linking e-commerce growth to changes in organizational characteristics are made, and evidence evaluated. Michael Ward (Chapter 5) distinguishes online from traditional retail markets, and asserts that lower consumer search costs lead to greater substitutability among homogeneous goods, but stronger brand preference for heterogeneous goods. Reduced delivery costs are argued to lead to rapid adoption of online retailing for travel and financial products, while higher delivery costs suggest more modest online retailing for groceries and sporting goods adoption. Robert Frieden (Chapter 6) addresses problems and opportunities presented by services provided via the Internet. After reviewing the pre-existing regulatory models that apply, the chapter considers whether a single legal and regulatory foundation exists to address the public policy issues raised by Internet-delivered communications and e-commerce. The argument concludes with suggestions on how governments should promote telecommunication and information processing infrastructure use as a conduit for services that promote education, rural development and commerce. Timothy Tardiff (Chapter 7) sees that trends in product innovation are producing complex service offerings and pricing structures. As a result, the prescription that price equals marginal cost in competitive markets is inadequate. To understand this outcome and related pricing patterns, cost structures of firms offering these services are considered and their pricing implications discussed. Examples of pricing trends for retail telecommunications services include the growth of pricing options in place of single undifferentiated prices, and the offering of 'one-stop shopping'. Finally, whether pricing responses are likely to be pro- or anti-competitive is discussed.

The following three chapters concern alternative technology platforms, their convergence and regulation, and the difficulties facing operators in making network investment decisions. Harald Gruber and Tommaso

Valletti (Chapter 8) show mobile telephony development is restricted by the available radio spectrum. Switching from analogue to digital transmission has somewhat relaxed the constraint. Further, extending cellular mobile telephony to higher frequency bands has allowed more subscribers to be accommodated. By relaxing the spectrum constraint more firms are able to supply service, and so improve service quality and price. Empirical studies find that digital technology and firm entry accelerate diffusion, but that national differences in diffusion persist. Joseph Pelton (Chapter 9) argues that satellite services growth is driven by price and technical competition, regulatory and trade reform, and advance in frequency allocation. However, satellite communications remain a complementary service and supplement to ground-based systems. Developing countries are a major growth opportunity for satellite operators by filling niches in their broadband fibre networks. He argues that geosynchronous satellite systems will dominate LEO satellite systems, and that stratospheric platforms will reduce LEO system opportunity further. Jerry Hausman (Chapter 10) regards advice on setting regulated prices is flawed when it does not recognize substantial industry fixed costs. Further, it is often claimed that economies of scale are such that regulated firms' costs are below those of entrants. Another problem arises when a regulated firm wants to decrease prices of services subject to competition. Most economists conclude that cost-based regulation leads to consumer harm and so regulatory attempts to set prices independent of demand does not make sense. Within this approach, why the failure to take account of sunk costs leads to a downward bias in regulated prices is discussed. Assumptions that network investment is fixed, not sunk, lead to error. Finally, which incumbent network elements should be subject to mandatory unbundling is considered.

Jorge Schement and Scott Forbes (Chapter 11) conclude Volume II with a discussion of universal service policy for the twenty-first century. They argue that basing policy solely on the short-term constraints faced by corporations and government misses an opportunity to build an equitable foundation for a global information age. Universal service should be sensitive to diverse population needs. The old concept of universal service simply represents an intention to wire the nation. It is proposed here that welfare is optimized when populations choose their access configuration.

Volume III contains a collection of chapters concerning the structure within which modern communications companies operate and evolve. This emerging structure is reflected in the transition from telecommunications to information social policy induced by service transformation. In this context, selected corporate experiences, international cooperation and coordination operations, and regional studies help describe the complex environment in which corporate strategies must develop and be applied. Eli

Noam (Chapter 1) describes the evolution of the telecommunications sector from incumbent monopoly status. He argues that incumbent carriers are in the midst of vertical diversification and horizontal expansion. Entrants with more efficient technology that has a greater modality impose these processes on incumbents. To maximize the gains from Schumpeterian creative destruction requires regulatory policy that ensures major companies become rivals and not partners.

The next three chapters concern the evolution of telecommunication policy to information social policy. Martin Fransman (Chapter 2) analyses the transformation from old to new telecommunications industry, and on to info-communications industry. He argues that the demise of the old telecommunications industry began with the opening of monopoly markets to competition, and the EU and WTO agreeing to liberalization. Further transformation to an info-communications industry was due to packet switching, IP and WWW technology. William Melody (Chapter 3) provides a review of telecommunications reform and information infrastructure development by focusing on traditional and network economy indicators. Traditional indicators include measures of network access and service, and competition in fixed network and mobile telephony markets. Information infrastructure development is measured by network investment, availability and use of new access technology, and Internet market development. Countries selected are leaders in global markets. Erik Bohlin (Chapter 4) considers that political and convergence dynamics will require telecommunications policy research to address a wider and more comprehensive set of questions. His chapter elaborates the nature of telecommunications policy research, addressing political inertia and its implications for policy, explaining the increased scope of telecommunications policy, and suggesting future research needs, including that of a sustainable information society.

The four chapters that follow describe the experience and visions of communications corporations. Peter Curwen (Chapter 5) argues that initial attempts to form global alliances floundered, even though 'New' Concert may have arisen from the ashes of its predecessor. Alliance formation is a slow process, partly because of the time required to obtain regulatory clearance, and because partners need time to ensure network architecture and standards are compatible. Time is a scarce resource in a rapidly changing industry. Furthermore, alliances consistently underestimate the difficulty of keeping relatively loose arrangements in place. Niall Levine, Douglas Pitt and David Lal (Chapter 6) analyse changes BT faced in the 1990s. They argue that the causes and effects of change are relevant to the disciplines of strategic planning, organizational behaviour, telecommunications management, economics and regulatory policy. They contend that BT faces permanent revolution in redefining its strategy and organizational practices to meet emerging market opportunities and environmental challenges posed by increasing competition and regulation. Martin Taschdjian (Chapter 7) considers that US WEST's evolution into international markets and broadband technology is typical of the telecommunications revolution. Its history of strategic initiatives, and the learning and adjustments made, offer insight into the issues of convergence, diversification, technology, competition, shareholder value and public policy. These lessons suggest AT&T's strategy of evolving from a long-distance carrier into a local broadband services company faced significant challenges. James Alleman and Lawrence Cole (Chapter 8) conclude that GTE purchased the long-distance telephone company SPC because it was already involved in telecommunications and other high-technology business, had an acceptable investment return, and an effective management in handling regulatory process. However, they claim the move was imprudent. Important mistakes include an inadequate understanding of interconnection charges and valuation analysis, and of a government imposed Consent Decree.

The seven chapters that appear next provide a complementary overview and analysis of national and international agencies that affect regulatory policy and international telecommunications aid activity. Richard Cawley (Chapter 9) provides insight into the substance and process of telecommunications regulation and policy in the EU in the latter part of the twentieth century. He considers an important message in the EU context is that regulatory policy can often not be designed from a first-best standpoint. It is constrained by the legal and institutional structures that are enshrined in successive EU treaties, and these in turn reflect evolving political, social and economic prerogatives and compromise between member states. Timothy Brennan (Chapter 10) argues that across the US, cable TV systems have begun to expand offerings beyond multi-channel point-to-multipoint video to include high-speed broadband Internet service. Recently, the US Federal Trade Commission, with the FCC, imposed open access requirements on Time Warner's cable systems as a condition for approving its acquisition by America On-Line. The merits of imposing such broadband access requirements are debatable. To ensure locality access rules are efficiency based, open access requirements should be imposed independently of merger occurrence. Tim Kelly (Chapter 11) asserts that ITRs, among the longest surviving international treaties, are regarded, especially in developing countries, as an expression of national sovereignty. The current ITRs, adopted in an era when old certainties were tumbling, remain essentially unchanged. For example, at the time of drafting there were only half a million Internet users globally. The current situation is one of atrophy in which the ITRs are slowly falling into neglect. Dimitri Ypsilanti (Chapter

12) considers that regulators are trying to interpret principles that are largely empty of detail. An associated difficulty is gaining consensus on the degree to which the details of policy should be the same, as opposed to the general frameworks. A means to ensure the continued convergence of regulation is through sharing experiences, and more open evaluation of regulator performance including benchmarking. Such an approach will ensure best practice regulation is more widely adopted and provide an impetus for convergence in policy and the implementation of regulation. Bruno Lanvin (Chapter 13) addresses the question as to whether it is too late to bridge the digital divide by stressing the main aspects of the present approach. That is, how do the attitudes of stakeholder groups differ from what they were two or three years ago? Why has the issue recently moved from an important problem to a matter of emergency? Finally, how are the nations of the G-8 gearing up to react to this challenge? Charles Kenny (Chapter 14) provides an early history of WBG involvement in the sector and describes change in the 1990s. In particular, the IFC undertook the role of investing in private telecommunications and the WBG moved toward supporting reform in policy and the development of regulatory institutions. After discussing developments in the late 1990s, including early reaction to evolution of the Internet, the increasing importance of universal access and the need for broadened strategy focus, the chapter examines ongoing efforts to reform the WBG strategy for information infrastructure. Claude Barfield and Steven Anderson (Chapter 15) indicate that 50 per cent of WTO members made no basic telecommunications commitment to the GATS. Further, they argue that ongoing negotiation should insist countries meet their 1997 commitments. Negotiations should also harvest more commitments from informal unilateral liberalization. An obvious omission involves basic telecommunications and technology convergence. Future negotiation must gain commitments concerning technology and anticompetitive practice by non-telecommunications entrants. Finally, the WTO basic telecommunications agreement needs to confront international pricing.

The last eight chapters in Volume III are concerned with regional issues. It is hoped that while the issues may be peculiar their lessons will be widely applicable. Andrea Kavanaugh (Chapter 16) considers that any assertion that the Internet favours a new pan-Arabism is ill founded and a more likely outcome is increased pan-elitism. Wealthy, well-educated and young Arab new media users are improving their lifestyles through consumerism, entertainment and realistic news coverage. Nor does the Internet suggest there will be new pan-Arabism, as elite Arabs are drifting farther from their less advantaged brethren. The gap between the haves and have-nots in the Arab Middle East is widening at a great pace. Mark Shadur, Kellie Caught and

René Kienzle (Chapter 17) argue that Australia and New Zealand have major differences in their telecommunications sectors as a result of deregulation. These differences have resulted in distinct market structures, competitor strategy and user outcomes. The chapter outlines change programmes for organizational design, work organization, employee involvement and SHRM. The question as to whether organizational changes have improved operational outcomes remains. Marcelo Resende (Chapter 18) traces the recent evolution of the Brazilian telecommunications sector, in particular the public provision of telecommunications and the system that replaced it. His principal focus is the emerging regulatory framework and challenges facing the sector. The period prior to privatization is examined through aggregate and firm-specific productivity indicators, and a description of the main events defining the institutional design is given. An examination of the post-privatization environment follows. Meheroo Jussawalla (Chapter 19) posits that as Chinese leadership encouraged the deployment of the latest IT and pursued development of global information infrastructure through the mid-1990s, the existing telephone network became obsolete and led to a planned broadband cable network. China now has to devise a future to better diffuse the information superhighway. Further, China needs to combine the goals of information infrastructure advancement with investment in primary education or face social unrest. Nicolas Curien and Dominique Bureau (Chapter 20) describe the change in regulatory oversight of the French telecommunications and electricity sectors. Among network industries, telecommunications and electricity are not uniquely affected by liberalization and market-oriented EU directives. Given the criteria of opening markets to competition and maintaining the public operators' dominant position in certain market segments, telecommunications and electricity are interesting sectors to study. Moreover sector comparisons are relevant as the public sector companies involved, France Télécom and EdF, are efficient and able to withstand international competition. Günter Knieps (Chapter 21) considers the German entry process and its relationship to EU network liberalization policy. New German telecommunications law tends toward over-regulation and a large regulatory base. Further, the potential for phasing out sector-specific regulation is sketched. A critical appraisal of the Review of the German Monopoly Commission is also provided. The analysis shows that after entry deregulation, regulation of market power is only justified in local telecommunications networks when they constitute monopolistic bottlenecks. Elsewhere actual and potential competition is sufficient to discipline the market. T.H. Chowdary (Chapter 22) observes that until recently the Indian telecommunications sector was characterized by millions of applicants waiting for network connection, unsatisfactory service delivery, inflated telephone tariffs – too high to generate funds for network improvement – inability to develop technology locally and poor communications for export-orientated business. Automation and computerization of telephone directory assistance, billing, complaint and fault registration, and control to improve quality of service were resisted. He describes how the move from centralized planning, liberalization, accession to the WTO and signing of the ITA have profoundly affected the provision of telecommunications and information services. Willem Hulsink and Andrew Davies (Chapter 23) argue that national governance regimes shape the internationalization approach of European PTOs. The strategic choice of PTOs, rivals and regulatory agencies determine the scope of market restructuring. The economics of innovation helps to explain innovation strategy. Occasionally industry is altered by strategic behaviour, disruptive technology and exogenous shocks, and so challenges the political and economic system. These situations create dissonance and technology may be replaced, and winning corporate strategy and industry dynamics transformed.

Acknowledgements

My primary obligation is to the authors of the chapters that comprise the *International Handbook of Telecommunications Economics*. I appreciate the authority, talent and imagination with which they prepared their drafts, the willingness to accept comment and efforts to meet deadlines.

Gary Madden
Curtin University of Technology

D 24
L 96

1.   On the subadditivity of cost functions

Russel J. Cooper, W. Erwin Diewert and Terence J. Wales

|U5|

INTRODUCTION

In the last decades of the twentieth century, the concatenation of the technological revolution in info-communications with the increasing financial capital mobility accompanying globalization led to some rethinking of the efficacy of competition policy, especially in the context of key industries which are the subject of major technological innovation, which provide important infrastructure both to other industries and to the community generally, and which have developed structures containing significant elements of monopoly provision and public provision of products and services. Since Baumol et al. (1982), it has become clear that in order to determine whether a multi-product regulated industry is a natural monopoly, it is helpful to estimate a multiple output cost function for the incumbent firm and then determine whether this estimated cost function is subadditive. Subadditivity may arise through economies of scale, economies of scope, or some combination of these. In the telecommunications area, the most important and influential studies that consider the subadditivity question in a multiple output setting are by Evans and Heckman (1983, 1984), who used data on the US Bell system. Evans and Heckman also pioneered an appropriate method for testing subadditivity in a multiple output setting. Diewert and Wales (1991) showed that the Evans and Heckman tests for subadditivity were flawed. Their estimated translog cost functions were improper in that they contained features incompatible with microeconomic theory, invalidating their use for examining the issue at hand. In particular, they exhibited negative marginal costs for a substantial fraction of their sample observations.

The purpose of this paper is to consider issues in the development of functional forms capable of addressing the subadditivity question in a theoretically valid context. We demonstrate our approach by re-examining the subadditivity of the Bell system cost function using a functional form

that is proper but nevertheless retains the advantages of a flexible functional form. As is now well known, it is not easy to obtain estimated cost
functions that satisfy the requirements imposed by microeconomic theory.
Thus a substantial part of the chapter is devoted to choosing an appropriate functional form for the firm's cost function. The next section
specifies toll (long-distance) and local (within some zone) outputs as produced by a separate production function. Then divisional costs are
brought together and a functional representation that allows for
economies of scale is presented. This functional form is then generalized
to permit economies of scope. Then a specification that is capable of examining economies of scale and scope jointly is developed. Further, the model
is generalized to allow for exogenous technical progress. Proofs of propositions developed to this point are contained in Appendix A. A version of
the Evans and Heckman subadditivity computations are presented and
empirical results are discussed. Supporting this discussion, the underlying
data are set out in Appendix B and details of the empirical results are contained in Appendix C. Appendices D and E contain results of subadditivity tests.

SPECIFICATION OF DIVISIONAL COST FUNCTIONS

Let $y \equiv (y_Z, y_T)$ denote a vector of nonnegative outputs that are produced
during a year and let $w \equiv (w_L, w_M, w_K)$ denote the vector of positive input
prices that the firm faces during that year. y_Z is the quantity of local calls
(zoned local output) produced, y_T the quantity of long-distance calls (toll
output) produced, w_L the price of labour, w_M the price of materials and w_K
the rental price of capital services. Let $C(y, w)$ denote the firm's cost function, the minimum cost of producing the output vector y given that the firm
faces the input price vector w. The goal of this section (and the next two
sections) is to specify an appropriate functional form for C. It is assumed
that each of the outputs can be produced by means of a separate divisional
production function and that there are separate divisional cost functions,
$C^Z(y_Z, w)$ for zoned local and $C^T(y_T, w)$ for toll output. Total cost is the
sum of the divisional costs, that is,

$$C(y_Z, y_T, w) = C^Z(y_Z, w) + C^T(y_T, w) \tag{1.1}$$

Consider the alternative specifications for the divisional cost functions:

$$C^n(y_n, w) \equiv w \cdot \beta^n + [w \cdot \phi^n + \tfrac{1}{2}(w \cdot \eta)^{-1} w \cdot \Gamma_n w] y_n, \quad n = Z, T \tag{1.2}$$

and

$$\tilde{C}^n(y_n, w, y_n{}^*) \equiv w\cdot\beta^n + [w\cdot\phi^n + \tfrac{1}{2}(w\cdot\eta)^{-1}w\cdot\Gamma_n w]H^n(y_n, y_n{}^*), \quad n = Z, T \ (1.3)$$

where $w\cdot\beta^n \equiv \Sigma_{i=L,M,K}w_i\beta_i^n$ is the inner product of the vectors $w \equiv (w_L, w_M, w_K)$ and $\beta^n \equiv (\beta_L^n, \beta_M^n, \beta_K^n)$ and $w\cdot\Gamma_n w \equiv \Sigma_{i=L,M,K}\Sigma_{j=L,M,K}w_i\gamma_{ij}^n w_j$.[1] The vectors β^n and $\phi^n \equiv (\phi_L^n, \phi_M^n, \phi_K^n)$, and symmetric matrices $\Gamma_n \equiv [\gamma_{ij}^n]$ which appear in (1.2) and (1.3) are to be statistically determined, as are certain parameters of the as yet unspecified functions $H^n(y_n, y_n{}^*)$ in (1.3). The vector of constants $\eta \equiv (\eta_L, \eta_M, \eta_K)$ is pre-specified. Specification (1.3) essentially represents a non-linear in output generalization of (1.2) to be developed to examine economies of scale.

The divisional cost functions (1.2) and (1.3) are variations of those introduced by Diewert and Wales (1987: 53).[2] To permit identification of the unknown parameters in the cost functions, some restrictions must be placed on the elements of the matrices Γ_n. Restrictions are also imposed on the form of the (yet to be fully specified) functions $H^n(y_n, y_n{}^*)$. Let $y^* \equiv (y_Z{}^*, y_T{}^*)$ be a positive pre-specified output vector and let $w^* \equiv (w_L{}^*, w_M{}^*, w_K{}^*)$ be a positive vector of pre-specified input prices. The reference point $(y_n{}^*, w^*)$ will serve as a point of approximation of the specified functions (1.2) and (1.3) to an unknown cost function satisfying the regularity conditions implied by economic theory. In addition, through the specification of the function H^n, the cost function (1.3) will be non-linear in output, in which case parameter estimates will be sensitive to the chosen reference point. An important aspect of this study concerns cost efficiency at substantially different levels of output. When comparing the costs of operating under these circumstances, it will be convenient to move the reference point for the flexible functional form approximation. For that reason, it is helpful to explicitly recognize the role of the reference point in (1.3); hence we write $\tilde{C}^n(y_n, w, y_n{}^*)$ in that case.

To be more specific, assume that $y_n{}^*$ is a relevant point for approximation of the functional form to an unknown regular function for values of y_n in some neighbourhood of $y_n{}^*$. This raises the issue, however, as to whether another point, say $y_n{}^{**}$, may not be more relevant as a point of approximation to the underlying unknown cost function for other values of y_n. Because of the dependence of $\tilde{C}^n$ on $y_n{}^*$, a dependence that is induced by the non-linearity of the functional form in output, the reference point $y_n{}^*$ therefore has an additional interpretation – as a kind of capacity indicator appropriate to the non-linearity in the technology. For example, the relevant flexible functional form to employ to approximate the true underlying function, when used in a comparison of the minimal cost of producing y_n with the minimal cost of producing $\frac{1}{2}y_n$, may be $\tilde{C}^n(y_n, w, y_n^*)$ compared to $\tilde{C}^n(\frac{1}{2}y_n, w, y_n^{**})$ where y_n^{**} may be more reasonably chosen close to $\frac{1}{2}y_n^*$ rather than chosen equal to y_n^*. Thus, a comparison of the optimal cost structure of, say, one large firm versus two smaller firms, when based on parameter estimates derived from a specification such as (1.3), may require a comparison for appropriately adjusted reference output levels.

The restrictions imposed on η, $H^Z(y_Z, y_Z^*)$, $H^T(y_T, y_T^*)$, Γ_Z and Γ_T are:[3]

$$w^* \cdot \eta = 1, \quad \eta >> 0_3 \tag{1.4}$$

$$\left.\begin{array}{l} H^n(0, y_n^*) = 0 \\[4pt] H^n(y_n^*, y_n^*) = y_n^* \\[4pt] \left.\dfrac{\partial H^n(y_n, y_n^*)}{\partial y_n}\right|_{y_n = y_n^*} = 1 \end{array}\right\} n = Z, T \tag{1.5}$$

$$\Gamma_n w^* = 0_3, \quad n = Z, T \tag{1.6}$$

The cost specification (1.2) splits naturally into fixed and variable cost components:

$$C^n(y_n, w) = C^{F,n}(w) + C^{V,n}(y_n, w) \tag{1.7}$$

where

$$C^{F,n}(w) = w \cdot \beta^n \tag{1.8}$$

$$C^{V,n}(y_n, w, y_n^*) = [w \cdot \phi^n + \tfrac{1}{2}(w \cdot \eta)^{-1} w \cdot \Gamma_n w] y_n \tag{1.9}$$

and using the restriction (1.5), the cost function (1.3) is split in similar fashion:

$$\tilde{C}^n(y_n, w, y_n^*) = C^{F,n}(w) + \tilde{C}^{V,n}(y_n, w, y_n^*) \tag{1.10}$$

where

$$\tilde{C}^{V,n}(y_n, w, y_n^*) = [w \cdot \phi^n + \tfrac{1}{2}(w \cdot \eta)^{-1} w \cdot \Gamma_n w] H^n(y_n, y_n^*) \tag{1.11}$$

Note that if $\beta^n = 0_3$ then there are no fixed costs and $C^n(y_n, w) = C^{V,n}(y_n, w)$ has the property of a homogeneous linear function:

$$C^n(y_n, w) = [\partial C^n(y_n, w)/\partial y_n] y_n \tag{1.12}$$

so that, in this case, dual to (1.2) is a divisional production function that exhibits constant returns to scale. Conversely, for (1.3) if $\beta'' = 0_3$ then $\tilde{C}^n(y_n, w, y_n^*) = \tilde{C}^{V,n}(y_n, w, y_n^*)$ has the property:

$$
\begin{aligned}
\tilde{C}^n(y_n, w, y_n^*) &= [w \cdot \phi^n + \tfrac{1}{2}(w \cdot \eta)^{-1} w \cdot \Gamma_n w] H^n(y_n, y_n^*) \\
&= [\tilde{C}^n(y_n^*, w, y_n^*)/y_n^*] H^n(y_n, y_n^*) \qquad (1.13) \\
&= [\partial \tilde{C}^n(y_n, w, y_n^*)/\partial y_n|_{y_n = y_n^*}] H^n(y_n, y_n^*)
\end{aligned}
$$

so that

$$
\tilde{C}^n(y_n^*, w, y_n^*) = \{[\partial \tilde{C}^n(y_n, w, y_n^*)/\partial y_n] y_n\}|_{y_n = y_n^*} \qquad (1.14)
$$

where the second property of H^n given in (1.5) is used, that is, $H^n(y_n^*, y_n^*) = y_n^*$ and the third property, that is, $\partial H^n(y_n, y_n^*)/\partial y_n|_{y_n = y_n^*} = 1$. Result (1.13) says that total cost in division n equals marginal cost of reference output y_n^* times $H^n(y_n, y_n^*)$. Result (1.14) says that the function $\tilde{C}^n$ acts 'locally linearly' at the reference point y_n^* in that, at that point total cost in division n equals marginal cost of reference output y_n^* times the reference output. That is, if $\beta'' = 0_3$ then (1.3) has the property that $\tilde{C}^n(y_n, w, y_n^*)$ approximates a homogeneous linear variable cost function in y_n at y_n^*.

In addition to this linear approximation property at y_n^*, it is also reasonable to assume that $H^n(y_n, y_n^*)$ is linearly homogeneous in (y_n, y_n^*); that is,

$$
H^n(\lambda y_n, \lambda y_n^*) = \lambda H^n(y_n, y_n^*) \qquad (1.15)
$$

so that, from (1.11), $\tilde{C}^{V,n}(y_n, w, y_n^*)$ is also linearly homogeneous in (y_n, y_n^*). The argument here is that if the technology is built to handle a reference level of output y_n^*, then costs specific to this reference output level are built into the fixed cost component.[4] Thus, although $H^n(y_n, y_n^*)$ and $\tilde{C}^{V,n}(y_n, w, y_n^*)$ are not necessarily linear in y_n, they are both linearly homogeneous in the pair (y_n, y_n^*) and, moreover, they both approximate linear functions in y_n at the point y_n^*. Under these conditions, if $\beta_n = 0_3$ then the divisional production function dual to (1.3) exhibits locally constant returns to scale at the reference point y_n^*.

Returning to the general case $\beta_n \neq 0_3$ and concentrating on (1.3) in which $H^n(y_n, y_n^*)$ is potentially non-linear in y_n, it may be noted that (1.3) contains (1.2) as a special case, that is, with $H^n(y_n, y_n^*) = y_n$ and $\partial^2 H^n(y_n, y_n^*)/\partial y_n^2 = 0$. In the general case, it is convenient to define a parameter:

$$
\left. \frac{\partial^2 H^n(y_n, y_n^*)}{\partial y_n^2} \right|_{y_n = y_n^*} = \theta_n, \quad n = Z, T \qquad (1.16)
$$

with $\theta_n = 0$ generating the linear case. It can be seen that each divisional cost function has three unknown β_i^n, three unknown ϕ_i^n, one unknown θ_n (at least, in the non-linear case), three unknown γ_{ij}^n (where the restrictions (1.6) and the symmetry of Γ_n are employed to calculate the number of unknown parameters in Γ_n), together with whatever unknown parameters may be contained in the H^n function in addition to the value of the second-derivative at the point of approximation. As long as H^n conforms to (1.5), there are enough parameters for $\tilde{C}^n(y_n, w, y_n^*)$ to be a flexible functional form, regardless of the precise specification of $H^n(y_n, y_n^*)$, as the following Proposition demonstrates.

Proposition 1: $\tilde{C}^n(y_n, w, y_n^*)$ defined by (1.3)–(1.6) can approximate an arbitrary twice-continuously differentiable divisional cost function $C^{n*}(y_n, w)$ at the reference point (y_n^*, w^*).[5]

It is desirable to use flexible functional forms in empirical work whenever possible since it reduces the risk of imposition of unwarranted restrictions on the cost functions. In the following section the divisional cost functions are combined into an overall function. The proposed specification for the H^n functions allows economies of scale to be examined. Particular attention is paid to the properness of the cost function.

OVERALL COST FUNCTION AND ECONOMIES OF SCALE

For (1.3) the marginal cost function $M\tilde{C}^n(y_n, w, y_n^*)$ for division n is obtained by differentiating $\tilde{C}^n(y_n, w, y_n^*)$ with respect to y_n:

$$M\tilde{C}^n(y_n, w, y_n^*) = \partial \tilde{C}^n(y_n, w, y_n^*)/\partial y_n$$
$$= [w \cdot \phi^n + \tfrac{1}{2}(w \cdot \eta)^{-1} w \cdot \Gamma_n w] \partial H^n(y_n, y_n^*)/\partial y_n. \quad (1.17)$$

The returns to scale measure for division n is:

$$R\tilde{S}^n(y_n, w, y_n^*) = [\partial \ln \tilde{C}^n(y_n, w, y_n^*)/\partial \ln y_n]^{-1} = \frac{\tilde{C}^n(y_n, w, y_n^*)}{M\tilde{C}^n(y_n, w, y_n^*)y_n} \quad (1.18)$$

where $M\tilde{C}^n(y_n, w, y_n^*)$ is defined by (1.17). If there are increasing (constant) returns to scale in division n at (y_n, w), then $R\tilde{S}^n(y_n, w, y_n^*)$ is greater than 1 (equal to 1). Note that at the point of approximation where (1.4)–(1.6) hold, $M\tilde{C}^n(y_n^*, w^*, y_n^*) = w^* \cdot \phi^n$ and $R\tilde{S}^n(y_n^*, w^*, y_n^*) = [w^* \cdot \beta^n + w^* \cdot \phi^n y_n^*]/[w^* \cdot \phi^n y_n^*]$. Hence, at (y_n^*, w^*), positivity of marginal cost in division n requires that $w^* \cdot \phi^n > 0$ and if, in addition, $w^* \cdot \beta^n > 0$, then returns to scale will be greater than 1. If each component of β^n is nonnegative, then β^n can be interpreted as a vector of fixed input requirements that give rise to the fixed costs $w \cdot \beta^n$.

Away from (y_n^*, w^*), (1.17) indicates that, for (1.3), positivity of marginal costs depends both on the input price factor $w \cdot \phi^n + \frac{1}{2}(w \cdot \eta)^{-1} w \cdot \Gamma_n w$ and the first-derivative of the function of output factor, that is, $\partial H^n(y_n, y_n^*)/\partial y_n$. For (1.2) and (1.3), positivity of the input price factor $w \cdot \phi^n + \frac{1}{2}(w \cdot \eta)^{-1} w \cdot \Gamma_n w$ cannot be guaranteed globally. The problem is that concavity of the cost function implies that Γ_n should be negative semi-definite, so that the second term of the input price factor is necessarily nonpositive if regularity is imposed.[6] Nevertheless, it would normally be possible to achieve positivity in the sample data range on unrestricted estimation of the parameters, with positive values for ϕ^n usually being of sufficient size that $w \cdot \phi^n > -\frac{1}{2}(w \cdot \eta)^{-1} w \cdot \Gamma_n w$ for a reasonable range of values of w. However, positive $\partial H^n(y_n, y_n^*)/\partial y_n$, even within the sample data range, is more problematic and is addressed later in this section through careful specification of the H^n function.

To ensure that the divisional cost functions are concave in input prices, following Diewert and Wales (1987: 52–53) this property is imposed without destroying the flexibility of the divisional cost functions by replacing the matrices Γ_n with:

$$\Gamma_n = -\Sigma_n \Sigma_n^{\#}, \quad n = Z, T \tag{1.19}$$

where $\Sigma_n \equiv [\sigma_{ij}^n]$ is a third-order lower triangular matrix (so that $\sigma_{ij}^n = 0$ if $i < j$) which, by virtue of (1.6), satisfies the following restrictions:

$$\Sigma_n^{\#} w^* = 0_3, \quad n = Z, T \tag{1.20}$$

where $\Sigma_n^{\#}$ denotes the transpose of Σ_n.

Using Shephard's (1953: 13) Lemma, the division n cost function is differentiated with respect to the components of the input vector w to generate the division's cost minimizing input demand vector $x^n \equiv (x_L^n, x_M^n, x_K^n)$. For the linear cost function (1.2), input demands are:

$$x^n = \beta^n + [\phi^n + (w \cdot \eta)^{-1} \Gamma_n w - \frac{1}{2}(w \cdot \eta)^{-2} w \cdot \Gamma_n w \eta] y_n, \quad n = Z, T \tag{1.21}$$

When the more general specification (1.3) is used, which is capable of investigating economies of scale for reasons other than the existence of fixed costs, the input demands are:[7]

$$x^n = \beta^n + [\phi^n + (w \cdot \eta)^{-1} \Gamma_n w - \frac{1}{2}(w \cdot \eta)^{-2} w \cdot \Gamma_n w \eta] H^n(y_n, y_n^*), \quad n = Z, T \tag{1.22}$$

As only the total divisional input demand vectors can be observed, adding the divisional demand vectors (1.21) or (1.22) gives the system of three aggregate input demand functions – for specification (1.2):

$$x = \beta + \phi^Z y_Z + \phi^T y_T + (w \cdot \eta)^{-1} \Gamma_Z w y_Z + (w \cdot \eta)^{-1} \Gamma_T w y_T$$
$$- \tfrac{1}{2}(w \cdot \eta)^{-2} w \cdot \Gamma_Z w \eta y_Z - \tfrac{1}{2}(w \cdot \eta)^{-2} w \cdot \Gamma_T w \eta y_T \tag{1.23}$$

while for specification (1.3):

$$x = \beta + \phi^Z H^Z(y_Z, y_Z^*) + \phi^T H^T(y_T, y_T^*)$$
$$+ (w \cdot \eta)^{-1} \Gamma_Z w H^Z(y_Z, y_Z^*) + (w \cdot \eta)^{-1} \Gamma_T w H^T(y_T, y_T^*) \tag{1.24}$$
$$- \tfrac{1}{2}(w \cdot \eta)^{-2} w \cdot \Gamma_Z w \eta H^Z(y_Z, y_Z^*) - \tfrac{1}{2}(w \cdot \eta)^{-2} w \cdot \Gamma_T w \eta H^T(y_T, y_T^*)$$

where we have defined $\beta = \beta^Z + \beta^T$, since we will not be able to identify the separate vectors β^Z and β^T by statistical techniques. Since aggregate input demand functions such as (1.23) or (1.24) will be used in place of the divisional demands (1.21) or (1.22), it is useful to record the overall cost functions. From (1.2) and (1.3) respectively, the definitions $y \equiv (y_Z, y_T)$ and $y^* \equiv (y_Z^*, y_T^*)$, together with the obvious definition of total cost over the divisions, these are:

$$C(y, w) = w \cdot \beta + [w \cdot \phi^Z + \tfrac{1}{2}(w \cdot \eta)^{-1} w \cdot \Gamma_Z w] y_Z$$
$$+ [w \cdot \phi^T + \tfrac{1}{2}(w \cdot \eta)^{-1} w \cdot \Gamma_T w] y_T \tag{1.25}$$

as the overall cost function corresponding to the divisional specification (1.2) and

$$\tilde{C}(y, w, y^*) = w \cdot \beta + [w \cdot \phi^Z + \tfrac{1}{2}(w \cdot \eta)^{-1} w \cdot \Gamma_Z w] H^Z(y_Z, y_Z^*)$$
$$+ [w \cdot \phi^T + \tfrac{1}{2}(w \cdot \eta)^{-1} w \cdot \Gamma_T w] H^T(y_T, y_T^*) \tag{1.26}$$

as the overall cost function corresponding to the divisional specification (1.3). Since (1.25) is a special case of (1.26) in which $H^n(y_n, y_n^*) = y_n$, $n = Z$, T, (1.26) is discussed in more detail, but (1.25) is retained as a useful special case on which to build initial variants of other generalizations in the next section.

Setting aside for the moment the details of the $H^n(y_n, y_n^*)$, $n = Z$, T functions required in (1.26) except for the proviso that they each contain (at least) a free parameter, θ_n (as required at a minimum for Proposition 1), specification (1.26) for the cost function contains the following parameters: the three parameters of $\beta \equiv (\beta_L, \beta_M, \beta_K)$, three parameters of $\phi^Z \equiv (\phi_L^Z, \phi_M^Z, \phi_K^Z)$, three parameters of $\phi^T \equiv (\phi_L^T, \phi_M^T, \phi_K^T)$, three free parameters within the matrix Γ_Z, which by virtue of (1.19) and (1.20) may be denoted σ_{LL}^Z, $\sigma_{ML}^Z, \sigma_{MM}^Z$, three free parameters within the matrix Γ_T, denoted σ_{LL}^T, $\sigma_{ML}^T, \sigma_{MM}^T$, and two parameters controlling the respective second derivatives of the $H^n(y_n, y_n^*)$, $n = Z, T$ functions evaluated at y_n^*, namely θ_Z and θ_T, or 17 parameters in all. To complete the specification of the divisional cost functions underlying (1.26) the output sub-functions $H^n(y_n, y_n^*)$, $n = Z, T$ are considered. The important issue is to provide a specification that allows returns to scale to be investigated empirically. This requires a potentially non-linear specification of cost in terms of output without creating excessive problems of negative marginal costs in estimation. For a given parameter, ε_n say, consider the specification:

$$H^n(y_n, y_n^*) = \left\{ \frac{\varepsilon_n y_n}{F_1^n(y,y^*)} + \frac{\varepsilon_n(y_n^* - y_n) + y_n^*}{F_2^n(y,y^*)} \right\} y_n \qquad (1.27)$$

where:

$$F_1^n(y, y^*) = \varepsilon_n y_n + y_n^* \qquad (1.28)$$

and

$$F_2^n(y, y^*) = \varepsilon_n[\varepsilon_n(y_n^* - y_n) + y_n^*] + y_n^* \qquad (1.29)$$

Note $\varepsilon_n = 0$ yields the linear case $H^n(y_n, y_n^*) = y_n$. It is easy to see that for (1.27), the first two properties in (1.5) and also (1.15) hold by construction, that is, $H^n(0, y_n^*) = 0$ and, since:

$$F_1^n(y_n^*, y_n^*) = F_2^n(y_n^*, y_n^*) = (\varepsilon_n + 1)y_n^*, \qquad (1.30)$$

$H^n(y_n^*, y_n^*) = y_n^*$. Moreover, $H^n(y_n, y_n^*)$ is linearly homogeneous in (y_n, y_n^*). To check the third property in (1.5), observe that specification (1.27) implies:

$$\frac{\partial H^n(y_n, y_n^*)}{\partial y_n} = \frac{\varepsilon_n y_n}{F_1^n(y_n, y_n^*)} + \frac{\varepsilon_n y_n y_n^*}{F_1^n(y_n, y_n^*)^2}$$

$$+ \left[\frac{\varepsilon_n(y_n^* - y_n) + y_n^*}{F_2^n(y_n, y_n^*)} - \frac{\varepsilon_n y_n y_n^*}{F_2^n(y_n, y_n^*)^2} \right] \qquad (1.31)$$

Given (1.30), it is straightforward to check from (1.31) that $\partial H^n(y_n, y_n^*)/\partial y_n|_{y_n = y_n^*} = 1$. It is also worth noting that:

$$\frac{\partial^2 H^n(y_n,y_n{}^*)}{\partial y_n{}^2} = \frac{\varepsilon_n}{F_1^n(y_n,y_n{}^*)} + \frac{\varepsilon_n y_n{}^* - \varepsilon_n^2 y_n}{F_1^n(y_n,y_n{}^*)^2} - \frac{2\varepsilon_n^2 y_n y_n{}^*}{F_1^n(y_n,y_n{}^*)^3}$$

$$- \left\{ \frac{\varepsilon_n}{F_2^n(y_n,y_n{}^*)} + \frac{\varepsilon_n y_n{}^* - \varepsilon_n^2[\varepsilon_n(y_n{}^* - y_n) + y_n{}^*]}{F_2^n(y_n,y_n{}^*)^2} + \frac{2\varepsilon_n^3 y_n y_n{}^*}{F_2^n(y_n,y_n{}^*)^3} \right\} \quad (1.32)$$

and recalling from (1.16) that $\theta_n \equiv \partial^2 H^n(y_n, y_n{}^*)/\partial y_n^2|_{y_n=y_n{}^*}$, then from (1.32) evaluated at $y_n = y_n{}^*$:

$$\theta_n = \frac{-2\varepsilon_n^2}{(1+\varepsilon_n)^2 y_n{}^*} \quad (1.33)$$

a parameter which is unrestricted, as required for Proposition 1, being dependent on the estimated value of the deeper parameter ε_n. Since (1.31) is positive (in fact, equal to 1) at $y_n = y_n{}^*$, it is likely that (1.31) will contribute to positive marginal costs in (1.17), at least for y_n sufficiently close to $y_n{}^*$. In fact, in free estimation of ε_n on the data set, choosing the approximate sample mid-point as the relevant point for $y_n{}^*$, it is found that this specification produces positive marginal costs over the entire sample space.

ECONOMIES OF JOINT PRODUCTION, SCOPE OR SHARED OVERHEADS

Thus far it has been assumed that outputs are produced by separate divisional production functions, that is, production is non-joint.[8] To obtain a model that is satisfactory for the telecommunications industry the possibility of joint production must be allowed. Baumol et al. (1988: 71) describe economies of scope as follows:

> In addition to economies deriving from the size or scale of a firm's operations (concepts at least intuitively familiar to most economists), there is also the possibility that cost savings may result from simultaneous production of several different outputs in a single enterprise, as contrasted with their production in isolation, each by its own specialized firm. That is, there may exist economies resulting from the *scope* of the firm's operations.

In this section, economies of joint production or economies of scope are introduced into the cost function model. The cost function (1.26) is unable to capture the cost reductions that might occur due to the divisions being able to share the services of some inputs. Cost reductions arising from shared inputs in the production of joint outputs are most naturally modelled by inclusion of cross-output terms in the variable cost component of the overall cost function. To model these economies of input sharing, it is useful to temporarily restrict the cost function specification that was postulated in the previous section. Initially, attention is paid to an augmentation of the variable cost component, $C^V(y, w)$ of (1.25), a specification that exhibits constant returns to scale. This allows an additional cross-output term to be added to the cost function, capturing possible economies of joint production initially in the context of a constant returns to scale specification. Assume overall variable cost $C^V(y_Z, y_T, w_L, w_M, w_K) \equiv C^V(y, w)$ is defined by:

$$C^V(y, w) = [w \cdot \phi^Z + \tfrac{1}{2}(w \cdot \eta)^{-1} w \cdot \Gamma_Z w] y_Z$$
$$+ [w \cdot \phi^T + \tfrac{1}{2}(w \cdot \eta)^{-1} w \cdot \Gamma_T w] y_T + \bar{Q}(y, w) \qquad (1.34)$$

where the cross-output term $\bar{Q}(y, w)$ is initially modelled as:

$$\bar{Q}(y, w) = \tfrac{1}{2}(y \cdot \zeta)^{-1} y \cdot [w_L A_L + w_M A_M + w_K A_K] y \qquad (1.35)$$

and where the additional parameters consist of (i) the elements of the second-order symmetric matrices $A_m \equiv [\alpha_{ij}^m]$ which satisfy the restrictions:[9]

$$A_m y^* = 0_2, \quad m = L, M, K \qquad (1.36)$$

and (ii) two normalizing parameters, given by the vector $\zeta = (\zeta_Z, \zeta_T) > 0_2$. From (1.34), (1.6) and the properties of $\bar{Q}$ as defined by (1.35), the following properties of $C^V(y, w)$ may be noted:[10]

$$C^V(0, w) = 0$$
$$C^V(y, w) \text{ is } HD1 \text{ in } y \equiv (y_Z, y_T)$$
$$C^V(y, w) \text{ is } HD1 \text{ in } w \equiv (w_L, w_M, w_K) \qquad (1.37)$$
$$C^V(y^*, w) = w \cdot \phi^Z y_Z^* + w \cdot \phi^T y_T^*$$

Given (1.36), the symmetric matrices A_m have rank 1. They can be written without loss of generality as:

$$A_m = \alpha_m \begin{bmatrix} y_T^*/y_Z^* & -1 \\ -1 & y_Z^*/y_T^* \end{bmatrix} \qquad (1.38)$$

so that each symmetric matrix A_m has one free parameter α_m. Defining $\alpha \equiv (\alpha_L, \alpha_M, \alpha_M)$, the cross-output term $\bar{Q}$ may be written as:

$$\bar{Q}(y, w) = \tfrac{1}{2} \frac{w \cdot \alpha}{y \cdot \zeta}(y_Z \ y_T) \begin{bmatrix} y_T^*/y_Z^* & -1 \\ -1 & y_Z^*/y_T^* \end{bmatrix} \begin{pmatrix} y_Z \\ y_T \end{pmatrix} \qquad (1.39)$$

Before discussing particular features of the specification consider the following preliminary result.

Proposition 2: $C^V(y, w)$ defined by (1.34), (1.6) and (1.39) can approximate to the second order an arbitrary twice continuously differentiable two-output, three-input cost function $C^*(y, w)$, which is linearly homogenous in y, at (y^*, w^*). This approximation can be accomplished when restrictions $\Gamma_Z = \Gamma_T$ and $A_L = A_M = A_K$ are imposed.

Thus C^V defined by (1.34) has some good approximation properties. However, there are also difficulties with the specification that manifest themselves in estimation. The linear homogeneity of $\bar{Q}$ in y, when $\bar{Q}$ is added to other linear terms in (1.34), creates the potential for an identification problem that increases the probability of obtaining insignificant and incorrectly signed parameter estimates. Additionally, the complexity of the specification of $\bar{Q}$ in y has the potential to produce negative marginal cost estimates. Correct signs on cross-output terms in $\bar{Q}$ are essential in testing for economies of joint production. Specification (1.34) is investigated in more detail in order to develop a modified specification that retains its attractive features but which is less prone to these problems.

Given (1.34) and (1.39), the vector of marginal costs is:

$$\begin{pmatrix} \dfrac{\partial C^V(y,w)}{\partial y_Z} \\[2ex] \dfrac{\partial C^V(y,w)}{\partial y_T} \end{pmatrix} = \begin{pmatrix} w \cdot \phi^Z + \tfrac{1}{2}(w \cdot \eta)^{-1} w \cdot \Gamma_{ZW} \\[2ex] w \cdot \phi^T + \tfrac{1}{2}(w \cdot \eta)^{-1} w \cdot \Gamma_{TW} \end{pmatrix}$$

$$- (y \cdot \zeta)^{-1} \bar{Q} \begin{pmatrix} \zeta_Z \\ \zeta_T \end{pmatrix} + \frac{w \cdot \alpha}{y \cdot \zeta} \begin{bmatrix} \dfrac{y_T^*}{y_Z^*} - 1 \\ -1 \quad \dfrac{y_Z^*}{y_T^*} \end{bmatrix} \begin{pmatrix} y_Z \\ y_T \end{pmatrix} \tag{1.40}$$

and positive marginal cost requires that, for zoned local output:

$$w \cdot \phi^Z > -\tfrac{1}{2}(w \cdot \eta)^{-1} w \cdot \Gamma_{ZW} + (y \cdot \zeta)^{-1} \bar{Q} \zeta_Z - \frac{w \cdot \alpha}{y \cdot \zeta} \left[\frac{y_T^*}{y_Z^*} y_Z - y_T \right] \tag{1.41}$$

while for toll output the condition for positive marginal cost is:

$$w \cdot \phi^T > -\tfrac{1}{2}(w \cdot \eta)^{-1} w \cdot \Gamma_{TW} + (y \cdot \zeta)^{-1} \bar{Q} \zeta_T - \frac{w \cdot \alpha}{y \cdot \zeta} \left[\frac{y_Z^*}{y_T^*} y_T - y_Z \right] \tag{1.42}$$

It may be noted that the normalizing factor $(y \cdot \zeta)^{-1}$ in (1.39), employed to maintain linear homogeneity, creates problems for the positivity of marginal costs through the second term in (1.41) and (1.42). In a later generalization in which linear homogeneity is not maintained, it will be possible to replace this normalizing factor by an alternative specification, obviating this problem.

In general, given a cost function $C^*(y_Z, y_T, w)$, outputs Z and T are defined to be *normal* at (y, w) provided that[11]

$$\partial^2 C^*(y, w)/\partial y_Z \partial y_T \leq 0. \tag{1.43}$$

Normality means the marginal cost of producing either output does not increase when the firm produces slightly more of the other output.

The matrix of second derivatives of (1.34) with respect to output is quite complex. As with the first derivatives, the situation is exacerbated by $(y \cdot \zeta)^{-1}$. For present purposes it is sufficient to examine the second-derivative matrix at (y^*, w^*). At this point, by Proposition 2, C^V can approximate an arbitrary regular underlying cost function that is linearly homogeneous in output. Given (1.36), the second-derivative matrix evaluated at (y^*, w^*), denoted $\nabla^2_{yy} C^V(y^*, w^*)$, is:

$$\nabla^2_{yy} C^V(y^*, w^*) = \frac{w^* \cdot \alpha}{y^* \cdot \zeta} \begin{bmatrix} \dfrac{y_T^*}{y_Z^*} & -1 \\ -1 & \dfrac{y_Z^*}{y_T^*} \end{bmatrix} \tag{1.44}$$

Since $\zeta > 0_2$, (1.44) indicates that $\nabla^2_{yy} C^V(y^*, w^*)$ is positive semi-definite and C^V is convex in y at the point (y^*, w^*) if and only if $w^* \cdot \alpha \geq 0$. For this functional form, at least at (y^*, w^*), the same condition that characterizes convexity of the cost function in output may also be employed to characterize normality at (y^*, w^*). Note that, at (y^*, w^*) where C^V approximates the true underlying cost function, the marginal cost cross-output sensitivity term may be obtained from (1.44) as:

$$\left. \frac{\partial^2 C^V(y, w^*)}{\partial y_Z \partial y_T} \right|_{y=y^*} = - \left(\frac{w^* \cdot \alpha}{y^* \cdot \zeta} \right) \tag{1.45}$$

Given that $y^* \cdot \zeta > 0$, then $w^* \cdot \alpha > 0$ is sufficient for marginal cross-output cost sensitivity (1.45) to be negative. In the case where outputs are normal and the cost function is convex in output, the overall technology will be said to exhibit *economies of shared production*. For this to apply under (1.34) at

(y^*, w^*), then from (1.44) it is apparent that at least one $\alpha_m > 0$, $m = L, M, K$ in this case. If $\alpha_m > 0$, then that input m exhibits economies of shared production at (y^*, w^*).[12] There is a relationship between technology that exhibits economies of shared production and technology that exhibits economies of scope, as we shall shortly explain.

Following Baumol et al. (1988: 17) and specializing their definition to the case of only two outputs, technology is *strictly subadditive* at $(y, w) \gg 0_5$ if for all points $y^1 \equiv (y_Z^1, y_T^1) > 0_2$ and $y^2 \equiv (y_Z^2, y_T^2) > 0_2$ satisfying $y^1 + y^2 = y$:

$$C(y, w) < C(y^1, w) + C(y^2, w) \tag{1.46}$$

where technology is represented by the cost function $C(y, w)$. Technology is *subadditive* at $(y, w) \gg 0_5$ if in the above definition, the strict inequality (1.46) is replaced by the weaker inequality:

$$C(y, w) \leq C(y^1, w) + C(y^2, w) \tag{1.47}$$

Again following Baumol et al. (1988: 71–2), technology exhibits *economies of scope* at $(y, w) \gg 0_5$ if

$$C(y, w) \equiv C(y_Z, y_T, w) < C(y_Z, 0, w) + C(0, y_T, w) \tag{1.48}$$

and the technology exhibits *weak economies of scope* at $(y, w) \gg 0_5$ if

$$C(y, w) \equiv C(y_Z, y_T, w) \leq C(y_Z, 0, w) + C(0, y_T, w) \tag{1.49}$$

The intuitive meaning of strict subadditivity is that it is cheaper for a single firm to serve the market than for two independent firms. It can be seen that economies of scope is a special case of strict subadditivity. From the viewpoint of public policy toward monopoly, the concept of strict subadditivity plays a key role: if the technology under consideration can be shown to be strictly subadditive for all (y, w) then, *ceteris paribus*, there exists a prima facie case that the industry should not be opened to competition, since this will only wastefully increase industry costs.[13]

Proposition 3: Let the technology of an industry be represented by $C^V(y, w)$ defined by (1.34), (1.6) and (1.39). If $\alpha \geq 0_3$ then the technology is convex in y and subadditive for all $(y, w) \gg 0_5$. If, in addition, outputs are normal and at least one input exhibits economies of shared production (so that at least one $\alpha_m > 0$) then the technology exhibits economies of scope at all points $(y, w) > 0_5$.

Thus our concept of economies of shared production appears to be closely related to the Baumol et al. concept of economies of scope.

The cost function used in this section is now generalized by adding back the terms $w \cdot \beta = w_L \beta_L + w_M \beta_M + w_K \beta_K$. Thus, define the overall cost function C as:

$$C(y, w) = w \cdot \beta + [w \cdot \phi^Z + \tfrac{1}{2}(w \cdot \eta)^{-1} w \cdot \Gamma_Z w] y_Z \qquad (1.50)$$

$$+ [w \cdot \phi^T + \tfrac{1}{2}(w \cdot \eta)^{-1} w \cdot \Gamma_T w] y_T + \tfrac{1}{2} \frac{w \cdot \alpha}{y \cdot \zeta} y \cdot \begin{bmatrix} y_T^*/y_Z^* & -1 \\ -1 & y_Z^*/y_T^* \end{bmatrix} y$$

As in the previous section, if β turns out to be nonnegative, then it can be interpreted as a vector of fixed input requirements which, when multiplied by the input price vector w, gives rise to the firm's fixed costs, $w \cdot \beta$. However, if one or more components of β are nonnegative, then care must be taken in extending the domain of definition of C defined by (1.50) to (y, w) that are far from the observed sample observations (y^t, w^t).

Proposition 4: Let the technology of the firm be represented by the cost function C defined by (1.50). If $\alpha \geq 0_3$ and $w \cdot \beta \geq 0$ then the technology is subadditive at (y, w) for all $y \gg 0_2$. If $\alpha \geq 0_3$ and $w \cdot \beta > 0$, then the technology is strictly subadditive and exhibits economies of scope at (y, w) for all $y > 0_2$.

Proposition 3 and Proposition 4 show that there are at least two ways in which strict subadditivity and economies of scope for C can occur: (i) an input can exhibit economies of shared production, that is, $\alpha \geq 0_3$ and $\alpha_m > 0$ for some input m or (ii) there are fixed costs at the vector of input prices w, that is, $w \cdot \beta > 0$. Of course, if say $\alpha_L > 0$ but at the same time $w \cdot \beta < 0$, then the inequalities go in opposite directions and it will not be possible to tell a priori whether C is subadditive at (y, w).

COMBINED ECONOMIES OF SCALE AND SCOPE

Another potential source of subadditivity is increasing returns to scale or diminishing costs that arise from the variable cost component of the model rather than simply from the presence of fixed costs. Thus, the model is now generalized to allow for the possible existence of divisional cost functions that exhibit non-constant returns to scale. Returning to the non-linear output specification $H^n(y_n, y_n^*)$ given by (1.27)–(1.29), define the vector $H(y, y^*) \equiv (H^Z(y_Z, y_Z^*), H^T(y_T, y_T^*))$. Referring to (1.34)–(1.35), the intention is to replace linear and quadratic terms in y with similar terms in

$H(y, y^*)$. In doing so advantage is taken of the fact that the cross-output term is no longer required to be linearly homogeneous in output to replace the empirically difficult scaling factor by one that is dependent on the reference level of output rather than the current output. Define fixed and variable parameter matrices $\Phi \equiv [\phi^Z, \phi^T]$ and $\Gamma^+(w) = [\Gamma_Z w, \Gamma_T w]$. Then the overall cost function may be defined as:[14]

$$\tilde{C}(y, w, y^*) = w \cdot \beta + w \cdot \Phi^+(w) \, H(y, y^*) + w \cdot \alpha \, Q(y, y^*) \qquad (1.51)$$

where

$$\Phi^+(w) \equiv \Phi + \tfrac{1}{2}(w \cdot \eta)^{-1} \Gamma^+(w), \qquad (1.52)$$

$$Q(y, y^*) = + \tfrac{1}{2} H(y, y^*) \cdot X(y^*) \, H(y, y^*)/J(y^*), \qquad (1.53)$$

$$X(y^*) = \begin{bmatrix} y_T^*/y_Z^* & -1 \\ -1 & y_Z^*/y_T^* \end{bmatrix}, \qquad (1.54)$$

and

$$J(y^*) = [(y_Z^*)^{\zeta_Z} + \psi_Z][(y_T^*)^{\zeta_T} + \psi_T], \; \zeta_Z + \zeta_T = 1. \qquad (1.55)$$

Noting that $H^n(0, y_n^*) = 0$, the variable cost component of (1.51) is written as:

$$\tilde{C}^V(y, w, y^*) = w \cdot \Phi^+(w) \, H(y, y^*) + w \cdot \alpha \, Q(y, y^*). \qquad (1.56)$$

The properties of $\tilde{C}^V(y, w, y^*)$ are:

$$\tilde{C}^V(0, w, y^*) = 0$$
$$\tilde{C}^V(y, w, y^*) \text{ is } HD1 \text{ in } w \qquad (1.57)$$
$$\tilde{C}^V(y^*, w, y^*) = w \cdot \Phi y^*$$

It is useful to record the first derivative vector and second derivative matrix of $\tilde{C}^V(y, w, y^*)$ with respect to the outputs as:[15]

$$\nabla_y \tilde{C}(y, w, y^*) = \begin{pmatrix} [w \cdot \phi^Z + \tfrac{1}{2}(w \cdot \eta)^{-1} w \cdot \Gamma_Z w] \nabla H^Z(y_Z, y_Z^*) \\ [w \cdot \phi^T + \tfrac{1}{2}(w \cdot \eta)^{-1} w \cdot \Gamma_T w] \nabla H^T(y_T, y_T^*) \end{pmatrix} \qquad (1.58)$$

$$+ \frac{w \cdot \alpha}{J(y^*)} \begin{bmatrix} \nabla H^Z(y_Z, y_Z^*) & 0 \\ 0 & \nabla H^T(y_T, y_T^*) \end{bmatrix} X(y^*) \, H(y, y^*)$$

$$\nabla^2_{yy}\tilde{C}(y,w,y^*) = \begin{bmatrix} \nabla^2 H^Z(y_Z,y_Z^*) & 0 \\ 0 & \nabla^2 H^T(y_T,y_T^*) \end{bmatrix}$$

$$\times \left\{ \begin{bmatrix} w\cdot\phi^Z + \frac{1}{2}(w\cdot\eta)^{-1}w\cdot\Gamma_Z w & 0 \\ 0 & w\cdot\phi^T + \frac{1}{2}(w\cdot\eta)^{-1}w\cdot\Gamma_T w \end{bmatrix} \right.$$

$$+ \frac{w\cdot\alpha}{J(y^*)}\begin{bmatrix} \frac{y_T^*}{y_Z^*} H^Z(y_Z,y_Z^*) - H^T(y_T,y_T^*)\frac{y_Z^*}{} & 0 \\ 0 & \left. \frac{y_T^*}{} H^T(y_T,y_T^*) - H^Z(y_Z,y_Z^*) \end{bmatrix} \right\}$$

$$+ \frac{w\cdot\alpha}{J(y^*)}\begin{bmatrix} \nabla H^Z(y_Z,y_Z^*) & 0 \\ 0 & \nabla H^T(y_T,y_T^*) \end{bmatrix}\begin{bmatrix} y_T^*/y_Z^* & -1 \\ -1 & y_Z^*/y_T^* \end{bmatrix}$$

$$\times \begin{bmatrix} \nabla H^Z(y_Z,y_Z^*) & 0 \\ 0 & \nabla H^T(y_T,y_T^*) \end{bmatrix} \tag{1.59}$$

From (1.58) the conditions for positivity of the marginal cost of zoned local output are:

$$\nabla H^Z(y_Z, y_Z^*) > 0 \tag{1.60}$$

and

$$w\cdot\phi^Z > -\frac{1}{2}(w\cdot\eta)^{-1}\, w\cdot\Gamma_Z w - \frac{w\cdot\alpha}{J(y^*)}\left[\frac{y_T^*}{y_Z^*}H^Z(y_Z,y_Z^*) - H^T(y_T,y_T^*)\right], \tag{1.61}$$

while for positive marginal cost of toll output, the conditions are:

$$\nabla H^T(y_T, y_T^*) > 0 \tag{1.62}$$

and

$$w\cdot\phi^T > -\frac{1}{2}(w\cdot\eta)^{-1}\, w\cdot\Gamma_T w - \frac{w\cdot\alpha}{J(y^*)}\left[\frac{y_Z^*}{y_T^*}H^T(y_T,y_T^*) - H^Z(y_Z,y_Z^*)\right]. \tag{1.63}$$

If (1.60)–(1.63) are satisfied the cost function $\tilde{C}^V(y, w, y^*)$ is proper. Conditional on (1.60)–(1.63) holding, it can then be observed from (1.59) that the following additional conditions are sufficient for positive semi-definiteness of $\nabla_{yy}^2 \tilde{C}^V(y, w, y^*)$ and hence for concavity of $\tilde{C}^V(y, w, y^*)$:[16]

$$\nabla^2 H^Z(y_Z, y_Z^*) \geq 0, \nabla^2 H^T(y_T, y_T^*) \geq 0, \text{ and } \alpha \geq 0_3 \qquad (1.64)$$

In view of the specification (1.27) for $H^n(y_n, y_n^*)$, $n = Z, T$, $\nabla H^n(y_n, y_n^*)$ and $\nabla^2 H^n(y_n, y_n^*)$ are as given by (1.31) and (1.32). The complex set of results and conditions (1.58)–(1.64) simplify considerably at $y_n = y_n^*$ since at this point $H^n(y_n^*, y_n^*), = y_n^*, \nabla H^n(y_n^*, y_n^*) = 1$ and $\nabla^2 H^n(y_n, y_n^*) = \theta_n$, where θ_n is defined by (1.33). Evaluating the results at $y_n = y_n^*$ provides:

$$\nabla_y \tilde{C}^V(y^*, w^*, y^*) = (w^* \cdot \phi^Z \quad w^* \cdot \phi^T)^\# \qquad (1.65)$$

$$\nabla_{yy}^2 \tilde{C}^V(y^*, w^*, y^*) = \begin{bmatrix} \theta_Z w \cdot \phi^Z & 0 \\ 0 & \theta_T w \cdot \phi^T \end{bmatrix} + \frac{w^* \cdot \alpha}{J(y^*)} \begin{bmatrix} y_T^*/y_Z^* & -1 \\ -1 & y_Z^*/y_T^* \end{bmatrix}$$
$$(1.66)$$

so that the conditions for positive marginal costs at (y^*, w^*) are:

$$\phi^Z \geq 0_3, \text{ with } \phi_m^Z > 0 \text{ for some } m = L, M, K$$
$$\text{and} \qquad (1.67)$$
$$\phi^T \geq 0_3, \text{ with } \phi_m^T > 0 \text{ for some } m = L, M, K$$

and the additional conditions for concavity of the cost function at (y^*, w^*) are:

$$\theta_Z \geq 0, \theta_T \geq 0 \text{ and } \alpha \geq 0_3. \qquad (1.68)$$

Moving from the linearly homogeneous specification (1.34) to the non-linear specification (1.56) requires a corresponding generalization of Proposition 2.

Proposition 5: $\tilde{C}^V(y, w, y^*)$ defined by (1.56), (1.6) and (1.52) can approximate to the second order an arbitrary twice continuously differentiable two-output, three-input cost function $C^*(y, w)$ at the point (y^*, w^*). This approximation can be accomplished even if the restrictions $\Gamma_Z = \Gamma_T$ and $A_L = A_M = A_K$ are imposed.

A non-linear generalization of the flexible functional form suitable for examining economies of scale and scope requires the extension of the definition of subadditivity and economies of scope to allow for adjustment of the reference vector in the non-linear flexible functional form approximation. Thus, when the technology is represented by a cost function

$\tilde{C}(y, w, y^*)$ the technology is said to be strictly subadditive at $(y, w, y^*) \gg 0_7$ if for all points $y^1 \equiv (y^1_Z, y^1_T) > 0_2$ and $y^2 \equiv (y^2_Z, y^2_T) > 0_2$ satisfying $y^1 + y^2 = y$, there exists a split of the reference vector y^* into $y^{1*} \equiv (y^{1*}_Z, y^{1*}_T) > 0_2$ and $y^{2*} \equiv (y^{2*}_Z, y^{2*}_T) > 0_2$, satisfying $y^{1*} + y^{2*} = y^*$, such that

$$\tilde{C}(y, w, y^*) < \tilde{C}(y^1, w, y^{1*}) + \tilde{C}(y^2, w, y^{2*}). \qquad (1.69)$$

The important point to note about (1.69) is that (strict) subadditivity is not defined for a common reference vector for the cost function approximations appearing in the definition. Equality of y^{1*} or y^{2*} with y^* is incompatible with the definition.

In a similar spirit, the technology is subadditive at $(y, w, y^*) \gg 0_7$ if, in the above definition, the strict inequality (1.69) is replaced by the weaker inequality:

$$\tilde{C}(y, w, y^*) \leq \tilde{C}(y^1, w, y^{1*}) + \tilde{C}(y^2, w, y^{2*}). \qquad (1.70)$$

The technology exhibits economies of scope at $(y, w, y^*) \gg 0_7$ if

$$\tilde{C}(y, w, y^*) \equiv \tilde{C}(y_Z, y_T, w, y_Z^*, y_T^*)$$
$$< \tilde{C}(y_Z, 0, w, y_Z^*, 0) + \tilde{C}(0, y_T, w, 0, y_T^*) \qquad (1.71)$$

and the technology exhibits weak economies of scope at $(y, w, y^*) \gg 0_7$ if

$$\tilde{C}(y, w, y^*) \equiv \tilde{C}(y_Z, y_T, w, y_Z^*, y_T^*)$$
$$\leq \tilde{C}(y_Z, 0, w, y_Z^*, 0) + \tilde{C}(0, y_T, w, 0, y_T^*). \qquad (1.72)$$

Proposition 6: Let the technology of the industry be represented by the cost function $\tilde{C}^V$ defined by (1.56), (1.6) and (1.52). If there exists $(y, w, y^*) \gg 0_7$ such that (1.60)–(1.64) hold for all $(\tilde{y}, w, \frac{1}{2}y^*)$, where $(w, \frac{1}{2}y^*)$ is given and $\tilde{y}$ is any output vector satisfying $0_2 < \tilde{y} \leq y$, then the technology is subadditive at (y, w, y^*).

Finally, the fixed cost component $w \cdot \beta$ is added back to the variable cost function $\tilde{C}^V(y, w, y^*)$ that was the object of study in Proposition 5 and Proposition 6. $\tilde{C}(y, w, y^*)$ is now as defined in (1.51). To match this general specification, it is straightforward to make amendments of Proposition 5 and Proposition 6, analogously to the way in which Proposition 4 generalized Proposition 3, to accommodate fixed costs. Before leaving this issue, however, it is appropriate to consider the implications of fixed costs more closely in the context of testing for subadditivity. A difficulty concerns whether fixed costs are sunk or recoverable on the sale of plant. In terms of specification, the sunkenness or recoverability of fixed costs has implications for the parameter vector β. If fixed costs are in part recoverable then β should be a function of the capacity of the plant, though not its operating level, so that divestiture of plant should entail a reduction in fixed costs for the remaining capacity. The situation is actually more complex. Even if some costs are sunk, a firm required to divest part of its operations may be able to make a case for compensation equivalent, for example, to the call value on the option to utilize the facility previously constructed. If these types of issues are ignored it would appear that on splitting a firm into two, say, fixed costs are doubled. At the other extreme, if all fixed costs attributable to the support of sold facility is recoverable fixed costs do not increase. The econometric difficulty in distinguishing these cases is that there is no observable explanator that can separate sunk from recoverable fixed costs.

If fixed costs are a relatively large component of costs, then treating β as a constant irrespective of the capacity of a plant will bias tests for subadditivity in favour of a single plant. In order to avoid this possible bias we have chosen to model the fixed cost vector β as dependent upon the reference level of output, y^*. This allows a different estimate of fixed costs to be determined for plants of fundamentally different capacity. However, because there is no within sample variation in y^*, parameters associated with β cannot be estimated independently of parameters in the variable cost component. The approach adopted here is to model the fixed cost component as dependent upon y^* in a similar manner to the way the variable cost component is modelled as dependent upon y, say in specification (1.56).

Define the '(linear component of the) average divisional cost at zero output (scaled by the reference output level)' function, $G^n(y^*)$ as:

$$G^n(y_n^*) \equiv \left[\left(\frac{H^n(y_n,y_n^*)}{y_n}\right)\Bigg|_{y_n=0}\right] y_n^* = \delta_n y_n^*, \quad n=Z, T \qquad (1.73)$$

where, from (1.27)–(1.29),

$$\delta_n = [\varepsilon_n + 1]/[\varepsilon_n^2 + \varepsilon_n + 1] \qquad (1.74)$$

Additionally, define a simplified variant of the cross-output term that captures the essential interactions in terms of the reference outputs:

$$V(y^*) = \delta_Z \delta_T y_Z^* y_T^*. \qquad (1.75)$$

Combine these terms using the parameters associated with the equivalent cost components in the variable cost function to obtain a composite factor specific parameter:

$$\bar{\beta}_m = \phi_m^Z \delta_Z y_Z^* + \phi_m^T \delta_T y_T^* + \alpha_m \delta_Z \delta_T y_Z^* y_T^*, \quad m = L, M, K \quad (1.76)$$

Specification (1.76) combines linear and interaction terms in the reference level of output with parameters estimated in the variable cost component of the cost function. Since factor demand equations have a free intercept that also contributes to fixed costs, they can be combined with (1.76) to provide an overall estimate of fixed costs that is dependent on the reference level of output and therefore may be adjusted according to empirical results on the parameter estimates when utilizing empirical results to test for sub-additivity.

Experimentation with the above type of form showed that results were poor when combined additively with free intercepts but better behaved when a multiplicative formulation was employed, where the criterion for good behaviour was the achievement of positive marginal costs over the sample. Accordingly, the specification ultimately employed for the 'fixed cost' component $w \cdot \beta$ of the cost function was:

$$\beta_m = \mu_m \bar{\beta}_m, \quad m = L, M, K \quad (1.77)$$

where μ_m, $m = L, M, K$ are freely estimated parameters and $\bar{\beta}_m$ is specified in (1.76).

SPECIFICATION OF TECHNICAL PROGRESS AND THE ESTIMATING FORM

The Bell System data set used by Evans and Heckman (1983) contains a measure of R&D expenditure which, in principle, could be relevant to the modelling of technical progress but which should be modelled as endogenous. However, because this analysis is conducted within the paradigm of an instantaneous (atemporal) cost-minimizing framework, the issue of endogenous technical progress will not be addressed. The specification opted for is the use of an exogenous technical progress variable (exponential time trend). Each factor input faces a different degree of technical progress and modifies the price of an effective unit of factor input. Letting the reference year, 1961, act as the base, the factor price adjustment terms are:

$$tech_m(t)=(1+\tau_m)^{1961-t}, \quad m=L, M, K, t=1947, \ldots, 1977 \quad (1.78)$$

where the parameters τ_m, $m=L, M, K$ denote the annual rate of factor price adjustment. Given (1.78), effective quality-adjusted factor prices are:

$$\tilde{w}_m(t)= tech_m(t)w_m, \quad m=L, M, K \qquad (1.79)$$

Given $\tilde{w}\equiv(\tilde{w}_L\ \tilde{w}_M\ \tilde{w}_K)$ with $\tilde{w}_m$ defined in (1.79), the final functional form for the overall cost function is:

$$\tilde{C}(y, w, y^*, t)=\tilde{w}\cdot\beta+\tilde{w}\cdot\Phi^+(\tilde{w})\ H(y, y^*)+\tilde{w}\cdot\alpha\ Q(y, y^*) \quad (1.80)$$

where the previously defined parameters and functional form specifications are summarized for $m=L, M, K$ and $n=Z, T$ as:

(i) $\beta\equiv(\beta_L, \beta_M, \beta_K)$ with
$$\beta_m = \mu_m\ [\phi_m^Z\delta_Z y_Z^*+\phi_m^T\delta_T y_T^*+\alpha_m\delta_Z\delta_T y_Z^* y_T^*],$$

(ii) $\Phi^+(\tilde{w})=[\phi^Z\ \phi^T]+\frac{1}{2}(\tilde{w}\cdot\eta)^{-1}[\Gamma_Z\tilde{w}\ \Gamma_T\tilde{w}]$ with $\phi^n\equiv(\phi_L^n\ \phi_M^n\ \phi_K^n)^{\#}$,

(iii) $\eta\equiv(\eta_L\ \eta_M\ \eta_K)$ and $\Gamma_n=-\Sigma_n\Sigma_n^{\#}$ where
$$\Sigma_n\equiv\begin{bmatrix} \sigma_{LL}^n & 0 & 0 \\ \sigma_{ML}^n & \sigma_{MM}^n & 0 \\ -(\sigma_{LL}^n+\sigma_{ML}^n) & -\sigma_{MM}^n & 0 \end{bmatrix},$$

(iv) $H(y, y^*)\equiv(H^Z(y_Z, y_Z^*)\ \ H^T(y_T, y_T^*))$ with
$$H^n(y_n, y_n^*)=\left\{\frac{\varepsilon_n y_n}{\varepsilon_n y_n+y_n^*}+\frac{\varepsilon_n(y_n^*-y_n)+y_n^*}{\varepsilon_n[\varepsilon_n(y_n^*-y_n)+y_n^*]+y_n^*}\right\}y_n,$$

(v) $\alpha\equiv(\alpha_L\ \alpha_M\ \alpha_K)$,

(vi) $Q(y, y^*)=\frac{1}{2}H(y, y^*)\cdot X(y^*)H(y, y^*)/J(y^*)$ where
$$X(y^*)=\begin{bmatrix} y_T^*/y_Z^* & -1 \\ -1 & y_Z^*/y_T^* \end{bmatrix}\text{ and}$$
$$J(y^*)=[(y_Z^*)^{\zeta_Z}+\psi_Z][(y_T^*)^{\zeta_T}+\psi_T], \quad \zeta_Z+\zeta_T=1$$

Three input demand functions, x_m, $m=L, M, K$, are obtained by differentiating (1.80) with respect to w_L, w_M and w_K, respectively. To reduce the effect of time trends in estimation the demand equations are written in share of total revenue form so that they can be estimated as a system of three similarly structured equations. A fourth revenue share equation is available by the adding up identity, but not explicitly estimated. Defining the factor input shares of revenue as $rs_m = w_m x_m / R$, $m = L, M, K$, the residual share is $rs_{\text{Profit}} = 1 - \sum_{m=L,M,K} rs_m$ and the estimating equations are:

$$rs_L = \frac{\tilde{w}_L}{R} \left\{ \begin{array}{l} \beta_L + [\phi_L^Z + \tfrac{1}{2}(\tilde{w} \cdot \eta)^{-1} \underline{\gamma}_L^Z \cdot \tilde{w} - (w \cdot \eta)^{-1} \eta_L \, S^Z(\tilde{w})] \, H^Z(y_Z, y_Z^*) \\ + [\phi_L^T + \tfrac{1}{2}(\tilde{w} \cdot \eta)^{-1} \gamma_L^T \cdot \tilde{w} - (w \cdot \eta)^{-1} \eta_L \, S^T(\tilde{w})] \, H^T(y_T, y_T^*) \\ + \alpha_L \, Q(y, y^*) \end{array} \right\} + u_L$$

$$rs_M = \frac{\tilde{w}_M}{R} \left\{ \begin{array}{l} \beta_M + [\phi_M^Z + \tfrac{1}{2}(\tilde{w} \cdot \eta)^{-1} \gamma_M^Z \cdot \tilde{w} - (w \cdot \eta)^{-1} \eta_M \, S^Z(\tilde{w})] \, H^Z(y_Z, y_Z^*) \\ + [\phi_m^T + \tfrac{1}{2}(\tilde{w} \cdot \eta)^{-1} \gamma_M^T \cdot \tilde{w} - (w \cdot \eta)^{-1} \eta_M \, S^T(\tilde{w})] \, H^T(y_T, y_T^*) \\ + \alpha_M \, Q(y, y^*) \end{array} \right\} + u_M$$

$$rs_K = \frac{\tilde{w}_K}{R} \left\{ \begin{array}{l} \beta_K + [\phi_K^Z + \tfrac{1}{2}(\tilde{w} \cdot \eta)^{-1} \gamma_K^Z \cdot \tilde{w} - (w \cdot \eta)^{-1} \eta_K \, S^Z(\tilde{w})] \, H^Z(y_Z, y_Z^*) \\ + [\phi_K^T + \tfrac{1}{2}(\tilde{w} \cdot \eta)^{-1} \gamma_K^T \cdot \tilde{w} - (w \cdot \eta)^{-1} \eta_K \, S^T(\tilde{w})] \, H^T(y_T, y_T^*) \\ + \alpha_K \, Q(y, y^*) \end{array} \right\} + u_K$$

$$(1.81)$$

where $\underline{\gamma}_m^n \equiv (\gamma_{mL}^n \ \gamma_{mM}^n \ \gamma_{mK}^n)$ denotes the mth row of Γ_n, which is a function of deeper parameters σ_{ij}^n in view of definition (iii) above, where $S^n(\tilde{w}) = \tfrac{1}{2}(\tilde{w} \cdot \eta)^{-1} \tilde{w} \cdot \Gamma_n \tilde{w}$, and where the u_m, $m = L, M, K$ are appended classical errors.

The factor-share equation system (1.81) is estimated by the maximum-likelihood routine contained in the SHAZAM package (White, 1978). Estimation results are provided in Appendix C. Before briefly describing the results the measures of technical progress and of subadditivity are described. For period t, technical progress as reported in Appendix C, Table 1C.11 under the column-heading *TECH*, is defined as:

$$TECH(t) \equiv -[\hat{C}(y(t), w(t), y^*, t) - \hat{C}(y(t), w(t), y^*, t-1)]/\hat{C}(y(t), w(t), y^*, t)$$

$$(1.82)$$

where $y(t) \equiv (y_Z(t), y_T(t))$ is the observed output quantity vector at time t, $w(t) \equiv (w_L(t), w_M(t), w_K(t))$ is the observed input price vector for period t and

$\hat{C}$ is the estimated cost function. From (1.82), $TECH(t)$ is equal to minus the period t (fitted) cost obtained by evaluating the period t-1 cost function $\hat{C}(y, w, y^*, t-1)$ at the quantity and price data for period t and this difference is divided by the period t fitted cost. Thus $TECH(t)$ is the percentage cost reduction due to (unexplained) technological progress going from period t-1 to period t. The fitted marginal costs $MC_n(t)$ for the outputs $n = Z, T$ by period are calculated by the rule:

$$MC_n(t) \equiv \partial \hat{C}(y(t), w(t), y^*, t)/\partial y_n, \quad n = Z, T \qquad (1.83)$$

and are reported in Table 1C.10. Further, price-marginal cost mark-ups are reported in that table. In Table 1C.11, traditional returns to scale measures for the sample period t, $RS(t)$, are provided, that is, the reciprocal of the sum of the cost elasticities with respect to the outputs. Using definition (1.83):

$$RS(t) = \hat{C}(y(t), w(t), y^*, t)/[MC_Z(t)\,y_z(t) + MC_T(t)\,y_T(t)]. \qquad (1.84)$$

CALCULATION OF SUBADDITIVITY MEASURES

In calculating subadditivity measures a variant of the method pioneered by Evans and Heckman (1983, 1984) is followed by adapting it to allow for splitting fixed costs and potentially very low output levels. Let $y(t) \equiv (y_Z(t)\ y_T(t))$ be the observed period t output vector. Assume further two firms A and B that have access to period t technology and divide the total industry output between them. The corresponding output vectors $y(A,t) \equiv (y_Z(A,t)\ y_T(A,t))$ and $y(B,t) \equiv (y_Z(B,t)\ y_T(B,t))$ satisfy the relations:

$$y(A,t) \geq \psi, \ y(B,t) \geq \psi, \ y(A,t) + y(B,t) = y(t) \qquad (1.85)$$

where $\psi \equiv (\psi_Z\ \psi_T)$ are minimal levels of output determined as parameters in the estimation. Reference output levels are split in the same manner between the firms. This ensures the point of approximation at which the estimated cost function is capable of representing an arbitrary regular cost function remains a point of relevance for the simulated firms. Since the reference output level serves as a measure of overall plant capacity, it ought to be related logically to fixed costs. Additionally, the estimating form allows limited freedom for fixed costs to be determined by parameters that are not linearly related to the reference output level. Splitting the reference output across firms can lead to a rise or fall of fixed costs in total relative to the single firm case. The outcome is empirically determined.

Based on these considerations, the reference output levels are assumed to satisfy relationships similar to (1.85), that is,

$$y^*(A,t) \geq \psi, \quad y^*(B,t) \geq \psi, \quad y^*(A,t) + y^*(B,t) = y^*(t) \qquad (1.86)$$

The period t subadditivity measure for firms A and B producing $y(A,t)$ and $y(B,t)$ is:

$$SUB(y^A, y^B, y^*, t) \equiv [\hat{C}(y(t), w(t), y^*(t), t) - \hat{C}_{sub}(y^A, y^B, y^*, t)]$$
$$/\hat{C}(y(t), w(t), y^*(t), t) \qquad (1.87)$$

where:

$$\hat{C}_{sub}(y^A, y^B, y^*, t) \equiv \hat{C}(y(A,t), w(t), y^*(A,t), t) + \hat{C}(y(B,t), w(t), y^*(B,t), t)$$
$$(1.88)$$

When it is cheaper for a single firm to produce total period t output vector $y(t)$ rather than two firms producing $y(A,t)$ and $y(B,t)$, then $SUB(y^A, y^B, y^*, t)$ is negative and the magnitude of $SUB(y^A, y^B, y^*, t)$ represents the percentage cost savings. When $SUB(y^A, y^B, y^*, t) \leq 0$ for all $y(A,t)$ and $y(B,t)$ satisfying the restrictions (1.85), then the cost function $\hat{C}(y, w, y^*, t)$ is effectively globally subadditive at $(y(t), w(t), y^*, (t))$.[17]

To find the points $y(A,t)$ and $y(B,t)$ that satisfy (1.85) it is necessary for $y(t) \geq 2\psi$. Since parameter estimates of ψ_Z and ψ_T are extremely small compared to the levels of y_Z and y_T for the sample period this is easily achieved. Points are calculated as follows:

$$y_Z(A,t) = \psi_Z + \lambda_Z[y_Z(t) - y_Z(1947)]$$
$$y_Z(B,t) = [y_Z(1947) - \psi_Z] + (1 - \lambda_Z)[y_Z(t) - y_Z(1947)]$$

$$(1.89)$$

$$y_T(A,t) = \psi_T + \lambda_T[y_T(t) - y_T(1947)]$$
$$y_T(B,t) = [y_T(1947) - \psi_T] + (1 - \lambda_T)[y_T(t) - y_T(1947)]$$

where $0 \leq \lambda_Z \leq 1$ and similarly $0 \leq \lambda_T \leq 1$. Appendix D and Appendix E report results of experiments for particular values of λ_Z and λ_T. Note, in simulating firm outputs using (1.89) for the period 1947–77, it is necessary because of the non-linearity of the cost function to choose specific reference output levels for the firms, levels that may be regarded as pointing to the capacity of the plants (though not indicating either plant's maximum capacity). Corresponding to the time-series pairs of (1.89) the fixed reference pairs are:

$$y_Z^*(A) = \psi_Z + \lambda_Z [y_Z^* - y_Z(1947)]$$
$$y_Z^*(B) = [y_Z(1947) - \psi_Z] + (1 - \lambda_Z)[y_Z^* - y_Z(1947)]$$

(1.90)

$$y_T^*(A) = \psi_T + \lambda_T [y_T^* - y_T(1947)]$$
$$y_T^*(B) = [y_T(1947) - \psi_T] + (1 - \lambda_T)[y_T^* - y_T(1947)]$$

Experiment 1 (Appendix D) sets $\lambda_Z = 0.2$ and $\lambda_T = 0.5$. This corresponds to A being a new entrant with minimal output and eventually capturing (at least, asymptotically) 50 per cent of the long-distance market. However, A makes a lesser impact in the local market, eventually gaining 20 per cent market share. Experiment 2 (Appendix E) has $\lambda_Z = 0.9$ and $\lambda_T = 0.75$ so that A, the incumbent, retains dominance in both markets. The entrant B starts off with a substantially higher level (ten times) starting value than the new entrant in Experiment 1 (there Firm A). The effect is that the entrant's growth prospects are substantially lower in Experiment 2. The simulation results are discussed in more detail following a description of the estimation results.

EMPIRICAL RESULTS

Evans and Heckman (1983, 1984) use data for the US Bell System constructed by Christensen et al. (1981) for 1947 through 1977. Data tabled in Diewert and Wales (1991) are supplemented by the aggregate output index data provided by Evans and Heckman (1983). These data are reproduced in Appendix B, Table 1B.1. In that table quantities of labour, capital and materials implied by calculating their values from the total cost, factor price and factor cost share data are exhibited. Because of the strong time-series trends in these quantities the demand equations are estimated in share form. As discussed earlier, the non-linear specification is sensitive to the units of measurement. Supplementary information is used to rescale output indices, generate implied price of output indices and construct revenue series. The resulting series for zoned local and toll call prices, local and toll revenue, local and toll output and total revenue are listed in Table 1B.2. The details of their construction are explained therein. Use is made of information given in Evans and Heckman (1983: 273) that indicates local call revenue reached 21 billion US dollars (USD) in 1979 while toll call revenue reached USD24 billion. Revenue series are constructed that, when examined for a linear trend for 1970–77, extrapolate to 1979 values. The procedure estimates price indices for the outputs in a two-step process and scales output indices to generate the results. Table 1B.2 also contains implied profits and factor shares as a proportion of total revenue. The factor shares of revenue are used as the dependent variables in the estimating equations. As expected, individually they exhibit substantially less trend than the quantity data, although there is an evident strong trend decline in profitability over the sample period.

Maximum likelihood estimates are reported in Table 1C.1. Overall, the estimating equations fit quite well, with R^2 values of 0.92, 0.96 and 0.99 for the labour, materials and capital factor share equations, respectively. However, the Durbin–Watson statistics indicate evidence of poorly behaved residuals, particularly the labour equation. In this non-linear share system it is not easy to determine what this poor result should be attributed to. Apparently better-behaved results from a statistical point of view under some alternative minor variations on the functional form could be obtained, for example, by modifying the scale factor on the quadratic-output term. However, such variations inevitably create problems with the economic quality of the results, leading to implied negative marginal costs. It was the over-riding importance of positive marginal costs that led to the final specification. The implied marginal costs, for the exhibited estimates, are positive not only throughout the sample period but also under simulation in the subadditivity tests. For this reason they are the preferred results even though there is some evidence of a pattern in the residuals that makes statistical hypothesis testing somewhat problematic.

While drawing attention to the positive marginal cost estimates in Table 1C.10, it should be pointed out the implied steady decline in the local call mark-up may be contrasted with the somewhat more erratic and substantially higher mark-up in the toll-call market. Also of relevance in assessing the results is the extent to which the non-linear in output specification matters. The estimates of ε_Z and ε_T are quite small but appear statistically significant, even if allowance is made for possible overestimation of the asymptotic t-statistics. Thus the departure from linearity appears small but significant. A better sense of the importance of this departure may be provided by Table 1C.4 and Table 1C.5. The structure of the $H^n(y_n, y_n^*)$ functions is clear from these tables. The effect of output on cost is slightly concavified at these parameter estimates. As Table 1C.6 shows, the quadratic terms add little by comparison. Table 1C.6 also shows technical progress is substantial in the case of labour. The apparent technical regress in materials and capital could have been avoided by constraining the coefficients τ_M and τ_K to zero. However, in view of the desire to construct and estimate a proper and flexible functional form, the parameters are left unconstrained wherever possible. Indeed, what is remarkable about the results is that the functional form specified does contain these economically desirable features, produced without undue constraint.[18]

Table 1C.7 reports some of the factor price related components of the cost function. First, $\tilde{w} \cdot \beta$ being positive may be interpreted as fixed costs. It is not clear whether these should be treated as sunk or recoverable. The costs are substantial and although a particular approach in tests for subadditivity (allocating them across firms) is followed it is possible, based on the evidence from the tests, that were the fixed costs replicated in their entirety for each firm then it would be sufficient to turn the test in favour of the cost effectiveness of a single firm. This points to the conclusion that a more detailed distinction between the types of fixed costs is crucial. Also apparent in Table 1C.7 is that the linear terms contributing to variable costs, $\tilde{w} \cdot \phi^Z$ and $\tilde{w} \cdot \phi^T$, are relatively more important than the quadratic terms. The second-order price terms S^Z and S^T are necessarily negative because of concavity of the cost function, and have the potential to exacerbate difficulty in obtaining positive marginal costs. However, for this specification their size and role is relatively unimportant. The same can be said for the quadratic-output overall price-scaling coefficient $\tilde{w} \cdot \alpha$. The negative sign, the overall outcome of a combination of positive and negative influences, works against the likelihood of subadditivity, but the effect is small. Tables 1C.8 and 1C.9 give some specific information on the factors. Table 1C.8 demonstrates the within sample predictive power of the model. Table 1C.9 shows that each factor makes a contribution to fixed costs. None of the intercepts are negative for this specification. The table also contains a measure of overall fixed costs constructed in several different ways as a check on calculations.

Table 1C.11 calculates *TECH*, a measure of technical progress based on the comparison of the cost of production with current technology compared with similar production but with prior technology. The measure is sensitive to the way in which technical progress is modelled. Although the freely estimated technological change parameters suggest regress for materials and capital versus progress for labour, the overall effect, as exhibited in *TECH*, is positive though possibly cyclical technical progress. In fact, the regress implied by the rising $tech_m$ terms for $m = M, K$ (Table 1C.6) is best interpreted as a correction for the strong downward trend in $tech_L$ since it is the relative dominance that matters for the construction of the overall *TECH* term. Finally, Table 1C.11 reports *RS*, a return to scale measure. Increasing returns to scale are evident throughout the sample period, though there is some evidence of a trend diminution in its degree.

Experiment 1 is concerned with an entrant, Firm A, ultimately capturing 50 per cent of the long-distance market while making smaller inroads into the local call market. The results are presented in Appendix D, beginning with an overall summary in Table 1D.1. In Column C actual costs are presented as a point of comparison for the cost figures. Column $\hat{C}$ gives the estimated total cost which is a good estimate of actual cost. Column $\hat{C}_{sub}$ represents the total costs of the firms, A and B, and is the relevant column for comparison with $\hat{C}$. The specific total costs of the individual firms are also given in the obvious columns. Column *SUB* represents the relevant cost differences as a proportion of the monopoly firm cost. The results clearly indicate the cost function is not subadditive and that it would have been more efficient to split the firm into two for 1947–77. However, this experiment result has embedded in it an assumption that the reference level of output can be split between the firms. In the estimated model, fixed costs are related to the reference level of output, though with an interaction term so not linearly. It is clear from the relative differences in the sizes of figures in columns $\hat{C}$ and $\hat{C}_{sub}$, and a comparison of these with the size of fixed costs, $\tilde{w} \cdot \beta$, say from Table 1C.11, that this result is, however, dependent upon the assumption implicit in this specification. Thus, the need for more detailed consideration of the exact nature of fixed costs.

Tables 1D.2 to 1D.5 give detailed results for A, and these are relevant to compare with Table 1D.6 to Table 1D.9, which give the equivalent results for B. It is interesting to compare the marginal costs of A (Table 1D.5) with those of B (Table 1D.9). The costs are remarkably close, even for cases where output levels are substantially different, as they are for local calls in this experiment, and also at least initially for toll calls (see Table 1D.2 and Table 1D.3 compared to Table 1D.6 and Table 1D.7). Finally, a potentially unrealistic feature of the presented results needs to be acknowledged. It should be pointed out that the experiment makes no allowance for possible output price change from competition, so the price mark-ups shown in Table 1D.5 (A) and Table 1D.9 (B) are effectively similar to those for the single firm estimated case, as shown in Table 1C.10.

Experiment 2, the results of which are reported in Appendix E, considers the case where an incumbent firm (A) manages to successfully keep the entrant, B, to a relatively minor portion of the market. The results demonstrate the robustness of the result on the commonality of marginal costs, despite overall cost conditions being very different for the firms. This result can be traced back to the near linearity of cost in output that was a feature of the estimated results. Comparison of Table 1E.4 and Table 1E.8 isolate another feature of relevance in these results, that is, the greater importance of technological progress as a cost-reducing device for the small firm. This can be seen from the *TECH* column. A glance at the estimated factor demand columns suggest the difference is due to the greater reliance on labour relative to capital for the small firm. In other respects, Experiment 2 concurs with Experiment 1. A lack of subadditivity is again apparent. As before, this is due to the split of the reference vector between the firms that leads to a substantial cost reduction for the small firm. While it is obvious the alternative assumption of full duplication of fixed costs would overturn this result, that would be unrealistic where firms are of such uneven sizes.

CONCLUSION

The concept of economies of scale and scope underlie the cost efficiency of large firms. The issues are of central importance in the development of appropriate policy to encourage best resource use. This is especially pertinent in the case of modern information-communication firms that play a key role in the new economy and whose cost structures have wide-ranging implications for business and consumers. Several functional form specifications that are capable of exhibiting economies of scale and scope within the flexible functional form framework have been examined, while coping with the problem of negative marginal costs. Using US Bell System data, cost function parameters are obtained by estimating a factor input demand system in share form. The estimates are fully consistent with positive marginal costs. The parameters are applied to tests for the subadditivity of the Bell System. Experimental outcomes and the interpretation of test results depend on the extent to which fixed costs are sunk or recoverable. Further, the associated non-linear cost function estimates are sensitive to the point of flexible functional form approximation. Fixed costs may well be linked to the overall capacity of plant, and a measure of this capacity may also be relevant when considering not only the point of approximation in flexible functional form estimation but also the reference level of capacity. The latter should receive attention when consideration is given to applying the estimated parameters to the cost functions of firms of very different size. These considerations lead to the conclusion that careful attention needs to be given to the precise nature of fixed costs in order to apply appropriate tests of subadditivity.

APPENDIX A

Proofs of Propositions

Proof of Proposition 1:   Since both $\tilde{C}^n(y_n, w, y_n*)$ and $C^{n*}(y_n, w)$ are linearly homogenous in the components of w, Euler's theorem on homogeneous functions implies the following restrictions on the derivatives of $\tilde{C}^n$ and C^{n*}:

$$\tilde{C}^n(y_n*, w*, y_n*) = w* \cdot \nabla_w \tilde{C}^n(y_n*, w*, y_n*)$$
$$C^{n*}(y_n*, w*) = w* \cdot \nabla_w C^{n*}(y_n*, w*) \tag{1.91}$$

$$\nabla^2_{ww} \tilde{C}^n(y_n*, w*, y_n*) w* = 0_3$$
$$\nabla^2_{ww} C^{n*}(y_n*, w*) w* = 0_3 \tag{1.92}$$

$$w* \cdot \nabla^2_{wy_n} \tilde{C}^n(y_n*, w*, y_n*) = \nabla_{y_n} \tilde{C}^n(y_n*, w*, y_n*)$$
$$w* \cdot \nabla^2_{wy_n} C^{n*}(y_n*, w*) = \nabla_{y_n} C^{n*}(y_n*, w*) \tag{1.93}$$

where $\nabla_w \tilde{C}^n(y_n*, w*) \equiv [\partial \tilde{C}^n(y_n*, w, y_n*)/\partial w_L \quad \partial \tilde{C}^n(y_n*, w, y_n*)/\partial w_M$
$\partial \tilde{C}^n(y_n*, w, y_n*)/\partial w_K]|_{w=w*}$ denotes the vector of first-order partial derivatives of $\tilde{C}^n$ with respect to the components of w; $\nabla_{y_n} \tilde{C}^n(y_n*, w*,$
$y_n*) \equiv \partial \tilde{C}^n(y_n, w*, y_n*)/\partial y_n|_{y_n=y_n*}$; $\nabla^2_{ww} \tilde{C}^n(y_n*, w*, y_n*)$ denotes the third-order matrix of second-order partial derivatives of $\tilde{C}^n$ with respect to the components of w, evaluated at $w*$, and so on.

In view of the restrictions (1.91)–(1.93), vectors β^n and ϕ^n, scalar θ_n and symmetric matrix Γ_n need to be found satisfying (1.6) such that the following hold:

$$\nabla_w \tilde{C}^n(y_n*, w*, y_n*) = \beta^n + \phi^n y_n* = \nabla_w C^{n*}(y_n*, w*); \tag{1.94}$$

$$\nabla^2_{ww} \tilde{C}^n(y_n*, w*, y_n*) = \Gamma_n y_n* = \nabla^2_{ww} C^{n*}(y_n*, w*); \tag{1.95}$$

$$\nabla^2_{y_n y_n} \tilde{C}^n(y_n*, w*, y_n*) = w* \cdot \phi^n \theta_n = \nabla^2_{y_n y_n} C^{n*}(y_n*, w*); \tag{1.96}$$

$$\nabla^2_{wy_n} \tilde{C}^n(y_n*, w*, y_n*) = \phi^n = \nabla^2_{wy_n} C^{n*}(y_n*, w*) \tag{1.97}$$

where the left-hand side (LHS) equalities are established, given (1.3), as follows. Use (1.5) and (1.6) to establish the LHS equality in (1.94); use (1.4) and (1.6) to establish the LHS equality in (1.95); use (1.5) and (1.6) to establish the LHS equality in (1.96); and use (1.5) and (1.6) to establish the LHS equality in (1.97). Now use the right-hand side (RHS) equality in (1.97) to solve for ϕ^n. Assume marginal cost is positive at (y_n^*, w^*), that is,

$$\partial C^{n*}(y_n, w^*)/\partial y_n|_{y_n=y_{n^*}} > 0. \tag{1.98}$$

Using (1.93), (1.97) and (1.98) provides:

$$w^* \cdot \phi^n = \partial C^{n*}(y_n, w^*)/\partial y_n|_{y_n=y_{n^*}} > 0. \tag{1.99}$$

Utilising (1.99), use the RHS in (1.96) to solve for θ_n. Finally, use the RHS equalities in (1.94) and (1.95) to solve for β^n and Γ_n, respectively. Using (1.92) and (1.95), $\Gamma_n w^* = 0_3$ so that Γ_n satisfies the restrictions (1.6). QED.

Proof of Proposition 2:   Since both $C^V(y, w)$ and $C^*(y, w)$ are linearly homogeneous in the components of w, the derivatives of these functions satisfy the following restrictions (see Diewert, 1974: 142–6 or Sakai 1974: 264):

$$C^V(y^*, w^*) = w^* \cdot \nabla_w C^V(y^*, w^*)$$
$$C^*(y^*, w^*) = w^* \cdot \nabla_w C^*(y^*, w^*) \tag{1.100}$$

$$\nabla^2_{ww} C^V(y^*, w^*) w^* = 0_3$$
$$\nabla^2_{ww} C^*(y^*, w^*) w^* = 0_3 \tag{1.101}$$

$$w^* \cdot \nabla^2_{wy} C^V(y^*, w^*) = \nabla_y C^V(y^*, w^*)$$
$$w^* \cdot \nabla^2_{wy} C^*(y^*, w^*) = \nabla_y C^*(y^*, w^*) \tag{1.102}$$

Since both $C^V(y, w)$ and $C^*(y, w)$ are linearly homogeneous in y, their derivatives satisfy the following restrictions (see Diewert, 1974: 142–6 or Sakai, 1974: 264):

$$C^V(y^*, w^*) = y^* \cdot \nabla_y C^V(y^*, w^*)$$
$$C^*(y^*, w^*) = y^* \cdot \nabla_y C^*(y^*, w^*) \tag{1.103}$$

$$\nabla^2_{yy} C^V(y^*, w^*) y^* = 0_2$$
$$\nabla^2_{yy} C^*(y^*, w^*) y^* = 0_2 \tag{1.104}$$

$$y^* \cdot \nabla^2_{yw} C^V(y^*, w^*) = \nabla_w C^V(y^*, w^*)$$
$$y^* \cdot \nabla^2_{yw} C^*(y^*, w^*) = \nabla_w C^*(y^*, w^*) \tag{1.105}$$

Let $A = A_L = A_M = A_K$ and $\Gamma = \Gamma_Z = \Gamma_T$. With the above restrictions, to establish the flexibility of $C^V(y, w)$ at (y^*, w^*) vectors ϕ^Z and ϕ^T, matrix Γ satisfying $\Gamma w^* = 0_3$ and matrix A satisfying $Ay^* = 0_2$ need be found such that the following hold:

$$\nabla^2_{wy} C^V(y^*, w^*) = [\phi^Z \; \phi^T] = \nabla^2_{wy} C^*(y_n^*, w^*) \qquad (1.106)$$

$$\nabla^2_{ww} C^V(y^*, w^*) = (y_Z^* + y_T^*) \, \Gamma = \nabla^2_{ww} C^*(y^*, w^*) \qquad (1.107)$$

$$\nabla^2_{yy} C^V(y^*, w^*) = (w_L^* + w_M^* + w_K^*) \, A = \nabla^2_{yy} C^*(y^*, w^*) \qquad (1.108)$$

Equations (1.106) determine ϕ^Z and ϕ^T, (1.107) determine Γ and (1.108) determine A. Using (1.107) and (1.101), gives $0_3 = \nabla^2_{ww} C^*(y^*, w^*)w^* = (y_Z^* + y_T^*) \, \Gamma w^*$ and hence, since $y_Z^* > 0$ and $y_T^* > 0$, $\Gamma w^* = 0_3$. Similarly, using (1.108) and (1.104), gives $0_2 = \nabla^2_{yy} C^*(y^*, w^*)y^* = (w_L^* + w_M^* + w_K^*) \, Ay^*$ and since $w_L + w_M + w_K > 0$, $Ay^* = 0$.

Since $C^*(y, w)$ is concave in w, $\nabla^2_{ww} C^*(y^*, w^*)$ is a negative semi-definite symmetric matrix. Hence, by using Lau's (1978: 427) Theorem (see also Diewert and Wales, 1987: 52–3 and 1988: 335–40), Γ can be represented as $-\Sigma\Sigma^{\#}$ where Σ is lower triangular. Since $\Gamma w^* = 0_3$, it must be $\Sigma^{\#} w^* = 0_3$. QED.

Proof of Proposition 3: Let $y \gg 0_2$, $w \gg 0_3$, $y^1 > 0_2$, $y^2 > 0_2$ and $y^1 + y^2 = y$. Since $\alpha_m \geq 0$, $m = L, M, K$, the matrices A_m defined by (1.35) are positive semi-definite. Then, using Theorem 10 in Diewert and Wales (1987: 66) it can be verified that for $C^V(y, w)$ defined by (1.31), $\nabla^2_{yy} C^V(y, w)$ is a positive semi-definite matrix. Hence, $C^V(y, w)$ is a convex function of y. Convexity of $C^V(y, w)$ in y implies:

$$C^V(\tfrac{1}{2}y^1 + \tfrac{1}{2}y^2, w) \leq \tfrac{1}{2}C^V(y^1, w) + \tfrac{1}{2}C^V(y^2, w) \qquad (1.109)$$

The linear homogeneity of $C^V(y, w)$ in y implies that

$$C^V(\tfrac{1}{2}(y^1 + y^2), w) = \tfrac{1}{2}C^V(y^1 + y^2) \qquad (1.110)$$

Combining (1.109) and (1.110) and multiplying by 2 yields:

$$C^V(y^1 + y^2, w) \leq C^V(y^1, w) + C^V(y^2, w) \qquad (1.111)$$

Thus, C^V is subadditive at (y, w).

If any $\alpha_m > 0$, then it can be verified that the inequality (1.109) is strict except when y^1 is proportional to y^2. If $y^1 = (y_Z, 0)$ and $y^2 = (0, y_T)$ then obviously y^1 and y^2 are not proportional and (1.111) becomes

$$C^V(y_Z + y_T, w) < C^V(y_Z, 0, w) + C^V(0, y_T, w). \qquad (1.112)$$

QED.

Proof of Proposition 4:   Let $y \gg 0_2$, $w \gg 0_3$, $y^1 > 0_2$, $y^2 > 0_2$ and $y^1 + y^2 = y$. From the proof of the previous Proposition, the inequality (1.111) is valid. If $w \cdot \beta \geq 0$ then

$$w \cdot \beta \leq w \cdot \beta + w \cdot \beta \qquad (1.113)$$

Adding the inequalities (1.111) and (1.113) and using (1.48) yields

$$C(y^1 + y^2, w) \leq C(y^1, w) + C(y^2, w) \qquad (1.114)$$

which establishes the subadditivity of C at (y, w). If $w \cdot \beta > 0$, then the inequality (1.113) becomes strict, as does (1.114) which establishes the strict subadditivity of C at (y, w) and the existence of economies of scope for C. QED.

Proof of Proposition 5:   Since both $\tilde{C}^V(y, w, y^*)$ and $C^*(y, w)$ are linearly homogeneous in the components of w, their derivatives satisfy the following:

$$\tilde{C}^V(y^*, w^*, y^*) = w^* \cdot \nabla_w \tilde{C}^V(y^*, w^*, y^*)$$
$$C^*(y^*, w^*) = w^* \cdot \nabla_w C^*(y^*, w^*) \qquad (1.115)$$

$$\nabla^2_{ww} \tilde{C}^V(y^*, w^*, y^*) w^* = 0_3$$
$$\nabla^2_{ww} C^*(y^*, w^*) w^* = 0_3 \qquad (1.116)$$

$$w^* \cdot \nabla^2_{wy} \tilde{C}^V(y^*, w^*, y^*) = \nabla_y \tilde{C}^V(y^*, w^*, y^*)$$
$$w^* \cdot \nabla^2_{wy} C^*(y^*, w^*) = \nabla_y C^*(y^*, w^*) \qquad (1.117)$$

Let $A = A_L = A_M = A_K$ and $\Gamma = \Gamma_Z = \Gamma_T$. In view of the above restrictions, to establish the flexibility of $\tilde{C}^V(y, w, y^*)$ at $(y^* \ w^*)$ requires vectors ϕ^Z and ϕ^T, matrix Γ satisfying $\Gamma w^* = 0_3$ and matrix A are found such that the following hold:

$$\nabla^2_{wy} \tilde{C}^V(y^*, w^*, y^*) = [\phi^Z \ \phi^T] = \nabla^2_{wy} C^*(y_n^*, w^*) \qquad (1.118)$$

$$\nabla^2_{ww} \tilde{C}^V(y^*, w^*, y^*) = (y_Z^* + y_T^*) \Gamma = \nabla^2_{ww} C^*(y^*, w^*) \qquad (1.119)$$

$$\nabla^2_{yy}\tilde{C}^V(y^*, w^*, y^*) = \begin{bmatrix} \theta_Z w \cdot \phi^Z & 0 \\ 0 & \theta_T w \cdot \phi^T \end{bmatrix} + \frac{\sum\limits_{m=L,M,K} w_m^*}{y^* \cdot \zeta} A = \nabla^2_{yy} C^*(y^*, w^*)$$
(1.120)

where

$$A = \bar{\alpha} \begin{bmatrix} y_T^*/y_Z^* & -1 \\ -1 & y_Z^*/y_T^* \end{bmatrix}.$$
(1.121)

Equation (1.118) determines ϕ^Z and ϕ^T, and (1.119) determines Γ. Using (1.119) and (1.116), gives $0_3 = \nabla^2_{ww} C^*(y^*, w^*)w^* = (y_Z^* + y_T^*) \Gamma w^*$ and hence, since $y_Z^* > 0$ and $y_T^* > 0$, $\Gamma w^* = 0_3$. Given (1.121), use either the (1, 2) or the (2, 1) equation in (1.120) to determine $\bar{\alpha}$ and hence A. If the outputs are normal then $\partial^2 C^*(y, w^*)/\partial y_Z \partial y_T|_{y=y^*} \leq 0$ and hence, given (1.121), $\bar{\alpha} \geq 0$. $Ay^* = 0$ by construction. Determine θ_Z by using the (1, 1) equation in (1.120) and determine θ_T by using the (2, 2) equation in (1.120). Since $C^*(y, w)$ is concave in w, $\nabla^2_{ww} C^*(y^*, w^*)$ is a negative semi-definite symmetric matrix. Following the argument in Proposition 2, Γ can be represented as $-\Sigma\Sigma^\#$, where Σ is lower triangular. Further, as $\Gamma w^* = 0_3$ then $\Sigma^\# w^* = 0_3$.

Consider now the additional possibility that $C^*(y, w)$ may be linearly homogeneous in y. Since (1.6) and (1.50) hold in the case of specification (1.51) for $\tilde{C}^V(y, w, y^*)$ and if $C^*(y, w)$ is assumed to be linearly homogeneous in y then their derivatives must also satisfy:

$$\tilde{C}^V(y^*, w^*, y^*) = y^* \cdot \nabla_y \tilde{C}^V(y^*, w^*, y^*)$$
$$C^*(y^*, w^*) = y^* \cdot \nabla_y C^*(y^*, w^*)$$
(1.122)

$$y^* \cdot \nabla^2_{yw} \tilde{C}^V(y^*, w^*, y^*) = \nabla_w \tilde{C}^V(y^*, w^*, y^*)$$
$$y^* \cdot \nabla^2_{yw} C^*(y^*, w^*) = \nabla_w C^*(y^*, w^*)$$
(1.123)

$$\nabla^2_{yy} \tilde{C}^V(y^*, w^*, y^*)y^* = 0_2$$
$$\nabla^2_{yy} C^*(y^*, w^*)y^* = 0_2$$
(1.124)

In view of these restrictions, to establish the ability of $\tilde{C}^V(y, w, y^*)$ to approximate a linearly homogeneous function at (y^*, w^*) it is necessary to demonstrate that there exist parameters for which the first equation in the sets (1.122)–(1.124) holds. The equalities are imposed through $\varepsilon_n = 0$, $n = Z, T$ in (1.49). This enforces $H^n(y_n, y_n^*) = y_n$ and $\theta_n = 0$, for $n = Z, T$. Alternatively, linear homogeneity may be tested for specification (1.50) by free estimation of the ε_n. QED.

Proof of Proposition 6: Let $w \gg 0_3$, $y^* \gg 0_2$ and consider the output vectors $y^1 > 0_2$, $y^2 > 0_2$ such that $y^1 + y^2 = y \gg 0$. Since (1.54)–(1.58) hold for $(\tilde{y}, w, \frac{1}{2}y^*)$ where $0_2 < \tilde{y} \leq y$, $\nabla^2_{yy} \tilde{C}^V(y, w, \frac{1}{2}y^*)$ defined by (1.53) is a positive semi-definite matrix. Hence, $\tilde{C}^V(y, w, \frac{1}{2}y^*)$ is a convex function of y. The convexity of $\tilde{C}^V(y, w, \frac{1}{2}y^*)$ in y implies:

$$\tilde{C}^V(\tfrac{1}{2}y^1 + \tfrac{1}{2}y^2, w, \tfrac{1}{2}y^*) \leq \tfrac{1}{2}\tilde{C}^V(y^1, w, \tfrac{1}{2}y^*) + \tfrac{1}{2}\tilde{C}^V(y^2, w, \tfrac{1}{2}y^*). \quad (1.125)$$

But by (1.51), $\tilde{C}^V(y, w, y^*)$ is linearly homogeneous in (y, y^*). Therefore:

$$\tilde{C}^V(\tfrac{1}{2}(y^1 + y^2), w, \tfrac{1}{2}y^*) = \tfrac{1}{2}\tilde{C}^V(y^1 + y^2, w, y^*). \quad (1.126)$$

Combining (1.109) and (1.110), and multiplying by 2 yields:

$$\tilde{C}^V(y^1 + y^2, w, y^*) \leq \tilde{C}^V(y^1, w, \tfrac{1}{2}y^*) + \tilde{C}^V(y^2, w, \tfrac{1}{2}y^*). \quad (1.127)$$

But $y^1 + y^2 = y$, so defining $y^{1*} = y^{2*} = \frac{1}{2}y^*$, under the conditions of the proposition $\tilde{C}^V$ implies a technology which is subadditive at (y, w, y^*) since it implies

$$\tilde{C}^V(y, w, y^*) \leq \tilde{C}^V(y^1, w, y^{1*}) + \tilde{C}^V(y^2, w, y^{2*}). \quad (1.128)$$

QED.

APPENDIX B

Table 1B.1   US Bell System data

Year	Cost	y_Z	y_T	P_K	P_L	P_M	R&D	cs_K	cs_L	y_{Agg}	x_K	x_L	x_M
1947	2550	0.410	0.346	0.500	0.536	0.670	0.580	0.396	0.496	0.372	2019	2363	411
1948	2994	0.458	0.372	0.559	0.582	0.751	0.555	0.404	0.483	0.411	2166	2483	449
1949	3291	0.487	0.383	0.574	0.610	0.745	0.553	0.419	0.471	0.433	2402	2543	483
1950	3563	0.520	0.416	0.618	0.632	0.765	0.570	0.441	0.454	0.466	2542	2558	491
1951	4047	0.556	0.466	0.700	0.669	0.816	0.596	0.453	0.442	0.508	2620	2674	517
1952	4616	0.592	0.501	0.795	0.710	0.829	0.621	0.467	0.432	0.545	2709	2808	566
1953	4935	0.625	0.523	0.809	0.734	0.844	0.639	0.464	0.436	0.573	2834	2931	581
1954	5258	0.657	0.550	0.813	0.761	0.856	0.651	0.466	0.429	0.604	3015	2960	647
1955	5770	0.703	0.619	0.861	0.807	0.876	0.662	0.478	0.414	0.661	3207	2962	708
1956	6305	0.757	0.684	0.880	0.811	0.905	0.680	0.476	0.411	0.721	3412	3192	788
1957	6351	0.804	0.740	0.820	0.848	0.939	0.714	0.471	0.414	0.773	3651	3097	777
1958	6788	0.842	0.777	0.873	0.851	0.953	0.768	0.508	0.389	0.811	3946	3099	740
1959	7334	0.897	0.863	0.911	0.920	0.974	0.839	0.520	0.373	0.880	4191	2976	801
1960	7912	0.953	0.935	0.957	0.960	0.991	0.919	0.531	0.361	0.945	4390	2974	862
1961	8516	1.000	1.000	1.000	1.000	1.000	1.000	0.544	0.346	1.000	4631	2947	938
1962	9018	1.054	1.082	1.015	1.036	1.020	1.085	0.551	0.340	1.067	4895	2955	968
1963	9508	1.111	1.175	1.008	1.074	1.034	1.190	0.551	0.334	1.140	5199	2952	1058
1964	10542	1.159	1.317	1.078	1.130	1.085	1.328	0.562	0.327	1.228	5499	3050	1075
1965	11207	1.228	1.474	1.061	1.171	1.107	1.500	0.553	0.329	1.336	5837	3150	1193
1966	11954	1.306	1.684	1.045	1.228	1.141	1.669	0.543	0.337	1.472	6213	3279	1257
1967	12710	1.383	1.843	1.041	1.297	1.174	1.868	0.541	0.341	1.585	6605	3337	1284
1968	13814	1.466	2.055	1.083	1.361	1.220	2.027	0.546	0.334	1.724	6964	3391	1357
1969	14940	1.559	2.334	1.046	1.494	1.283	2.163	0.514	0.358	1.899	7343	3579	1490
1970	16516	1.639	2.537	1.049	1.624	1.352	2.284	0.498	0.371	2.032	7841	3776	1596
1971	17951	1.710	2.698	1.041	1.804	1.420	2.400	0.483	0.383	2.141	8334	3811	1691
1972	20161	1.805	2.969	1.092	2.062	1.477	2.521	0.480	0.391	2.312	8856	3818	1773
1973	21190	1.912	3.316	1.003	2.263	1.562	2.655	0.446	0.414	2.521	9412	3880	1899
1974	23168	2.008	3.605	1.001	2.516	1.741	2.805	0.434	0.425	2.698	10046	3911	1877
1975	27376	2.075	3.864	1.189	2.855	1.913	2.972	0.462	0.406	2.842	10628	3894	1891
1976	31304	2.173	4.244	1.328	3.219	2.014	3.151	0.470	0.395	3.054	11075	3841	2100
1977	34745	2.292	4.685	1.419	3.407	2.129	3.334	0.467	0.393	3.304	11434	4003	2289

Table 1B.2　Constructed US Bell System-compatible data

										Revenue Shares			
Year	p_Z	p_T	R_Z	R_T	y_Z	y_T	R	C	Profit	rs_L	rs_M	rs_K	rs_{Profit}
1947	0.815	1.183	2456	2253	3013	1905	4710	2550	2159	0.269	0.059	0.214	0.458
1948	0.915	1.084	3078	2217	3364	2046	5296	2994	2301	0.273	0.064	0.229	0.435
1949	0.956	1.043	3422	2197	3578	2106	5620	3291	2329	0.276	0.064	0.246	0.414
1950	0.960	1.040	3668	2378	3821	2288	6046	3563	2483	0.267	0.062	0.260	0.411
1951	0.949	1.050	3876	2688	4082	2560	6565	4047	2517	0.273	0.064	0.280	0.384
1952	0.964	1.035	4191	2853	4346	2756	7045	4616	2429	0.283	0.067	0.306	0.345
1953	0.992	1.008	4553	2897	4589	2875	7451	4935	2516	0.289	0.066	0.308	0.338
1954	1.006	0.994	4854	3007	4825	3025	7862	5258	2603	0.287	0.070	0.312	0.331
1955	0.995	1.005	5140	3423	5164	3407	8563	5770	2793	0.279	0.072	0.322	0.326
1956	1.012	0.989	5622	3719	5558	3762	9342	6305	3036	0.277	0.076	0.322	0.325
1957	1.021	0.979	6030	3985	5904	4071	10016	6351	3664	0.262	0.073	0.299	0.366
1958	1.043	0.957	6457	4089	6188	4272	10546	6788	3758	0.250	0.067	0.327	0.356
1959	1.020	0.980	6720	4651	6588	4746	11372	7334	4037	0.241	0.069	0.336	0.355
1960	1.044	0.957	7309	4922	7003	5144	12231	7912	4319	0.233	0.070	0.344	0.353
1961	1.000	1.000	7348	5501	7348	5501	12849	8516	4332	0.229	0.073	0.360	0.337
1962	1.058	0.943	8194	5613	7745	5954	13807	9018	4789	0.222	0.072	0.360	0.347
1963	1.116	0.886	9107	5721	8161	6461	14828	9508	5320	0.214	0.074	0.354	0.359
1964	1.144	0.858	9740	6218	8517	7245	15958	10542	5415	0.216	0.073	0.372	0.339
1965	1.136	0.866	10254	7019	9025	8110	17273	11207	6066	0.214	0.076	0.359	0.351
1966	1.132	0.870	10863	8058	9597	9265	18921	11954	6967	0.213	0.076	0.343	0.368
1967	1.128	0.874	11460	8858	10163	10136	20319	12710	7608	0.213	0.074	0.338	0.374
1968	1.128	0.874	12143	9880	10770	11305	22024	13814	8210	0.210	0.075	0.343	0.373
1969	1.128	0.874	12916	11221	11453	12841	24138	14940	9197	0.222	0.079	0.318	0.381
1970	1.129	0.873	13597	12177	12043	13955	25774	16516	9257	0.238	0.084	0.319	0.359
1971	1.130	0.871	14198	12930	12562	14840	27129	17951	9177	0.253	0.089	0.320	0.338
1972	1.133	0.868	15027	14183	13260	16334	29211	20161	9050	0.270	0.090	0.331	0.310
1973	1.137	0.865	15974	15776	14050	18243	31750	21190	10559	0.277	0.093	0.297	0.333
1974	1.140	0.862	16819	17089	14754	19832	33908	23168	10740	0.290	0.096	0.297	0.317
1975	1.147	0.855	17491	18171	15249	21257	35662	27376	8285	0.312	0.101	0.355	0.232
1976	1.153	0.848	18417	19810	15968	23349	38228	31304	6924	0.324	0.111	0.385	0.181
1977	1.158	0.844	19492	21758	16838	25770	41250	34745	6505	0.331	0.118	0.393	0.158

Explanatory Notes for Constructed Data

Initially assume price relatives satisfy $\tilde{p}_Z y_Z^{index} + \tilde{p}_T y_T^{index} = y_{Agg}^{index}$ and $\tilde{p}_Z + \tilde{p}_T = 1$.

Construct initial price relatives as: $\tilde{p}_Z(t) = \begin{cases} \dfrac{y_{Agg}^{index}(t) - y_T^{index}(t)}{y_Z^{index}(t) - y_T^{index}(t)}, & t \neq 1961 \\ [\tilde{p}_Z(1960) + \tilde{p}_Z(1962)]/2, & t = 1961 \end{cases}$ and

$\tilde{p}_T(t) = 1 - \tilde{p}_Z(t)$. Construct initial price indices as scaled variants of the initial price relatives $p_n^{(1)} = \tilde{p}_n/\tilde{p}_n(1961)$, $n = Z,T$. Hence construct revenue indices as $R_n^{index} = p_n^{(1)} y_n^{index}$, $n = Z, T$. Next, extrapolate 1970–1977 trend in revenue indices to 1979. From linear regressions of R_n^{index} against time: $\hat{R}_Z^{index} = 1.693 + 0.106$ $(t - 1970) \Rightarrow \hat{R}_Z^{index}(1979) = 2.648$ and $\hat{R}_T^{index} = 2.310 + 0.269 \, (t - 1970) \Rightarrow \hat{R}_T^{index}(1979) = 4.731$. Evans and Heckman (1983: 273) suggest $R_Z(1979) = 21\,000$, $R_T(1979) = 24\,000$. Hence construct $R_n^{(0)} = R_n^{index} \left[\dfrac{R_n(1979)}{\hat{R}_n^{index}(1979)} \right]$, $n = Z, T$, and

$y_n^{(1)} = R_n^{(0)}/p_n^{(1)}$, $n = Z, T$. Construct initial total revenue: $R^{(0)} = R_Z^{(0)} R_T^{(0)}$ and hence initial total revenue index: $R^{index}(t) = R^{(0)}(t)/R^{(0)}(1961)$. Thence construct initial aggregate price index: $p_{Agg}^{(1)}(t) = R^{index}(t)/y_{Agg}^{index}(t)$. Next, examine the relationship between $p_Z^{(1)}$, $p_T^{(1)}$ and $p_{Agg}^{(1)}$ by regression and modify initial price relative assumption based on this. The regression for 1947–1977 implies $\hat{p}_{Agg}^{(1)} = 0.490197 \, p_Z^{(1)} + 0.496212$ $p_T^{(1)}$. Replace initial price relative assumption by: $\tilde{p}_Z^{(2)} y_Z^{index} + \tilde{p}_T^{(2)} y_T^{index} = p_{Agg}^{(1)} y_{Agg}^{index}$ and $0.490197 \, \tilde{p}_Z^{(2)} + 0.496212 \, \tilde{p}_T^{(2)} = p_{Agg}^{(1)}$, implying

$$\frac{\tilde{p}_Z^{(2)}}{p_{Agg}^{(1)}} = \frac{y_{Agg}^{index} - \left[\dfrac{0.490197 + 0.496212}{2(0.496212)} \right] y_T^{index}}{y_Z^{index} - \left[\dfrac{0.490197}{0.496212} \right] y_T^{index}} \quad \text{and} \quad \frac{\tilde{p}_T^{(2)}}{p_{Agg}^{(1)}} = \frac{0.490197 + 0.496211}{2(0.496212)} - \frac{\tilde{p}_Z^{(2)}}{p_{Agg}^{(1)}}.$$

Construct revised price indices as $p_n^{(2)} = [\tilde{p}_n^{(2)}/p_{Agg}^{(1)}]/[(\tilde{p}_n^{(2)}/p_{Agg}^{(1)}) \, (1961)]$, $n = Z, T$. Revised revenue measures are: $R_n^{(1)} = p_n^{(2)} y_n^{(1)}$, $n = Z, T$. Check extrapolated values of revised revenue measures: From linear regressions of $R_n^{(1)}$ against time: $\hat{R}_Z^{(1)} = 14500.71 + 907.08 \, (t - 1970) \Rightarrow \hat{R}_Z^{(1)}(1979) = 22,664.46$ and $\hat{R}_T^{(1)} = 10795.39 + 1259.72$ $(t - 1970) \Rightarrow \hat{R}_T^{(1)}(1979) = 22,132.82$. Given $R_Z(1979) = 21,000$ and $R_T(1979) = 24,000$, construct $R_n^{(2)} = R_n^{(1)} \left[\dfrac{R_n(1979)}{\hat{R}_n^{(1)}(1979)} \right]$, $n = Z, T$. The finally constructed output measures are: $y_n^{(2)} = R_n^{(2)}/p_n^{(2)}$, $n = Z, T$. Table B-2 contains $p_Z^{(2)}$, $p_T^{(2)}$, $R_Z^{(2)}$, $R_T^{(2)}$, $y_Z^{(2)}$, $y_T^{(2)}$, $R^{(2)} \equiv R_Z^{(2)} + R_T^{(2)}$, C, Profit $\equiv R^{(2)} - C$, and revenue shares defined as $rs_m \equiv w_m x_m / R^{(2)}$, $m = L, M, K$ and $rs_{Profit} = 1 - \displaystyle\sum_{m=L,M,K} rs_m$.

APPENDIX C

Table 1C.1   Estimation results

Parameter	Estimate	t-ratio
μ_L	0.0117	1.1134
μ_M	−0.0067	−3.0307
μ_K	−0.0247	−3.3706
ϕ_L^Z	0.0309	7.0595
ϕ_M^Z	0.0865	13.5520
ϕ_K^Z	0.5078	9.8011
ϕ_L^T	0.1402	2.6774
ϕ_M^T	0.0458	6.6340
ϕ_K^T	−0.0885	−4.3121
α_L	0.0001	2.3462
α_M	−0.0002	−8.6123
α_K	−0.0002	−22.2780
σ_{LL}^Z	−0.1649	−7.4026
σ_{ML}^Z	0.2244	1.2046
σ_{MM}^Z	0.0239	0.9280
σ_{LL}^T	−0.2274	−9.6105
σ_{ML}^T	0.1906	7.6473
σ_{MM}^T	0.1584	8.1163
ε_Z	0.1286	6.3881
ε_T	0.0536	4.2167
$\sqrt{(1/\zeta_Z) - 1}$	19.3300	17.7150
$\sqrt{\psi_Z}$	3.1993	2.9019
$\sqrt{\psi_T}$	2.2150	1.9119
τ_L	0.0572	15.8910
τ_M	−0.0104	−6.8492
τ_K	−0.0241	−4.5808

Table 1C.1 *(cont.)*

Implied parameters

δ_Z	δ_T	ζ_Z	ζ_T	ψ_Z	ψ_T
0.986	0.997	0.003	0.997	10.236	4.906
Γ_Z	Γ_T				
−0.0272	0.0037	0.0235	−0.0517	0.0434	0.0084
0.0037	−0.0011	−0.0026	0.0434	−0.0614	0.0181
0.0235	−0.0026	−0.0209	0.0084	0.0181	−0.0265
θ_Z	θ_T	β_L	β_M	β_K	
−0.000003	−0.0000009	73.364	55.488	1326.786	

Table 1C.2   Parameters, formulas and nomenclature

$\delta_n = (\varepsilon_n + 1)/(\varepsilon_n^2 + \varepsilon_n + 1)$, $n = Z, T$, where ε_Z and ε_T are freely estimated

$\zeta_T = 1 - \zeta_Z$, where ζ_Z is estimated subject to $0 < \zeta_Z < 1$

$$\Gamma_n = -\Sigma_n \Sigma_n^{\#} \text{ where } \Sigma_n = \begin{bmatrix} \sigma_{LL}^n & 0 & 0 \\ \sigma_{ML}^n & \sigma_{MM}^n & 0 \\ -(\sigma_{LL}^n + \sigma_{ML}^n) & -\sigma_{MM}^n & 0 \end{bmatrix}, n = Z, T \text{ where}$$

$\sigma_{LL}^n, \sigma_{ML}^n, \sigma_{MM}^n, n = Z, T$ are freely estimated

$\beta_m = \mu_m [\phi_m^Z \delta_Z y_Z^* + \phi_m^T \delta_T y_T^* + \alpha_m \delta_Z \delta_T y_Z^* y_T^*]$,
$\mu_m, \phi_m^n, \alpha_m, n = Z, T, m = L, M, K$ are freely estimated

$\theta_n = -2\varepsilon_n^2/[(1 + \varepsilon_n)^2 \, y_n^*]$, $n = Z, T$, second-derivatives of non-linear output functions at reference output level

$$H^n(y_n, y_n^*) = \left\{ \frac{\varepsilon_n y_n}{\varepsilon_n y_n + y_n^*} + \frac{\varepsilon_n(y_n^* - y_n) + y_n^*}{\varepsilon_n[\varepsilon_n(y_n^* - y_n) + y_n^*] + y_n^*} \right\} y_n, n = Z, T,$$

non-linear output functions

$Q(y, y^*) = \frac{1}{2}H(y, y^*) \cdot X(y^*)H(y, y^*)/J(y^*) \equiv Q$, quadratic output term in cost function

$$X(y^*) = \begin{bmatrix} y_T^*/y_Z^* & -1 \\ -1 & y_Z^*/y_T^* \end{bmatrix} \text{ with } X(y^*)y^* = 0, \text{ negative semi-definite matrix}$$

constrained to rank 1 at reference output

$J(y^*) = [(y_Z^*)^{\zeta_Z} + \psi_Z][(y_T^*)^{\zeta_T} + \psi_T]$, $\zeta_Z + \zeta_T = 1$, normalizing term *HD1* in y^* if $\psi_Z = \psi_T = 0$

$\tilde{w}_m = tech_m \, w_m$ where $tech_m = (1 + \tau_m)^{1961 - year}$, $m = L, M, K$, factor price adjustment for technical progress

$S^n(\tilde{w}) = \frac{1}{2}(\tilde{w} \cdot \eta)^{-1}\tilde{w} \cdot \Gamma_n \tilde{w} \equiv S^n$, $n = Z, T$, quadratic price terms in cost function

---

Table 1C.3   Estimating equations, underlying cost function and components

Cost function

$$\tilde{C}(y, \tilde{w}, y^*) = \tilde{w} \cdot \beta + [\tilde{w} \cdot \phi^Z + S^Z(\tilde{w})] H^Z(y_Z, y_Z^*) + [\tilde{w} \cdot \phi^T + S^T(\tilde{w})] H^T(y_T, y_T^*)$$
$$+ \tilde{w} \cdot \alpha Q(y, y^*)$$

where $\tilde{w} \cdot \beta \equiv \displaystyle\sum_{m=L,M,K} \tilde{w}_m \beta_m \equiv \tilde{C}^F(\tilde{w}, y^*)$, $\tilde{w} \cdot \phi^n \equiv \displaystyle\sum_{m=L,M,K} \tilde{w}_m \phi_m^n, n = Z, T$

and $\tilde{w} \cdot \alpha \equiv \displaystyle\sum_{m=L,M,K} \tilde{w}_m \alpha_m$

Estimating equations

$$rs_L = \frac{\tilde{w}_L}{R} \left\{ \begin{array}{l} \beta_L + [\phi_L^Z + \frac{1}{2}(\tilde{w} \cdot \eta)^{-1} \underline{\gamma}_L^Z \cdot \tilde{w} - (w \cdot \eta)^{-1} \eta_L S^Z(\tilde{w})] H^Z(y_Z, y_Z^*) \\ + [\phi_L^T + \frac{1}{2}(\tilde{w} \cdot \eta)^{-1} \underline{\gamma}_L^T \cdot \tilde{w} - (w \cdot \eta)^{-1} \eta_L S^T(\tilde{w})] H^T(y_T, y_T^*) + \alpha_L Q(y, y^*) \end{array} \right\}$$

Labour demand equation as share of total revenue

$$rs_M = \frac{\tilde{w}_M}{R} \left\{ \begin{array}{l} \beta_M + [\phi_M^Z + \frac{1}{2}(\tilde{w} \cdot \eta)^{-1} \underline{\gamma}_M^Z \cdot \tilde{w} - (w \cdot \eta)^{-1} \eta_M S^Z(\tilde{w})] H^Z(y_Z, y_Z^*) \\ + [\phi_m^T + \frac{1}{2}(\tilde{w} \cdot \eta)^{-1} \underline{\gamma}_M^T \cdot \tilde{w} - (w \cdot \eta)^{-1} \eta_M S^T(\tilde{w})] H^T(y_T, y_T^*) \\ + \alpha_M Q(y, y^*)\} \end{array} \right\}$$

Materials demand equation as share of total revenue

$$rs_K = \frac{\tilde{w}_K}{R} \left\{ \begin{array}{l} \beta_K + [\phi_K^Z + \frac{1}{2}(\tilde{w} \cdot \eta)^{-1} \underline{\gamma}_K^Z \cdot \tilde{w} - (w \cdot \eta)^{-1} \eta_K S^Z(\tilde{w})] H^Z(y_Z, y_Z^*) \\ + [\phi_K^T + \frac{1}{2}(\tilde{w} \cdot \eta)^{-1} \underline{\gamma}_K^T \cdot \tilde{w} - (w \cdot \eta)^{-1} \eta_K S^T(\tilde{w})] H^T(y_T, y_T^*) + \alpha_T Q(y, y^*) \end{array} \right\}$$

Capital demand equation as share of total revenue

where $rs_M = w_m x_m / R, m = L, M, K$

mth factor share as a proportion of total revenue

$$R \equiv \sum_{n=Z,T} p_n y_n$$

Definition of total revenue

$\underline{\gamma}_M^n \equiv (\gamma_{mL}^n \, \gamma_{mM}^n \, \gamma_{mK}^n), n = Z, T, m = L, M, K$

m^{th} row of $\Gamma_n, n = Z, T, m = L, M, K$

Table 1C.4   Local output and nonlinear output functions

Year	y_Z	$H^Z(y_Z, y_Z{}^*)$	$\partial H^Z/\partial y_Z$	$\partial^2 H^Z/\partial y_Z{}^2$
1947	3013	2999	1.002	0.000002
1948	3364	3350	1.002	0.000002
1949	3578	3565	1.003	0.000001
1950	3821	3809	1.003	0.000001
1951	4082	4071	1.004	0.000001
1952	4346	4336	1.004	0.000000
1953	4589	4580	1.004	0.000000
1954	4825	4817	1.004	−0.000000
1955	5164	5158	1.004	−0.000000
1956	5558	5554	1.004	−0.000001
1957	5904	5901	1.003	−0.000001
1958	6188	6186	1.003	−0.000001
1959	6588	6587	1.002	−0.000002
1960	7003	7003	1.001	−0.000003
1961	7348	7348	1.000	−0.000003
1962	7745	7745	0.998	−0.000004
1963	8161	8160	0.996	−0.000004
1964	8517	8514	0.994	−0.000005
1965	9025	9019	0.992	−0.000005
1966	9597	9586	0.988	−0.000006
1967	10164	10147	0.985	−0.000006
1968	10771	10741	0.980	−0.000007
1969	11455	11409	0.975	−0.000008
1970	12046	11983	0.970	−0.000008
1971	12562	12485	0.965	−0.000009
1972	13260	13156	0.958	−0.000010
1973	14050	13911	0.950	−0.000010
1974	14754	14576	0.942	−0.000011
1975	15249	15042	0.936	−0.000012
1976	15968	15712	0.927	−0.000012
1977	16838	16514	0.916	−0.000013

Table 1C.5   Toll output and nonlinear output functions

Year	y_T	$H^T(y_T, y_T{}^*)$	$\partial H^T/\partial y_T$	$\partial^2 H^T/\partial y_T{}^2$
1947	1905	1903	1.000	0.0000008
1948	2046	2044	1.000	0.0000008
1949	2106	2104	1.000	0.0000008
1950	2288	2285	1.000	0.0000007
1951	2560	2558	1.000	0.0000005
1952	2756	2755	1.001	0.0000004
1953	2875	2873	1.001	0.0000004
1954	3025	3024	1.001	0.0000003
1955	3407	3406	1.001	0.0000001
1956	3762	3761	1.001	−0.0000000
1957	4071	4070	1.001	−0.0000000
1958	4272	4271	1.001	−0.0000000
1959	4746	4745	1.001	−0.0000000
1960	5144	5144	1.000	−0.0000000
1961	5501	5501	1.000	−0.0000000
1962	5954	5953	0.999	−0.0000001
1963	6461	6460	0.998	−0.0000001
1964	7245	7244	0.997	−0.0000001
1965	8110	8106	0.995	−0.0000002
1966	9265	9255	0.993	−0.0000002
1967	10137	10119	0.990	−0.0000003
1968	11305	11274	0.986	−0.0000003
1969	12841	12786	0.980	−0.0000004
1970	13955	13876	0.976	−0.0000004
1971	14840	14738	0.972	−0.0000004
1972	16334	16184	0.964	−0.0000005
1973	18243	18015	0.953	−0.0000006
1974	19832	19522	0.943	−0.0000006
1975	21257	20861	0.934	−0.0000006
1976	23349	22799	0.918	−0.0000007
1977	25770	25001	0.899	−0.0000008

Table 1C.6   Quadratic output terms, technical progress by factor

Year	$\alpha_L Q$	$\alpha_M Q$	$\alpha_K Q$	$tech_L$	$tech_M$	$tech_K$
1947	0.000	0.000	0.000	2.18	0.86	0.71
1948	0.000	−0.001	−0.001	2.06	0.87	0.72
1949	0.000	−0.001	−0.001	1.95	0.88	0.74
1950	0.000	−0.001	−0.001	1.84	0.89	0.76
1951	0.000	−0.001	−0.001	1.74	0.90	0.78
1952	0.000	−0.001	−0.001	1.65	0.91	0.80
1953	0.000	−0.001	−0.001	1.56	0.91	0.82
1954	0.000	−0.001	−0.001	1.47	0.92	0.84
1955	0.000	−0.001	0.000	1.39	0.93	0.86
1956	0.000	0.000	0.000	1.32	0.94	0.88
1957	0.000	0.000	0.000	1.24	0.95	0.90
1958	0.000	0.000	0.000	1.18	0.96	0.92
1959	0.000	0.000	0.000	1.11	0.97	0.95
1960	0.000	0.000	0.000	1.05	0.99	0.97
1961	0.000	0.000	0.000	1.00	1.00	1.00
1962	0.000	0.000	0.000	0.94	1.01	1.02
1963	0.000	0.000	0.000	0.89	1.02	1.05
1964	0.001	−0.002	−0.002	0.84	1.03	1.07
1965	0.002	−0.005	−0.004	0.80	1.04	1.10
1966	0.004	−0.011	−0.010	0.75	1.05	1.13
1967	0.006	−0.016	−0.015	0.71	1.06	1.15
1968	0.010	−0.027	−0.025	0.67	1.07	1.18
1969	0.016	−0.046	−0.043	0.64	1.08	1.21
1970	0.022	−0.061	−0.057	0.60	1.09	1.24
1971	0.026	−0.074	−0.069	0.57	1.11	1.27
1972	0.036	−0.102	−0.095	0.54	1.12	1.30
1973	0.053	−0.147	−0.137	0.51	1.13	1.34
1974	0.067	−0.189	−0.176	0.48	1.14	1.37
1975	0.084	−0.235	−0.219	0.45	1.15	1.40
1976	0.111	−0.311	−0.289	0.43	1.17	1.44
1977	0.145	−0.408	−0.379	0.41	1.18	1.48

Table 1C.7   Linear and quadratic price components of cost

Year	$\tilde{w}\cdot\beta$	$\tilde{w}\cdot\phi^Z$	S^Z	$\tilde{w}\cdot\phi^T$	S^T	$\tilde{w}\cdot\alpha$
1947	588	0.59	−0.013	0.15	−0.016	0.000
1948	664	0.63	−0.011	0.16	−0.014	0.000
1949	692	0.64	−0.010	0.15	−0.013	0.000
1950	750	0.65	−0.008	0.15	−0.010	0.000
1951	854	0.70	−0.006	0.14	−0.008	0.000
1952	974	0.75	−0.004	0.14	−0.006	0.000
1953	1009	0.75	−0.004	0.13	−0.005	0.000
1954	1035	0.76	−0.003	0.13	−0.004	0.000
1955	1114	0.79	−0.002	0.13	−0.003	0.000
1956	1160	0.80	−0.001	0.12	−0.002	0.000
1957	1114	0.78	−0.001	0.12	−0.001	0.000
1958	1201	0.80	0.000	0.11	0.000	0.000
1959	1278	0.84	0.000	0.11	0.000	0.000
1960	1368	0.87	0.000	0.10	0.000	0.000
1961	1455	0.90	0.000	0.09	0.000	0.000
1962	1508	0.92	0.000	0.09	0.000	0.000
1963	1533	0.92	0.000	0.08	0.000	0.000
1964	1671	0.98	0.000	0.08	−0.001	0.000
1965	1685	0.98	−0.001	0.08	−0.001	0.000
1966	1701	0.99	−0.001	0.08	−0.002	0.000
1967	1735	1.00	−0.001	0.08	−0.002	0.000
1968	1845	1.05	−0.002	0.07	−0.003	−0.001
1969	1834	1.06	−0.001	0.08	−0.004	−0.001
1970	1888	1.09	−0.001	0.09	−0.005	−0.001
1971	1925	1.13	−0.001	0.10	−0.006	−0.001
1972	2068	1.21	−0.001	0.10	−0.005	−0.001
1973	1967	1.19	−0.001	0.12	−0.007	−0.001
1974	2024	1.24	−0.001	0.14	−0.012	−0.001
1975	2439	1.44	−0.002	0.13	−0.013	−0.001
1976	2773	1.60	−0.003	0.13	−0.013	−0.001
1977	3025	1.71	−0.004	0.12	−0.016	−0.001

Table 1C.8   Actual and estimated factor demand

Year	x_L	$\hat{x}_L$	x_M	$\hat{x}_M$	x_K	$\hat{x}_K$
1947	2363	2428	411	421	2019	2027
1948	2483	2562	449	451	2166	2195
1949	2543	2567	484	474	2402	2324
1950	2558	2620	491	502	2542	2454
1951	2674	2702	517	533	2620	2585
1952	2808	2753	566	567	2709	2724
1953	2931	2759	581	592	2834	2877
1954	2960	2758	647	619	3015	3034
1955	2962	2838	708	670	3207	3220
1956	3192	2940	788	710	3412	3434
1957	3097	2967	777	748	3651	3669
1958	3099	2988	740	774	3946	3848
1959	2976	3046	801	837	4191	4089
1960	2974	3102	862	893	4390	4342
1961	2947	3117	938	943	4631	4571
1962	2955	3151	968	995	4895	4836
1963	2952	3181	1058	1051	5199	5120
1964	3050	3237	1075	1112	5499	5338
1965	3150	3314	1193	1183	5837	5662
1966	3279	3422	1257	1265	6213	6012
1967	3337	3477	1284	1339	6605	6397
1968	3391	3574	1357	1426	6964	6768
1969	3579	3674	1490	1514	7343	7222
1970	3776	3709	1596	1579	7841	7661
1971	3811	3681	1691	1641	8334	8112
1972	3818	3708	1773	1795	8856	8579
1973	3880	3788	1899	1838	9412	9225
1974	3911	3849	1877	1821	10046	9862
1975	3894	3842	1891	1944	10628	10154
1976	3841	3870	2100	2136	11075	10555
1977	4003	3967	2289	2239	11434	11094

Table 1C.9   Estimated fixed cost factor

Year	$\tilde{w}_L\beta_L$	$\tilde{w}_M\beta_M$	$\tilde{w}_K\beta_K$	$\hat{C}^F$	$\equiv \tilde{w}\cdot\beta$
1947	85.6	32.0	470.9	588	588
1948	88.1	36.3	539.8	664	664
1949	87.2	36.4	568.6	692	692
1950	85.4	37.8	627.0	750	750
1951	85.6	40.7	727.9	854	854
1952	85.9	41.8	846.8	974	974
1953	84.0	43.0	882.4	1009	1009
1954	82.4	44.1	908.9	1035	1035
1955	82.6	45.6	986.3	1114	1114
1956	78.5	47.6	1033.8	1160	1160
1957	77.7	49.9	986.7	1114	1114
1958	73.7	51.2	1076.5	1201	1201
1959	75.4	52.9	1150.5	1278	1278
1960	74.4	54.3	1239.5	1368	1368
1961	73.3	55.4	1326.7	1455	1455
1962	71.9	57.1	1379.4	1508	1508
1963	70.4	58.5	1404.7	1533	1533
1964	70.1	62.1	1538.8	1671	1671
1965	68.7	64.0	1552.6	1685	1685
1966	68.2	66.7	1566.0	1701	1701
1967	68.1	69.3	1598.3	1735	1735
1968	67.6	72.8	1705.0	1845	1845
1969	70.2	77.4	1686.6	1834	1834
1970	72.1	82.4	1733.4	1888	1888
1971	75.8	87.5	1762.1	1925	1925
1972	82.0	91.9	1894.2	2068	2068
1973	85.1	98.3	1783.6	1967	1967
1974	89.5	110.7	1823.8	2024	2024
1975	96.0	123.0	2220.6	2439	2439
1976	102.4	130.8	2540.2	2773	2773
1977	102.5	139.8	2782.6	3025	3025

Table 1C.10   Output price, estimated marginal cost, markup

Year	p_Z	p_T	MC_Z	MC_T	$\dfrac{p_Z}{MC_Z}$	$\dfrac{p_T}{MC_T}$
1947	0.81	1.18	0.58	0.14	1.40	8.29
1948	0.91	1.08	0.62	0.14	1.46	7.28
1949	0.95	1.04	0.63	0.14	1.50	7.12
1950	0.96	1.04	0.65	0.14	1.46	7.28
1951	0.94	1.05	0.70	0.14	1.35	7.43
1952	0.96	1.03	0.75	0.13	1.28	7.59
1953	0.99	1.00	0.75	0.13	1.30	7.59
1954	1.00	0.99	0.76	0.13	1.31	7.66
1955	0.99	1.00	0.79	0.12	1.24	7.91
1956	1.01	0.98	0.80	0.11	1.25	8.30
1957	1.02	0.97	0.78	0.12	1.30	7.97
1958	1.04	0.95	0.80	0.11	1.29	8.61
1959	1.02	0.98	0.84	0.11	1.21	8.83
1960	1.04	0.95	0.87	0.10	1.19	9.15
1961	1.00	1.00	0.90	0.09	1.10	10.24
1962	1.05	0.943	0.91	0.09	1.15	10.18
1963	1.11	0.88	0.92	0.08	1.20	9.93
1964	1.14	0.858	0.97	0.08	1.17	10.48
1965	1.13	0.86	0.97	0.07	1.16	10.90
1966	1.13	0.87	0.97	0.07	1.15	11.03
1967	1.12	0.87	0.99	0.07	1.13	11.22
1968	1.12	0.87	1.02	0.07	1.09	12.24
1969	1.12	0.87	1.03	0.01	1.00	10.90
1970	1.12	0.87	1.06	0.08	1.06	10.47
1971	1.13	0.87	1.09	0.09	1.03	9.54
1972	1.13	0.86	1.16	0.09	0.97	8.91
1973	1.13	0.86	1.13	0.11	1.00	7.72
1974	1.14	0.86	1.17	0.12	0.97	7.08
1975	1.14	0.85	1.35	0.11	0.84	7.38
1976	1.15	0.84	1.49	0.11	0.77	7.60
1977	1.15	0.84	1.56	0.09	0.73	8.56

Table 1C.11   Estimated cost, technical progress, returns to scale

Year	$\tilde{w} \cdot \beta$	C	$\hat{C}$	$\hat{C}_{tech(-1)}$	*TECH*	*RS*
1947	588	2550	2596	2643	0.018	1.28
1948	664	2994	3058	3110	0.017	1.27
1949	692	3291	3254	3307	0.017	1.26
1950	750	3563	3556	3611	0.015	1.25
1951	854	4047	4054	4109	0.014	1.25
1952	974	4616	4589	4644	0.012	1.26
1953	1009	4935	4852	4907	0.011	1.25
1954	1035	5258	5096	5151	0.011	1.24
1955	1114	5770	5647	5705	0.011	1.23
1956	1160	6305	6050	6107	0.009	1.23
1957	1114	6351	6229	6293	0.011	1.21
1958	1201	6788	6640	6697	0.009	1.21
1959	1278	7334	7341	7403	0.008	1.20
1960	1368	7912	8020	8081	0.008	1.20
1961	1455	8516	8632	8691	0.007	1.20
1962	1508	9018	9188	9246	0.006	1.19
1963	1533	9508	9667	9726	0.006	1.19
1964	1671	10542	10618	10676	0.005	1.19
1965	1685	11207	11201	11264	0.006	1.18
1966	1701	11954	11928	12002	0.006	1.17
1967	1735	12710	12739	12820	0.006	1.17
1968	1845	13814	13935	14019	0.006	1.17
1969	1834	14940	14986	15098	0.007	1.16
1970	1888	16516	16196	16325	0.008	1.16
1971	1925	17951	17414	17567	0.009	1.15
1972	2068	20161	19664	19848	0.009	1.15
1973	1967	21190	20700	20937	0.011	1.15
1974	2024	23168	22728	23011	0.012	1.15
1975	2439	27376	26766	27064	0.011	1.15
1976	2773	31304	30778	31109	0.011	1.16
1977	3025	34745	34032	34376	0.011	1.17

APPENDIX D

Table 1D.1   Experiment 1 subadditivity test

Year	C	$\hat{C}$	$\hat{C}_{sub}$	$\hat{C}_A$	$\hat{C}_B$	SUB
1947	2550	2596	2291	20	2270	0.117
1948	2994	3058	2710	76	2633	0.114
1949	3291	3254	2889	108	2780	0.112
1950	3563	3556	3157	154	3002	0.112
1951	4047	4054	3593	219	3374	0.114
1952	4616	4589	4058	282	3776	0.116
1953	4935	4852	4300	328	3971	0.114
1954	5258	5096	4527	374	4153	0.112
1955	5770	5647	5033	464	4568	0.109
1956	6305	6050	5408	545	4862	0.106
1957	6351	6229	5613	612	5001	0.099
1958	6788	6640	5972	669	5303	0.101
1959	7334	7341	6628	787	5840	0.097
1960	7912	8020	7254	896	6357	0.095
1961	8516	8632	7815	990	6824	0.095
1962	9018	9188	8339	1091	7248	0.092
1963	9508	9667	8803	1189	7614	0.089
1964	10542	10618	9674	1332	8341	0.089
1965	11207	11201	10248	1462	8786	0.085
1966	11954	11928	10966	1626	9339	0.081
1967	12710	12739	11756	1789	9966	0.077
1968	13814	13935	12886	1991	10895	0.075
1969	14940	14986	13942	2252	11697	0.070
1970	16516	16196	15120	2497	12622	0.066
1971	17951	17414	16315	2759	13555	0.063
1972	20161	19664	18477	3191	15286	0.060
1973	21190	20700	19566	3542	16024	0.055
1974	23168	22728	21554	3996	17558	0.052
1975	27376	26766	25339	4611	20728	0.053
1976	31304	30778	29139	5278	23860	0.053
1977	34745	34032	32223	5778	26444	0.053

Reference Output Levels

Firm A

$$y_Z^*(A) = \psi_Z + 0.2\,[y_Z^* - y_Z(1947)] = \;\; 877.1$$
$$y_T^*(A) = \psi_T + 0.5\,[y_T^* - y_T(1947)] = 1802.6$$

Firm B

$$y_Z^*(B) = [y_Z(1947) - \psi_Z] + 0.8\,[y_Z^* - y_Z(1947)] = 6471.0$$
$$y_T^*(B) = [y_T(1947) - \psi_T] + 0.5\,[y_T^* - y_T(1947)] = 3698.5$$

Output Profiles

Firm A

$$y_Z(A, t) = \psi_Z + 0.2\,[y_Z(t) - y_Z(1947)]$$
$$y_T(A, t) = \psi_T + 0.5\,[y_T(t) - y_T(1947)]$$

Firm B

$$y_Z(B, t) = [y_Z(1947) - \psi_Z] + 0.8\,[y_Z(t) - y_Z(1947)]$$
$$y_T(B, t) = [y_T(1947) - \psi_T] + 0.5\,[y_T(t) - y_T(1947)]$$

Table 1D.2   Experiment 1 local calls, Firm A

Year	y_Z	$H^Z(y_Z, y_Z{}^*)$	$\partial H^Z/\partial y_Z$	$\partial^2 H^Z/\partial y_Z{}^2$
1947	10	10	0.986	0.000068
1948	80	79	0.990	0.000058
1949	123	121	0.993	0.000052
1950	171	170	0.995	0.000046
1951	224	222	0.997	0.000040
1952	276	274	0.999	0.000033
1953	325	323	1.001	0.000027
1954	372	370	1.002	0.000022
1955	440	438	1.003	0.000014
1956	519	518	1.004	0.000005
1957	588	587	1.004	−0.000001
1958	645	644	1.004	−0.000007
1959	725	724	1.003	−0.000015
1960	808	808	1.001	−0.000023
1961	877	877	1.000	−0.000029
1962	956	956	0.997	−0.000036
1963	1039	1039	0.994	−0.000043
1964	1110	1109	0.990	−0.000049
1965	1212	1210	0.985	−0.000058
1966	1326	1322	0.978	−0.000066
1967	1440	1432	0.969	−0.000075
1968	1561	1550	0.960	−0.000083
1969	1698	1680	0.948	−0.000092
1970	1816	1791	0.936	−0.000100
1971	1919	1888	0.926	−0.000106
1972	2059	2016	0.910	−0.000115
1973	2217	2159	0.891	−0.000124
1974	2358	2283	0.873	−0.000131
1975	2457	2369	0.860	−0.000136
1976	2601	2491	0.840	−0.000143
1977	2775	2635	0.814	−0.000151

Table 1D.3   Experiment 1 toll calls, Firm A

Year	y_T	$H^T(y_T, y_T^*)$	$\partial H^T/\partial y_T$	$\partial^2 H^T/\partial y_T^2$
1947	4	4	0.997	0.000006
1948	75	75	0.997	0.000005
1949	105	105	0.997	0.000005
1950	196	195	0.998	0.000005
1951	332	331	0.999	0.000004
1952	430	429	0.999	0.000003
1953	489	489	0.999	0.000003
1954	564	564	0.999	0.000003
1955	755	755	1.000	0.000002
1956	933	932	1.000	0.000001
1957	1087	1087	1.000	0.000000
1958	1188	1187	1.000	0.000000
1959	1425	1424	1.000	−0.000001
1960	1624	1624	1.000	−0.000002
1961	1802	1802	1.000	−0.000002
1962	2029	2028	0.999	−0.000003
1963	2282	2282	0.998	−0.000004
1964	2674	2673	0.995	−0.000006
1965	3107	3103	0.992	−0.000008
1966	3684	3675	0.987	−0.000010
1967	4120	4103	0.982	−0.000012
1968	4704	4675	0.974	−0.000014
1969	5472	5419	0.962	−0.000016
1970	6029	5952	0.952	−0.000018
1971	6472	6372	0.944	−0.000019
1972	7219	7071	0.928	−0.000021
1973	8173	7947	0.906	−0.000024
1974	8968	8660	0.886	−0.000026
1975	9680	9285	0.867	−0.000027
1976	10726	10176	0.836	−0.000029
1977	11937	11167	0.799	−0.000032

Table 1D.4   Experiment 1 predicted cost, Firm A

Year	$\tilde{w} \cdot \beta$	$\hat{x}_L$	$\hat{x}_M$	$\hat{x}_K$	$\hat{C}$	TECH
1947	14.4	23.7	2.3	13.3	20.9	0.026
1948	15.4	78.4	12.4	37.7	76.1	0.026
1949	15.6	104.4	17.9	54.1	108.1	0.025
1950	16.1	143.7	27.3	69.8	154.8	0.025
1951	17.2	191.5	39.1	84.1	19.0	0.025
1952	18.5	228.0	49.3	100.3	282.4	0.024
1953	18.7	251.4	56.0	118.9	328.0	0.024
1954	18.8	274.1	63.6	136.7	374.3	0.023
1955	19.7	323.1	79.8	155.4	464.4	0.024
1956	19.8	373.3	92.9	179.9	545.1	0.023
1957	19.3	406.3	103.6	207.5	612.1	0.024
1958	19.8	429.4	110.6	227.1	669.0	0.022
1959	20.8	471.2	129.6	250.6	787.9	0.023
1960	21.6	507.4	145.5	277.5	896.9	0.022
1961	22.4	531.5	159.9	299.5	990.9	0.022
1962	22.8	562.5	174.7	325.1	1091.1	0.022
1963	23.0	593.0	190.5	352.3	1189.1	0.022
1964	24.4	637.6	210.3	356.3	1332.6	0.022
1965	24.5	686.4	231.4	378.6	1462.0	0.023
1966	24.6	748.8	256.0	396.4	1626.1	0.024
1967	25.0	787.0	275.7	427.8	1789.6	0.025
1968	26.1	840.9	300.8	443.5	1991.4	0.025
1969	26.3	901.1	323.8	469.0	2252.4	0.027
1970	27.1	932.8	338.0	501.2	2497.7	0.028
1971	27.9	940.8	350.5	542.4	2759.8	0.028
1972	30.0	966.8	400.1	555.5	3191.0	0.029
1973	29.3	1017.3	397.6	616.6	3542.3	0.031
1974	30.5	1058.3	369.9	688.6	3996.2	0.032
1975	35.5	1069.5	409.8	650.6	4611.1	0.032
1976	39.7	1091.6	474.1	609.8	5278.7	0.033
1977	42.4	1137.4	494.4	598.9	5778.3	0.033

Table 1D.5   Experiment 1 marginal cost, markup, Firm A

Year	p_Z	p_T	MC_Z	MC_T	$\dfrac{p_Z}{MC_Z}$	$\dfrac{p_T}{MC_T}$
1947	0.81	1.18	0.57	0.14	1.42	8.32
1948	0.91	1.08	0.61	0.14	1.48	7.30
1949	0.95	1.04	0.62	0.14	1.52	7.14
1950	0.96	1.04	0.64	0.14	1.48	7.30
1951	0.94	1.05	0.69	0.14	1.36	7.44
1952	0.96	1.03	0.74	0.13	1.29	7.60
1953	0.99	1.00	0.75	0.13	1.31	7.60
1954	1.00	0.99	0.76	0.13	1.31	7.67
1955	0.99	1.00	0.79	0.12	1.24	7.91
1956	1.01	0.98	0.80	0.11	1.25	8.30
1957	1.02	0.97	0.78	0.12	1.30	7.97
1958	1.04	0.95	0.80	0.11	1.29	8.61
1959	1.02	0.98	0.84	0.11	1.21	8.83
1960	1.04	0.95	0.87	0.10	1.19	9.15
1961	1.00	1.00	0.90	0.09	1.10	10.27
1962	1.05	0.94	0.91	0.09	1.15	10.14
1963	1.11	0.88	0.92	0.08	1.21	9.94
1964	1.14	0.85	0.97	0.08	1.17	10.50
1965	1.13	0.86	0.96	0.07	1.17	10.94
1966	1.13	0.87	0.96	0.07	1.16	11.10
1967	1.12	0.87	0.97	0.07	1.15	11.31
1968	1.12	0.87	1.00	0.07	1.11	12.39
1969	1.12	0.87	1.00	0.07	1.12	11.11
1970	1.12	0.87	1.02	0.08	1.10	10.72
1971	1.13	0.87	1.04	0.08	1.08	9.82
1972	1.13	0.86	1.10	0.09	1.02	9.25
1973	1.13	0.86	1.06	0.10	1.06	8.11
1974	1.14	0.86	1.09	0.11	1.04	7.53
1975	1.14	0.85	1.24	0.10	0.92	7.95
1976	1.15	0.84	1.35	0.10	0.85	8.33
1977	1.15	0.84	1.39	0.08	0.83	9.63

Table 1D.6   Experiment 1 local calls, Firm B

Year	y_Z	$H^Z(y_Z, y_Z{}^*)$	$\partial H^Z/\partial y_Z$	$\partial^2 H^Z/\partial y_Z{}^2$
1947	3003	2991	1.003	0.000002
1948	3283	3273	1.003	0.000001
1949	3455	3445	1.003	0.000001
1950	3649	3640	1.004	0.000001
1951	3858	3850	1.004	0.000000
1952	4069	4062	1.004	0.000000
1953	4263	4257	1.004	−0.000000
1954	4452	4447	1.004	−0.000000
1955	4724	4719	1.004	−0.000000
1956	5039	5036	1.003	−0.000001
1957	5316	5314	1.003	−0.000002
1958	5543	5542	1.002	−0.000002
1959	5863	5862	1.002	−0.000002
1960	6195	6195	1.001	−0.000003
1961	6471	6471	1.000	−0.000004
1962	6789	6788	0.998	−0.000004
1963	7121	7120	0.997	−0.000005
1964	7406	7404	0.995	−0.000005
1965	7812	7808	0.993	−0.000006
1966	8270	8262	0.990	−0.000006
1967	8723	8710	0.986	−0.000007
1968	9208	9188	0.983	−0.000008
1969	9755	9724	0.978	−0.000009
1970	10227	10185	0.973	−0.000009
1971	10642	10588	0.969	−0.000010
1972	11200	11128	0.963	−0.000010
1973	11832	11735	0.956	−0.000011
1974	12395	12271	0.949	−0.000012
1975	12792	12647	0.944	−0.000012
1976	13367	13188	0.937	−0.000013
1977	14063	13837	0.927	−0.000014

Table 1D.7   Experiment 1 toll calls, Firm B

Year	y_T	$H^T(y_T, y_T{}^*)$	$\partial H^T/\partial y_T$	$\partial^2 H^T/\partial y_T{}^2$
1947	1900	1899	1.000	0.000000
1948	1971	1970	1.000	0.000000
1949	2001	2000	1.000	0.000000
1950	2091	2090	1.000	0.000000
1951	2228	2227	1.000	0.000000
1952	2326	2325	1.000	0.000000
1953	2385	2384	1.000	0.000000
1954	2460	2460	1.000	-0.000000
1955	2651	2651	1.000	-0.000000
1956	2829	2828	1.000	-0.000000
1957	2983	2983	1.000	-0.000000
1958	3084	3083	1.000	-0.000000
1959	3321	3320	1.000	-0.000000
1960	3520	3520	1.000	-0.000001
1961	3698	3698	1.000	-0.000001
1962	3924	3924	0.999	-0.000001
1963	4178	4178	0.999	-0.000001
1964	4570	4570	0.998	-0.000002
1965	5003	5001	0.997	-0.000002
1966	5580	5577	0.995	-0.000003
1967	6016	6010	0.993	-0.000003
1968	6600	6590	0.991	-0.000004
1969	7368	7350	0.988	-0.000005
1970	7925	7900	0.985	-0.000005
1971	8368	8335	0.982	-0.000005
1972	9115	9068	0.977	-0.000006
1973	10692	9998	0.971	-0.000007
1974	10863	10767	0.965	-0.000007
1975	11576	11453	0.959	-0.000008
1976	12622	12452	0.950	-0.000009
1977	13833	13595	0.938	-0.000010

Table 1D.8   Experiment 1 predicted cost, components, Firm B

Year	$\tilde{w} \cdot \beta$	$\hat{x}_L$	$\hat{x}_M$	$\hat{x}_K$	$\hat{C}$	*TECH*
1947	267	2375	399	1460	2270	0.023
1948	299	2456	419	1589	2633	0.022
1949	310	2436	437	1688	2780	0.021
1950	333	2451	454	1788	3002	0.019
1951	375	2486	474	1890	3374	0.018
1952	424	2501	497	1997	3776	0.016
1953	438	2486	515	2117	3971	0.015
1954	448	2463	534	2239	4153	0.014
1955	480	2495	569	2390	4568	0.013
1956	497	2548	596	2563	4862	0.012
1957	479	2543	623	2754	5001	0.013
1958	513	2542	641	2896	5303	0.011
1959	545	2559	686	3095	5840	0.010
1960	581	2580	725	3303	6357	0.009
1961	616	2571	761	3491	6824	0.008
1962	637	2575	798	3711	7248	0.007
1963	647	2575	838	3949	7614	0.007
1964	702	2587	878	4142	8341	0.006
1965	708	2615	928	4423	8786	0.006
1966	714	2662	985	4733	9339	0.006
1967	729	2679	1038	5065	9966	0.006
1968	773	2722	1100	5397	10894	0.005
1969	770	2761	1164	5803	11690	0.006
1970	793	2764	1215	6186	12622	0.007
1971	810	2728	1263	6571	13555	0.007
1972	870	2727	1367	7000	15286	0.008
1973	832	2755	1411	7559	16024	0.009
1974	858	2773	1421	8097	17558	0.010
1975	1029	2752	1503	8401	20728	0.009
1976	1166	2755	1629	8817	23860	0.008
1977	1269	2801	1709	9341	26444	0.007

Table 1D.9   Experiment 1 marginal cost, markup, Firm B

Year	p_Z	p_T	MC_Z	MC_T	$\dfrac{p_Z}{MC_Z}$	$\dfrac{p_T}{MC_T}$
1947	0.815	1.183	0.581	0.143	1.404	8.29
1948	0.915	1.084	0.626	0.149	1.462	7.27
1949	0.956	1.043	0.635	0.146	1.507	7.12
1950	0.960	1.040	0.654	0.143	1.468	7.28
1951	0.949	1.050	0.700	0.141	1.356	7.42
1952	0.964	1.035	0.751	0.136	1.285	7.59
1953	0.992	1.008	0.759	0.133	1.307	7.59
1954	1.006	0.994	0.765	0.130	1.315	7.66
1955	0.995	1.005	0.798	0.127	1.246	7.91
1956	1.012	0.989	0.803	0.119	1.259	8.30
1957	1.021	0.979	0.785	0.123	1.302	7.97
1958	1.043	0.957	0.805	0.111	1.296	8.61
1959	1.020	0.980	0.842	0.111	1.211	8.84
1960	1.044	0.957	0.874	0.105	1.194	9.15
1961	1.000	1.000	0.904	0.098	1.107	10.24
1962	1.058	0.943	0.919	0.093	1.151	10.17
1963	1.116	0.885	0.923	0.089	1.209	9.93
1964	1.144	0.858	0.977	0.082	1.171	10.48
1965	1.136	0.866	0.977	0.079	1.163	10.89
1966	1.132	0.870	0.981	0.079	1.154	11.00
1967	1.128	0.874	0.993	0.078	1.135	11.18
1968	1.128	0.874	1.032	0.072	1.093	12.18
1969	1.128	0.874	1.038	0.081	1.086	10.82
1970	1.129	0.873	1.067	0.084	1.059	10.37
1971	1.130	0.871	1.096	0.092	1.032	9.44
1972	1.133	0.868	1.169	0.099	0.969	8.78
1973	1.137	0.865	1.143	0.114	0.995	7.57
1974	1.140	0.862	1.185	0.124	0.962	6.92
1975	1.147	0.855	1.366	0.119	0.840	7.19
1976	1.153	0.848	1.505	0.115	0.766	7.34
1977	1.158	0.844	1.588	0.103	0.729	8.21

APPENDIX E

Table 1E.1   Experiment 2 subadditivity test

Year	C	$\hat{C}$	$\hat{C}_{sub}$	$\hat{C}_A$	$\hat{C}_B$	SUB
1947	2550	2596	2413	2349	63	0.070
1948	2994	3058	2849	2755	94	0.068
1949	3291	3254	3035	2925	110	0.067
1950	3563	3556	3317	3183	134	0.067
1951	4047	4054	3779	3610	169	0.068
1952	4616	4589	4272	4069	203	0.069
1953	4935	4852	4522	4296	226	0.068
1954	5258	5096	4756	4507	249	0.067
1955	5770	5647	5280	4984	296	0.065
1956	6305	6050	5667	5331	335	0.063
1957	6351	6229	5861	5493	368	0.059
1958	6788	6640	6241	5845	396	0.060
1959	7334	7341	6916	6457	458	0.058
1960	7912	8020	7563	7049	513	0.057
1961	8516	8632	8145	7583	561	0.056
1962	9018	9188	8681	8069	612	0.055
1963	9508	9667	9151	8490	661	0.053
1964	10542	10618	10055	9319	735	0.053
1965	11207	11201	10632	9832	799	0.051
1966	11954	11928	11354	10472	882	0.048
1967	12710	12739	12153	11187	965	0.046
1968	13814	13935	13311	12242	1069	0.045
1969	14940	14986	14366	13163	1202	0.041
1970	16516	16196	15557	14229	1328	0.039
1971	17951	17414	16763	15299	1464	0.037
1972	20161	19664	18963	17275	1687	0.036
1973	21190	20700	20034	18167	1867	0.032
1974	23168	22728	22042	19940	2102	0.030
1975	27376	26766	25934	23508	2426	0.031
1976	31304	30778	29828	27052	2776	0.031
1977	34745	34032	32990	29950	3040	0.031

Reference Output Levels

Firm A

$$y_Z^*(A) = [y_Z(1947) - 10\psi_Z] + 0.9\,[y_Z^* - y_Z(1947)] = 6812$$
$$y_T^*(A) = [y_T(1947) - 10\psi_T] + 0.75\,[y_T^* - y_T(1947)] = 4553$$

Firm B

$$y_Z^*(B) = 10\psi_Z + 0.1\,[y_Z^* - y_Z(1947)] = 535$$
$$y_T^*(B) = 10\psi_T + 0.25\,[y_T^* - y_T(1947)] = 947$$

Output Profiles

Firm A

$$y_Z(A, t) = [y_Z(1947) - 10\psi_Z] + 0.9\,[y_Z(t) - y_Z(1947)]$$
$$y_T(A, t) = [y_T(1947) - 10\psi_T] + 0.75\,[y_T(t) - y_T(1947)]$$

Firm B

$$y_Z(B, t) = 10\psi_Z + 0.1\,[y_Z(t) - y_Z(1947)]$$
$$y_T(B, t) = 10\psi_T + 0.25\,[y_T(t) - y_T(1947)]$$

Table 1E.2   Experiment 2 local calls, Firm A

Year	y_Z	$H^Z(y_Z, y_Z{}^*)$	$\partial H^Z/\partial y_Z$	$\partial^2 H^Z/\partial y_Z{}^2$
1947	2911	2898	1.002	0.000002
1948	3226	3214	1.003	0.000002
1949	3419	3408	1.003	0.000001
1950	3638	3627	1.003	0.000001
1951	3873	3863	1.004	0.000001
1952	4110	4102	1.004	0.000000
1953	4329	4321	1.004	0.000000
1954	4541	4535	1.004	−0.000000
1955	4847	4842	1.004	−0.000000
1956	5201	5197	1.004	−0.000000
1957	5513	5510	1.003	−0.000000
1958	5769	5767	1.003	−0.000000
1959	6128	6127	1.002	−0.000000
1960	6502	6502	1.001	−0.000000
1961	6812	6812	1.000	−0.000000
1962	7170	7170	0.998	−0.000000
1963	7544	7543	0.996	−0.000000
1964	7864	7862	0.995	−0.000000
1965	8321	8316	0.992	−0.000000
1966	8836	8826	0.989	−0.000000
1967	9346	9329	0.985	−0.000000
1968	9892	9867	0.981	−0.000000
1969	10507	10469	0.976	−0.000000
1970	11038	10986	0.971	−0.000000
1971	11504	11438	0.966	−0.000001
1972	12133	12043	0.960	−0.000001
1973	12844	12724	0.952	−0.000001
1974	13477	13324	0.944	−0.000001
1975	13923	13745	0.939	−0.000001
1976	14570	14349	0.930	−0.000001
1977	15353	1507435	0.919	−0.000001

Table 1E.3   Experiment 2 toll calls, Firm A

Year	y_T	$H^T(y_T, y_T{}^*)$	$\partial H^T/\partial y_T$	$\partial^2 H^T/\partial y_T{}^2$
1947	1856	1854	1.000	0.000000
1948	1962	1960	1.000	0.000000
1949	2007	2005	1.000	0.000000
1950	2143	2141	1.000	0.000000
1951	2348	2346	1.000	0.000000
1952	2495	2493	1.000	0.000000
1953	2584	2582	1.000	0.000000
1954	2696	2695	1.000	0.000000
1955	2982	2982	1.000	0.000000
1956	3249	3248	1.000	−0.000000
1957	3480	3480	1.000	−0.000000
1958	3631	3631	1.000	−0.000000
1959	3986	3986	1.000	−0.000000
1960	4285	4285	1.000	−0.000000
1961	4553	4553	1.000	−0.000001
1962	4892	4892	0.999	−0.000001
1963	5273	5272	0.998	−0.000001
1964	5861	5860	0.997	−0.000002
1965	6510	6507	0.996	−0.000002
1966	7376	7369	0.994	−0.000003
1967	8030	8018	0.991	−0.000003
1968	8906	8886	0.988	−0.000003
1969	10058	10022	0.983	−0.000004
1970	10894	10842	0.979	−0.000005
1971	11557	11492	0.976	−0.000005
1972	12678	12582	0.969	−0.000006
1973	14110	13964	0.960	−0.000006
1974	15301	15103	0.952	−0.000007
1975	16370	16117	0.943	−0.000007
1976	17939	17588	0.931	−0.000008
1977	19755	19264	0.914	−0.000009

Traditional telecommunications networks

Table 1E.4   Experiment 2 predicted cost, component, Firm A

Year	$\tilde{w}\cdot\beta$	$\hat{x}_L$	$\hat{x}_M$	$\hat{x}_K$	$\hat{C}$	TECH
1947	407	2328	398	1673	2349	0.021
1948	458	2439	422	1820	2755	0.019
1949	477	2434	443	1932	2925	0.019
1950	515	2472	466	2046	3183	0.017
1951	584	2533	491	2161	3610	0.016
1952	664	2568	520	2283	4070	0.014
1953	687	2567	541	2417	4296	0.013
1954	704	2557	564	2555	4507	0.013
1955	757	2616	607	2721	4984	0.012
1956	787	2696	641	2914	5331	0.011
1957	756	2709	673	3124	5493	0.012
1958	814	2721	695	3283	5845	0.010
1959	865	2760	749	3501	6457	0.009
1960	925	2801	797	3730	7049	0.008
1961	982	2806	840	3936	7583	0.008
1962	1017	2826	884	4176	8069	0.007
1963	1034	2843	932	4435	8490	0.007
1964	1125	2879	983	4638	9319	0.006
1965	1135	2933	1043	4939	9832	0.006
1966	1145	3012	1113	5266	10472	0.006
1967	1169	3050	1176	5622	11187	0.006
1968	1242	3121	1251	5970	12242	0.006
1969	1235	3192	1327	6397	13163	0.007
1970	1272	3212	1385	6805	14229	0.008
1971	1297	3181	1439	7220	15299	0.008
1972	1394	3196	1569	7664	17275	0.009
1973	1328	3251	1612	8263	18167	0.011
1974	1367	3292	1608	8847	19940	0.012
1975	1645	3279	1711	9141	23508	0.010
1976	1868	3295	1870	9546	27052	0.010
1977	2036	3367	1961	10073	29950	0.009

Table 1E.5   Experiment 2 marginal cost, markup, Firm A

Year	p_Z	p_T	MC_Z	MC_T	$\dfrac{p_Z}{MC_Z}$	$\dfrac{p_T}{MC_T}$
1947	0.815	1.183	0.580	0.143	1.405	8.29
1948	0.915	1.084	0.626	0.149	1.463	7.28
1949	0.956	1.043	0.634	0.146	1.507	7.12
1950	0.960	1.040	0.654	0.143	1.468	7.28
1951	0.949	1.050	0.700	0.141	1.356	7.43
1952	0.964	1.035	0.750	0.136	1.285	7.59
1953	0.992	1.008	0.759	0.133	1.307	7.59
1954	1.006	0.994	0.765	0.130	1.315	7.66
1955	0.995	1.005	0.798	0.127	1.246	7.91
1956	1.012	0.989	0.803	0.119	1.259	8.30
1957	1.021	0.979	0.785	0.123	1.301	7.97
1958	1.043	0.957	0.805	0.111	1.296	8.61
1959	1.020	0.980	0.842	0.111	1.211	8.83
1960	1.044	0.957	0.874	0.105	1.194	9.15
1961	1.000	1.000	0.904	0.098	1.107	10.24
1962	1.058	0.943	0.919	0.093	1.151	10.18
1963	1.116	0.885	0.923	0.089	1.209	9.93
1964	1.144	0.858	0.976	0.082	1.171	10.48
1965	1.136	0.866	0.976	0.079	1.164	10.90
1966	1.132	0.870	0.980	0.079	1.155	11.02
1967	1.128	0.874	0.992	0.078	1.137	11.20
1968	1.128	0.874	1.030	0.072	1.094	12.22
1969	1.128	0.874	1.036	0.080	1.089	10.87
1970	1.129	0.873	1.064	0.084	1.061	10.43
1971	1.130	0.871	1.092	0.092	1.035	9.51
1972	1.133	0.868	1.165	0.098	0.973	8.86
1973	1.137	0.865	1.137	0.113	0.999	7.66
1974	1.140	0.862	1.179	0.123	0.967	7.02
1975	1.147	0.855	1.357	0.117	0.845	7.31
1976	1.153	0.848	1.495	0.113	0.772	7.50
1977	1.158	0.844	1.575	0.100	0.735	8.42

Table 1E.6   Experiment 2 zoned local call, Firm B

Year	y_Z	$H^Z(y_Z, y_Z{}^*)$	$\partial H^Z/\partial y_Z$	$\partial^2 H^Z/\partial y_Z{}^2$
1947	102	101	0.995	0.000077
1948	137	136	0.997	0.000065
1949	158	157	0.999	0.000058
1950	183	182	1.000	0.000050
1951	209	208	1.001	0.000042
1952	235	234	1.002	0.000033
1953	259	258	1.003	0.000026
1954	283	282	1.003	0.000019
1955	317	316	1.004	0.000009
1956	356	356	1.004	−0.000001
1957	391	391	1.004	−0.000011
1958	419	419	1.003	−0.000018
1959	459	459	1.002	−0.000023
1960	501	501	1.001	−0.000039
1961	535	535	1.000	−0.000048
1962	575	575	0.997	−0.000058
1963	617	616	0.995	−0.000067
1964	652	652	0.992	−0.000075
1965	703	702	0.988	−0.000087
1966	760	759	0.983	−0.000099
1967	817	814	0.977	−0.000110
1968	877	873	0.970	−0.000122
1969	946	939	0.961	−0.000135
1970	1005	996	0.953	−0.000146
1971	1057	1045	0.945	−0.000155
1972	1126	1110	0.933	−0.000167
1973	1206	1184	0.920	−0.000182
1974	1276	1248	0.907	−0.000191
1975	1325	1293	0.897	−0.000198
1976	1397	1357	0.882	−0.000209
1977	1484	1433	0.864	−0.000221

Table 1E.7   Experiment 2 toll call, Firm B

Year	y_T	$H^T(y_T, y_T{}^*)$	$\partial H^T/\partial y_T$	$\partial^2 H^T/\partial y_T^2$
1947	49	48	0.997	0.000010
1948	84	84	0.998	0.000010
1949	99	99	0.998	0.000009
1950	144	144	0.998	0.000008
1951	212	212	0.999	0.000007
1952	261	261	0.999	0.000006
1953	291	291	0.999	0.000006
1954	329	328	1.000	0.000005
1955	424	424	1.000	0.000003
1956	513	512	1.000	0.000002
1957	590	590	1.000	0.000000
1958	640	640	1.000	−0.000000
1959	759	759	1.000	−0.000002
1960	858	858	1.000	−0.000003
1961	947	947	1.000	−0.000005
1962	1061	1061	0.999	−0.000007
1963	1187	1187	0.998	−0.000009
1964	1384	1383	0.996	−0.000012
1965	1600	1598	0.993	−0.000015
1966	1889	1884	0.988	−0.000019
1967	2106	2099	0.983	−0.000022
1968	2399	2385	0.976	−0.000026
1969	2783	2758	0.965	−0.000030
1970	3061	3025	0.956	−0.000033
1971	3282	3236	0.948	−0.000036
1972	3656	3588	0.934	−0.000040
1973	4133	4029	0.914	−0.000044
1974	4530	4388	0.895	−0.000048
1975	4887	4704	0.877	−0.000051
1976	5409	5156	0.849	−0.000055
1977	6015	5660	0.815	−0.000059

Table 1E.8   Experiment 2 predicted cost, component, Firm B

Year	$\tilde{w}\cdot\beta$	$\hat{x}_L$	$\hat{x}_M$	$\hat{x}_K$	$\hat{C}$	TECH
1947	−2	81	11	23	63	0.034
1948	−2	106	16	36	94	0.031
1949	−3	116	19	44	110	0.030
1950	−4	133	23	52	134	0.028
1951	−5	154	29	59	169	0.028
1952	−7	170	34	67	203	0.026
1953	−8	179	38	77	226	0.025
1954	−8	188	42	86	249	0.025
1955	−9	210	50	95	296	0.025
1956	−10	233	56	108	335	0.024
1957	−9	247	61	122	368	0.024
1958	−11	257	65	132	396	0.023
1959	−12	276	74	144	458	0.023
1960	−13	292	82	158	513	0.022
1961	−14	302	90	169	561	0.022
1962	−15	316	97	182	612	0.022
1963	−16	329	105	196	661	0.022
1964	−18	350	115	198	735	0.022
1965	−18	373	125	210	799	0.023
1966	−18	403	138	219	882	0.024
1967	−19	421	147	236	965	0.024
1968	−20	446	160	244	1069	0.025
1969	−20	475	172	258	1202	0.027
1970	−20	490	179	275	1328	0.027
1971	−20	493	185	297	1464	0.028
1972	−22	505	210	305	1687	0.029
1973	−20	530	209	338	1867	0.031
1974	−21	550	195	376	2102	0.032
1975	−26	555	216	358	2426	0.031
1976	−30	566	249	340	2776	0.032
1977	−33	589	259	337	3040	0.032

Table 1E.9 Experiment 2 marginal cost, markup, Firm B

Year	p_Z	p_T	MC_Z	MC_T	$\dfrac{p_Z}{MC_Z}$	$\dfrac{p_T}{MC_T}$
1947	0.815	1.183	0.576	0.142	1.41	8.31
1948	0.915	1.084	0.622	0.149	1.47	7.29
1949	0.956	1.043	0.632	0.146	1.51	7.14
1950	0.960	1.040	0.651	0.142	1.47	7.29
1951	0.949	1.050	0.698	0.141	1.35	7.44
1952	0.964	1.035	0.749	0.136	1.28	7.60
1953	0.992	1.008	0.758	0.133	1.30	7.60
1954	1.006	0.994	0.764	0.130	1.31	7.67
1955	0.995	1.005	0.798	0.127	1.24	7.91
1956	1.012	0.989	0.804	0.119	1.25	8.30
1957	1.021	0.979	0.785	0.123	1.30	7.97
1958	1.043	0.957	0.806	0.111	1.29	8.61
1959	1.020	0.980	0.843	0.111	1.21	8.83
1960	1.044	0.957	0.874	0.105	1.19	9.15
1961	1.000	1.000	0.904	0.098	1.10	10.27
1962	1.058	0.943	0.918	0.093	1.15	10.18
1963	1.116	0.885	0.922	0.089	1.21	9.94
1964	1.144	0.858	0.974	0.082	1.17	10.50
1965	1.136	0.866	0.972	0.079	1.16	10.93
1966	1.132	0.870	0.974	0.078	1.16	11.09
1967	1.128	0.874	0.983	0.077	1.14	11.30
1968	1.128	0.874	1.018	0.071	1.10	12.37
1969	1.128	0.874	1.020	0.079	1.10	11.08
1970	1.129	0.873	1.044	0.082	1.08	10.68
1971	1.130	0.871	1.068	0.089	1.05	9.78
1972	1.133	0.868	1.133	0.094	1.00	9.19
1973	1.137	0.865	1.099	0.107	1.03	8.05
1974	1.140	0.862	1.132	0.115	1.00	7.46
1975	1.147	0.855	1.297	0.109	0.88	7.85
1976	1.153	0.848	1.418	0.103	0.81	8.21
1977	1.158	0.844	1.479	0.089	0.78	9.45

NOTES

1. In the case of the quadratic form, although w is initially defined as a row vector, to avoid notational clutter a transposition sign is not used on w when forming the matrix product $\Gamma_n w$. A similar convention is maintained throughout.
2. Initially, Diewert and Wales (1987, p. 52) call (1.2) the Symmetric Generalized McFadden cost function. Later Diewert and Wales (1988) shorten the name to the Normalized Quadratic cost function. The specification (1.3) varies in that the output variable attached to the linear- and quadratic-price terms are modified.
3. The notation: 0_3 is a vector of zeros of dimension 3, $x \gg 0_3$ means each component of x is greater than zero, $x \geq 0_3$ means each component of x is nonnegative and $x > 0_3$ means $x \geq 0_3$ but $x \neq 0_3$.
4. This also suggests that in (1.10) $C^{F,n}(w)$ may be more precisely denoted $\tilde{C}^{F,n}(w, y^*)$.
5. Proofs of propositions are found in Appendix A.
6. The definition of the cost function as a minimum implies the concavity property (Shephard, 1953; Samuelson, 1953; Diewert, 1974, p. 156). Other regularity conditions include linear homogeneity in input prices and monotonicity in input prices. It is difficult to specify analytical functions that are both flexible and globally regular. A compromise is to impose flexibility at a relevant point (as confirmed by Proposition 1), homogeneity (evident in (1.3) above) and global concavity (see (1.19)) and to search for empirical implementations that generate monotonicity in the sample data.
7. Although (1.3) allows the examination of economies of scale through non-linearity in y_n, the restrictions (1.5) and (1.15) limit this potential. Later, the functional form is further generalized.
8. See, for example, Woodland (1982, pp. 134–5).
9. Strictly speaking, these restrictions imply $\bar{Q}(y, w)$ and $C^V(y, w)$ are functions of y^*. This dependence is ignored for notional clarity here and taken up later when, in the non-linear specification, the reference vector y^* has greater relevance.
10. *HD1* denotes homogeneous of degree 1, for example, the second property of (1.37) is equivalent to $C^V(\lambda y_Z, \lambda y_T, w) = \lambda C^V(y_Z, y_T, w)$ for $\lambda \geq 0$. The property is a consequence of the property holding for $\bar{Q}(y, w)$.
11. See Sakai (1974, p. 272). McFadden (1978, p. 48) terms two outputs complements if $\partial^2 C^*(y, w)/\partial y_Z \partial y_T < 0$.
12. It is possible for two inputs to exhibit economies of shared production when the overall technology does not. Further, the condition $\alpha \geq 0_3$, necessary for economies of shared production, also guarantees that Q is nonnegative for this specification. This exacerbates the likelihood of nonpositive marginal costs because of the presence of Q in the second term in (1.41) and (1.42). This problem is addressed in a later generalization of the functional form.
13. The qualification is important as, for example, lack of competition may have other consequences, such as encouraging a natural monopoly to not undertake costly innovation. This could have consequences for future costs of industries dependent on monopoly provider inputs. This may be especially important in the context of increasingly global competition in downstream industry. Conversely, R&D may be subject to a subadditive cost structure. The analysis of the cost conditions of an industry is a partial equilibrium one. Nevertheless it is an important ingredient in a more complete analysis.
14. As noted in the initial discussion of (1.3), because of the non-linearity in output it is useful to recognize the dependence of this specification on the reference vector y^*.
15. In the case of the H^n functions, $\nabla H^n(y_n, y_n^*)$ and $\nabla^2 H^n(y_n, y_n^*)$ denote the first- and second-derivatives of $H^n(y_n, y_n^*)$ with respect to y_n. For $\tilde{C}^V$ ∇_y and ∇^2_{yy} are used to distinguish output from input price derivatives.
16. Assume the normalization $J(y^*)$ is chosen such that $J(y^*) > 0$, for example, by setting $\psi \equiv (\psi_Z, \psi_T) > 0_2$.
17. 'Effective' indicates the analysis is for output above a certain minimum, as specified in (1.85).

18. Certain parameters in the scale factor on the quadratic term are constrained to allow an economic interpretation of ψ_Z and ψ_T as minimum output levels. The usual regularity condition of concavity in factor prices is imposed.

REFERENCES

Baumol, W.J., Panzar, C. and Willig, R.D. (1988), *Contestable Markets and the Theory of Industry Structure*, Revised Edition, New York: Harcourt Brace Jovanovich.

Christensen, L.R., Christensen, D.C. and Schoech, P.E. (1981), 'Total factor productivity in the Bell System', Christensen Associates, Madison, Wisconsin, September.

Diewert, W.E. (1973), 'Functional forms for profit and transformation functions', *Journal of Economic Theory*, **6**, 284–316.

Diewert, W.E. (1974), 'Applications of duality theory', in Intriligator, M.D. and Kendrick, D.A. (eds.), *Frontiers of Quantitative Economics: Volume II*, Amsterdam: North-Holland, 106–71.

Diewert, W.E. and Wales, T.J. (1987), 'Flexible functional forms and global curvature conditions', *Econometrica*, **55**, 43–68.

Diewert, W.E. and Wales, T.J. (1988), 'A normalized quadratic semiflexible functional form', *Journal of Econometrics*, **37**, 327–42.

Diewert, W.E. and Wales, T.J. (1991), 'Multiproduct cost function estimation and subadditivity tests: A critique of the Evans and Heckman research on the US Bell System', Discussion Paper, Department of Economics, University of British Columbia, Vancouver, Canada, April.

Evans, D.S. and Heckman, J.J. (1983), 'Multiproduct cost function estimates and natural monopoly tests for the Bell System', in Evans, D.S. (ed.), *Breaking Up Bell*, Amsterdam: North-Holland, 253–82.

Evans, D.S. and Heckman, J.J. (1984), 'A test for subadditivity of the cost function with an application to the Bell System', *American Economic Review*, **74**, 615–23.

Lau, L.J. (1978), 'Testing and imposing monotonicity, convexity and quasiconvexity constraints' in Fuss, M. and McFadden, D. (eds), *Production Economics: A Dual Approach to Theory and Applications, Volume 1*, Amsterdam, North-Holland, 409–53.

McFadden, D. (1978), 'Cost, revenue and profit functions', in Fuss, M. and McFadden, D. (eds), *Production Economics: A Dual Approach to Theory and Applications, Volume 1*, Amsterdam, North-Holland, 3–109.

Sakai, Y. (1974), 'Substitution and expansion effects in production theory: the case of joint production', *Journal of Economic Theory*, **9**, 255–74.

Samuelson, P.A. (1953), 'Prices of factors and goods in general equilibrium', *Review of Economic Studies*, **21**, 1–20.

Shephard, R.W. (1953), *Cost and Production Functions*, Princeton, Princeton University Press.

White, K.J. (1978), 'A general computer program for econometric methods – SHAZAM', *Econometrica*, **46**, 239–40.

Woodland, A.D. (1982), *International Trade and Resource Allocation*, Amsterdam: North-Holland.

024
L96

2.   Cost function issues and estimation

Yale M. Braunstein and Grant Coble-Neal

INTRODUCTION

Since the ratification in 1997 of the World Trade Organization agreement on basic telephone services, more than 1000 facilities-based international carriers have become operational in a global market (TeleGeography, 1999). The response from large carriers is to establish global service platforms by horizontal and vertical integration, and a changing web of international alliances (Braunstein et al., 1999). These changes have resulted from fundamental shifts in regulatory environment and the presence of economies of scale and scope. The bundling of voice, data and video services provides substantial scope for innovation and cost saving. However, horizontal and vertical integration has seen a reduction of competition in global markets. United States (US) and European Union regulators have permitted mergers in the belief that larger scale in telecommunications is in the public interest. Clearly, reliable estimate of telecommunications carrier cost structures is essential for decision-making. Much recent research builds on studies of telecommunication costs published in the 1960s and 1970s.[1] Development in the estimation of production technology follows from the work of Arrow et al. (1961) and Nerlove (1963) that relax restrictive factor substitution and returns to scale assumptions implied by Cobb–Douglas technology. Early-1970s developments include duality theory by Diewert (1971) and the use of flexible functional forms by Diewert (1971) and Christensen et al. (1971). The subsequent development of economies of scope by Panzar and Willig (1975) and the notion of subadditivity by Baumol (1977) provided further impetus to applied research.

Typical applied studies by Dobell et al. (1972) and Davis et al. (1972) provide estimates for two-factor Cobb–Douglas production functions with value-added output and Hicks-neutral technology change for Bell Canada and the Bell System. Vinod (1975, 1976) used a Translog function to estimate two-output, two-input production. Subsequent analysis introduces the use of the dual to simultaneously estimate Translog cost and factor cost share equations.[2] Fuss and Waverman (1977) estimate a Translog cost function for the Bell Canada System, while Christensen et al. (1981, 1983) provide estimates of the System's economies of scale. Fuss and Waverman (1981) generalise the Translog to a three-output, three-input model. The more general form substitutes a Box–Cox transformation for natural logarithms of output and technology change. This generalised form nests the Translog. Cost and cost-share equations are estimated simultaneously using the seemingly unrelated regression estimation technique proposed by Zellner (1962). Since cost share equations sum to unity a cost-share equation is dropped to ensure non-singularity of the variance–covariance matrix. Alternative estimation procedures employed are three-stage least squares and full-information maximum likelihood (Kiss and Lefebvre, 1987).

This chapter reviews issues related to cost function estimation in the telecommunications industry. The next section summarises concern associated with specification of functional forms. Then problems with data and approaches to addressing them are discussed. The following section examines the treatment of technical change. Estimation and testing of economic theory appears before an overview and summary.

FUNCTIONAL FORMS

Diewert (1971) and Christensen et al. (1971) introduced the Generalised Leontief and the Translog functions as flexible alternatives to the Cobb–Douglas and Leontief production functions.[3] Subsequent research extended the application of flexible econometric techniques with particular emphasis on the imposition of theoretical restrictions. To be proper and consistent with the general propositions of duality theory, cost functions are described by positive output quantities and input prices. Further, a cost function is strictly quasi-concave in input prices, that is, a continuous increase in the price of an input will induce a reduction in its use in favour of cheaper alternatives (Coelli et al., 1999). Subadditivity highlights the elusiveness of estimating a proper cost function whilst retaining sufficient flexibility to measure but not prejudge scale, scope and subadditivity, for example, Evans and Heckman (1983), Diewert and Wales (1991a, 1991b), Fuss (1992), Röller (1990) and Shin and Ying (1992). Kiss and Lefebvre (1985) report that the estimated economic properties of the Bell System are sensitive to transformations of local output. Moreover, tests of second-order output parameters cannot reject the hypothesis that they are equal to zero. Röller (1990) argues parameter estimates obtained with the Translog function are not robust due to a 'Flip-Flop' property. This property refers to the degeneration of the Translog close to zero output. Thus, sign changes of the second-order parameters on input prices may lead to fundamentally different cost structures. Similarly, Charnes et al. (1988), show that by using a goal/programming-constrained regression the Translog function, data and subadditivity test used by Evans and Heckman yield diametrically opposite results. To overcome this shortcoming, Röller (1990) utilises a Generalised CES-Quadratic cost function. The function is quadratic in outputs and imposes restrictions required to produce a proper cost function while allowing flexible estimation of output. This approach explicitly confronts the trade-off between flexibility and properness. Using this approach Röller finds economies of scale and scope, and cost complementarity.

Early studies of telecommunications, particularly of the US Bell System, aggregate diverse telecommunications service outputs into a single measure, estimate single-product cost functions and determine whether there are scale economies. However, as noted by Evans and Heckman (1983), these models are based on an aggregate measure of output that is valid only under highly restrictive assumptions that are not tested. Assumptions of particular concern are separability and non-jointness. Separability implies that outputs can be aggregated so that estimates of total cost are invariant to the use of aggregate of sub-aggregate measures of output while non-jointness implies the cost of producing several outputs is equal to the sum of the costs of producing the outputs separately. When production technology is non-joint there are no economies of scope.

Despite achieving results using a cost function that conforms to theory, there is no clear trade-off between flexibility and properness of a production function. In particular, Pulley and Braunstein (1992) point out that the CES-Quadratic model imposes strong separability between outputs and inputs. Multiplicative separability means that factor demand equations are comprised of, and are fully described by, an output function multiplied by an input function. This is a direct result of the assumption of profit maximisation (or cost minimisation). With multiplicatively separable forms, input ratios and cost shares are independent of output levels. Moreover, input demand elasticities with respect to outputs are equal and independent of input prices. Pulley and Braunstein (1992) propose the Composite Cost function that combines the Translog Quadratic input structure with a quadratic structure for outputs. This general parsimonious design has the advantage of nesting the standard and Generalised Translog forms and Röller's (1990) Generalised CES-Quadratic form. Replacing the logarithmic transformation of individual outputs with a Box–Cox transformation averts the problem of non-robustness of the input structure. The separability problem is overcome by including a price and output interaction term. An appeal of the Composite cost function is its combination of a quadratic output structure, desirable for estimating multi-product firms with an input structure that is easily restricted to be linearly homogenous in input prices. The general composite cost function is:

$$C^{(\phi)} =$$

$$\left\{ \begin{array}{l} \exp\left[\alpha_0 + \sum \alpha_i q_i^{(\pi)} + \frac{1}{2}\sum\sum \alpha_{ij} q_i^{(\pi)} q_j^{(\pi)} + \sum\sum \delta_{ik} q_i^{(\pi)} \ln r_k\right]^{(\tau)} \\ \cdot \exp\left[\beta_0 + \sum \beta_k \ln r_k + \frac{1}{2}\sum\sum \beta_{kl} \ln r_k + \sum\sum \mu_{ik} q_i^{(\pi)} \ln r_k\right] \end{array} \right\}^{(\phi)}$$

(2.1)

where,

$$C^{(\phi)} = \frac{(C^{(\phi)} - 1)}{\phi},$$

(2.2)

$$q_i^{(\pi)} = \frac{(q_i^{(\pi)} - 1)}{\pi} \text{ and}$$

(2.3)

$$\exp[\cdot] = \frac{(\exp[\cdot]^{(\tau)} - 1)}{\tau}.$$

(2.4)

Equations (2.2) through (2.4) hold for $\phi \neq 0$; if $\phi = 0$ then $y(\phi) = \ln\phi$. q_i ($i = 1, 2, ..., m$) are outputs and r_k ($k = 1, 2, ..., n$) are input prices. Equations (2.2), (2.3) and (2.4) represent the Box–Cox transformation. In general, this function allows the Composite cost function to flexibly select the logarithmic form when ϕ, π, $\tau = 0$.[4] Pulley and Braunstein (1992) utilise the Box–Cox transformation to develop a general specification. By restricting the transformation parameters (see Table 2.1) the standard and generalised forms of the Translog, Separable Quadratic and Composite specifications are given.

Table 2.1 Parameter restrictions and specifications

Specification	ϕ	π	τ	Other
Standard Translog	0	0	1	
Generalised Translog	0	–	1	δ_{ik}, $\mu_{ik} = 0$ for all i, k
Separable Quadratic	–	1	0	
Composite	–	1	0	

Source: Braunstein and Pulley (1998).

An advantage of the Composite cost function is its more parsimonious form, thereby increasing the degrees of freedom relative to other functional forms. Other restrictions such as those required for linear homogeneity are easily applied.[5] Further, the transform-both-sides application of the Box–Cox transformation preserves the composite structure, thus the results are easy to interpret. The quadratic structure means the composite structure has the ability to model cost behaviour near zero outputs, thus providing a viable alternative to the Translog and Generalised Translog forms.

DATA ISSUES

Cost function estimation typically requires the decomposition of annual revenue and expenditure data into separate price and quantity indexes. When wage and capital price data are well defined and readily available, quantity indexes can be calculated. Hence, in developing output and capital quantity indexes data assumptions form an integral part of the maintained hypothesis. For capital stocks, assumptions include a linear factor price frontier so that capital goods of the same vintage are homogenous, with the latest vintage more efficient, the marginal labour product across vintages are equal, and re-switching and capital reversals are absent (Nadiri, 1982). While these assumptions are restrictive the paucity of available data demands pragmatism. It is useful to consider a telecommunications network as comprised of several components providing voice and data services at local (regional areas) and long-distance, on both public and private lines, via exchange facilities. Attached to exchanges are factors of production such as labour skills, a variety of capital and materials, ranging from plant and machinery to buildings and motor vehicles. This suggests that analysing telecommunications cost structures and estimating cost functions requires data on inputs and outputs for local and toll exchanges and possibly on access lines. With this kind of micro-data, synergies, scale and scope economies arising from joint production could be explicitly tested. With data on cross-sections of carriers with different scales of operation and configurations, it would be possible to test for efficiency and determine optimal network size. Moreover, the addition of time-series on such micro-data would allow the existence of cost complementarity between local and long-distance exchanges to be tested. Unfortunately, such data are rarely available, forcing researchers to estimate aggregate cost functions.

Where appropriate, aggregation permits different quantities of factors to be compared, allowing analysis of the changing proportions of factors, for example, capital and labour, through time. However, inappropriate aggregation can lead to incorrect inferences about the production process, that is, the aggregation problem. Aspects of the aggregation problem include identifying appropriate factor classification, constructing quantity indexes and identifying appropriate production sequence classifications. Identifying appropriate factor classifications is important because, for example, most firms produce a variety of goods of different quality. Moreover, the inputs used can also vary in quality and productivity. That is, substantial quality differences between inputs lead to substantial productivity differences in the input quantity required to produce a fixed output. Failure to distinguish between inputs leads to incorrect inferences about average productivity and impedes analysis of the impact of changes in factors on productivity. Another aspect of the aggregation problem is developing comparable quantity measures across factors – the index number problem. This problem is overcome by finding a characteristic across items contained in a common category. A convenient measure is the value of a factor. Cost function estimation, however, requires the construction of price indices for inputs and quantity indices for outputs. Commonly used methods are Laspeyres and Paasche indexes. Laspeyres price and quantity indexes, respectively, are:

$$P_{st}^L = \sum_{i=1}^N \frac{p_{it}}{p_{is}} \times \frac{p_{is}q_{is}}{\sum_{i=1}^N p_{is}q_{is}}$$

and

$$Q_{st}^L = \sum_{i=1}^N \frac{q_{it}}{q_{is}} \times \frac{p_{is}q_{is}}{\sum_{i=1}^N p_{is}q_{is}}, \qquad (2.5)$$

where p is the price of commodity i at time t relative to the price of i at base period s. Thus, the Laspeyres price index is the summation of the relative prices weighted by the relative value shares of the commodities corresponding to a fixed base historical date. The corresponding Paasche indexes are:

$$P_{st}^P = \sum_{i=1}^N \frac{p_{it}}{p_{is}} \times \frac{p_{it}q_{it}}{\sum_{i=1}^N p_{it}q_{it}}$$

and

$$Q_{st}^P = \sum_{i=1}^N \frac{q_{it}}{q_{is}} \times \frac{p_{it}q_{it}}{\sum_{i=1}^N p_{it}q_{it}}, \qquad (2.6)$$

which uses the current values of each commodity as base period price weights. Since both the Laspeyres and Paasche methods use relative-value measures, they have the advantage of eliminating the dependence of quantity on the measurement unit used. Both indexes are calculated from the weighted average ratio of current-period quantity to base-year quantity. In both methods the sum of the weights equals one.

The index procedures are simple to implement when both price and quantity measures are known. Typical firm data available are financial statements containing annual labour, current and non-current investment expenditure. Independent price data, such as average hourly wage rates by occupation or industry are generally available from national statistical agencies, along with various proxy measures for the rental price of capital. While persons employed (labour quantity) can be readily observed, measuring capital employed is not so clear. An indirect approach utilises the definition of expenditure:

$$V_{st} = P_{st} \times Q_{st}. \tag{2.7}$$

Thus, given expenditure and price (2.7) is rearranged to solve for quantity:

$$Q_{st} = \frac{V_{st}}{P_{st}} = \frac{\displaystyle\sum_{i=1}^{N} p_{it}q_{it}}{\displaystyle\sum_{i=1}^{N} p_{is}q_{is}} \times \frac{1}{P_{st}} = \frac{\displaystyle\sum_{i=1}^{N} p_{it}q_{it}}{P_{st}} \times \frac{1}{\displaystyle\sum_{i=1}^{N} p_{is}q_{is}} \tag{2.8}$$

Equation (2.8) states that the quantity of capital can be derived as the ratio of current value of investment at constant period s prices to period s investment expenditure. An interesting feature of this decomposition is that, rearranging (2.8), the index becomes:

$$Q_{st}P_{st} = \frac{\displaystyle\sum_{i=1}^{N} p_{it}q_{it}}{\displaystyle\sum_{i=1}^{N} p_{is}q_{is}}, \tag{2.9}$$

which is exactly the Paasche index. However, realisation of the Laspeyres and Paasche index duality raises an issue of potential divergence of direct and indirect quantity indexes. Since the Laspeyres index uses base period quantities and Paasche uses current period quantities, the measures only coincide when relative prices are stable. Should relative prices display considerable variation the indexes define an upper and lower bound of the true quantity. Unfortunately, the tendency for relative prices to diverge through time means that only indexes using the same base period are directly comparable (Heathfield and Wibe, 1987). An alternative method, suggested by Diewert (1976) that overcomes this time reversal problem is Fisher's Ideal index. This index is constructed by calculating the geometric mean of the Laspeyre and the Paasche indices. The Fisher Ideal price and quantity indices are:

$$FP=[P_{st}^{L}\times P_{st}^{P}]^{\frac{1}{2}}$$

and

$$FQ=[Q_{st}^{L}\times Q_{st}^{P}]^{\frac{1}{2}}. \qquad (2.10)$$

In addition, Diewert shows the Fisher Ideal index direct and indirect quantity measures are identical. Thus, this index always yields an accurate measure.[6] Fisher's Ideal index is termed ideal because it 'corresponds to a functional form which is capable of providing a second-order approximation to an arbitrary twice-differentiable function' (Diewert, 1976: 117).[7] Moreover, Diewert (1983) demonstrates that, when the underlying cost function is quadratic, the Fisher Ideal Index corresponds exactly. A popular alternative to the Fisher Ideal index is the Törnqvist (Divisia) formula, which is a weighted geometric average of output and price relativities. The Törnqvist price and quantity indexes, in log-change form, are:

$$\ln P_{st}=\sum_{i=1}^{N}\left(\frac{p_{is}q_{is}/\sum_{i=1}^{N}p_{is}q_{is}+p_{it}q_{it}/\sum_{i=1}^{N}p_{it}q_{it}}{2}\right)[\ln p_{it}-\ln p_{is}] \quad (2.11)$$

and

$$\ln Q_{st}=\sum_{i=1}^{N}\left(\frac{p_{is}q_{is}/\sum_{i=1}^{N}p_{is}q_{is}+p_{it}q_{it}/\sum_{i=1}^{N}p_{it}q_{it}}{2}\right)[\ln q_{it}-\ln q_{is}]. \quad (2.12)$$

The popularity of the Törnqvist index is due to its theoretical correspondence to the Translog cost function. However, Eichorn and Voeller (1983) show that, unlike the Fisher Ideal index, the Törnqvist index does not satisfy either the factor or time reversal tests. A further shortcoming of the index is its inability to accommodate zero values.

TECHNICAL CHANGE

The impact of technology, from an economic perspective, is analysed in terms of how producers respond to changes in economic variables that impact on their profitability, for example, factor price changes. Formally, technology is defined as the collection of available techniques of combining inputs to produce outputs (Dixit, 1976). Technology can be described by several means including an isoquant map, production function or cost function. Clearly, in responding to changes such as a reduction in the use of a production input in favour of a cheaper alternative, producer behaviour must be constrained by the technology, which determines the substitution possibilities between production inputs. Thus, as outlined in Jorgenson (2000), the goal of empirical research is to determine the nature of substitution between inputs, the character of differences in technology, and the roles of economies of scale and scope. The responses of demand and supply to changes in prices, technology, economies of scale, capital vintage and human capital characterise the behaviour of producers. Measures of substitution are specified in terms of the impact of price change on demand and supply, while measures of technical change are specified in terms of their impact on technology change. A flexible means of modelling such response is given by Shephard (1970), who demonstrates that all information about the underlying technology is contained in the cost function. Thus, research has concentrated on cost function estimation because it yields direct parameter estimates of the pattern and degree of substitution between factors without imposing restrictions on returns to scale.[8]

While the form of current technology largely constrains short-run production response, in the long run technology is flexible. Thus, it is necessary to provide a means of modelling exogenous technology change. Hicks (1932) introduced the notion of biased technical change as a measure of the share of an input in the value of output to a change in technology. A positive bias indicates an increase in the demand for an input as technology improves, while a negative bias indicates a reduction in the use of the input. Technical change can be neutral in that price change elicits no response from the producer. Hicks-neutrality implies the shape of the isoquants, which describe the substitution possibilities, is invariant to cost function shifts, that is, the marginal rate of technical substitution is independent of cost function shifts. The use of time as a technology index implies that technical change is related to shifts of the cost function through time. For the

Translog cost function, the average rates of technical change between discrete time periods are the difference between the growth rates of outputs and the weighted average growth of inputs. The use of time trends in modelling exogenous technical change is attractive when long-run technical change and productivity growth are mainly determined by capital, but may be too restrictive when there are changes in regulatory structures, privatisation or deregulation that impact on technical change. Flexible functional forms, such as the Translog or Generalised Leontief, provide the ability to allow technical change to increase at non-constant rates, be non-neutral and scale augmenting. In this context, flexible cost functions incorporate Hicks neutrality as a testable hypothesis since Hicks-neutral advance production implies the time-price cross-product (interaction terms) parameters are zero. However, Hicks's response to technical change imposes arbitrary restrictions on producer behaviour. To provide more flexibility in substitution Christensen et al. (1971, 1973) introduced share elasticities as an alternative substitution measure.

While technical change may be analysed in terms of its interaction with price and output change, further insight can be gained by analysing the sources of technical change in terms of both embodied and disembodied technical change. Disembodied technical change analyses the impact of innovative processes that cannot be solely attributed to an input, that is, labour or capital. Innovation is often attributed to changes in organisation structure such as reordering production processes or inventory reductions. By contrast, embodied technical change recognises that new labour and capital inputs reflect new knowledge, enabling greater production efficiency. In the case of capital-embodied technical change new capital is more efficient than old vintages. In this sense, the efficiency of production depends on the history of investment, whereas differences in labour quality reflect variations in human capital and are often proxied by shifts in the occupation shares, for example, Sung (1998). Hence, the ratio of the number of managers to workers can be used to compare differences in labour quality across firms or time. Measuring embodied technical change requires the use of efficiency units. An efficiency unit is an analytical device used to separate increases in output as a result of embodied technical change from increases in output due to increased input. The approach assumes changes in efficiency are specified in terms of geometrically declining weights, that is,

$$EK_t = e^{-\alpha t}K_t$$

and

$$EL_t = e^{-\beta t}L_t \qquad (2.13)$$

where *EK* and *EL* represent capital and labour inputs, respectively in efficiency units, *t* denotes time and α, β measure the constant percentage change in efficiency through time. When parameter constancy is too restrictive, linear or quadratic terms can be specified. The quadratic specification, for example, is:

$$EK_t = \exp\left(\alpha_1 t + \alpha_2 \frac{1}{2} t^2\right) K_t$$

and

$$EL_t = \exp\left(\beta_1 t + \beta_2 \frac{1}{2} t^2\right) L_t. \qquad (2.14)$$

To estimate the efficiency parameters in, say, (2.13) it must be embedded in the cost function. Gollop and Roberts (1981) show that since total input expenditure is defined as the product between the unit cost multiplied by the quantity of the input used, then total labour and capital expenditure is:

$$P_k K = P_k \frac{EK}{e^{-\alpha t}} = \frac{P_k}{e^{-\alpha t}} = EK$$

and

$$P_L L = P_L \frac{EL}{e^{-\beta t}} = \frac{P_L}{e^{-\beta t}} EL \qquad (2.15)$$

where P_k is the rental price of capital and P_L the wage rate. Equation (2.15) shows the use of efficiency units *EK* and *EL* as quantity measures implies the efficiency prices $\dfrac{P_K}{e^{-\alpha_i t}}$ and $\dfrac{P_L}{e^{-\alpha_L t}}$, respectively. Thus input-augmenting technical change can be incorporated in the Translog cost function as:

$$\ln C = \alpha_0 + \sum_i \beta_i \ln p_i - \sum_i \beta_i \alpha_i T + \sum_k \beta_k \ln Q_k + \frac{1}{2} \sum_i \sum_j \gamma_{ij} \ln p_i \ln p_j$$

$$-\frac{1}{2} \sum_i \sum_j \gamma_{ij} \alpha_i \ln p_j T + \frac{1}{2} \sum_i \sum_k \gamma_{ik} \ln p_i \ln Q_k - \frac{1}{2} \sum_i \sum_k \gamma_{ik} \alpha_i \ln Q_k T$$

$$+\frac{1}{2} \sum_k \sum_l \gamma_{kl} \ln Q_k \ln Q_l + \frac{1}{2} \gamma_{TT} T^2, \qquad (2.16)$$

where the α_i correspond to the embodied efficiency parameters ($i = L, K$). Imposing homogeneity and symmetry restrictions on the estimating equations permits the identification of the parameters ($\alpha_i, \beta_i, \gamma_{ij}, \gamma_{ik}$ and γ_{TT}) in (2.16).

While use of flexible cost functions permits considerable generalisation of technical change, the specification of linear or quadratic time trends means technical change is restricted to a smooth, slowly changing time path. When technical change is highly variable, simple time trends are likely to be poor proxies. In response, Baltagi and Griffin (1988) argue for the specification of a general index model that does not impose any functional structure on the behaviour of the technical change. By using pooled data on firms for an industry, Baltagi and Griffin show a completely general technical change index is achieved by estimating a Translog cost function using time-specific dummy variables across firms. Equating the index model to a General Translog function identifies relevant technical change variables including the index. An advantage of this specification is that technical change is allowed to exhibit an erratic pattern of change. However, panel data are not available, and Diewert and Wales (1987) propose a system of producer supply and demand functions with separate quadratic time trends. Quadratic spline functions provide an efficient method of allowing model parameters to vary across sub-samples of data while maintaining continuity at the joins. Radical changes in technology require a function capable of permitting the effects of technology to vary through time. A specification of flexible technological progress function is provided by:

$$\text{Tech} = [a_0 + b_0(t - t_0) + c_0(t - t_0)^2] \cdot d_0 + [a_1 + b_1(t - t_1) + c_1(t - t_1)^2] \cdot d_1$$

$$+ \sum_{i=2}^{n} [a_i + b_i(t - t_i) + c_i(t - t_i)^2] \cdot d_i \qquad (2.17)$$

where t is a time trend, t_i are specific fixed periods (knots), the dummy variable $d_0 = 1$ when $t_0 < t \leq t_1$ and zero otherwise, $d_1 = 1$ when $t_1 < t \leq t_2$ and zero otherwise, $d_2 = 1$ when $t_2 < t \leq t_3$ and zero otherwise, and the a_i, b_i and c_i are parameters to be estimated. To ensure the technology function meets at the knots and is continuous to at least the first derivative, it is necessary to constrain (2.17). Following Suits et al. (1978), the Tech function is specified:

$$\text{Tech} = a_0 + b_0(t - t_0) + c_0(t - t_0)^2 + (c_1 - c_0)(t - t_1)^2 \cdot D_1 + (c_2 - c_1)(t - t_2)^2 \cdot D_2$$

$$+ \sum_{i=3}^{n} (c_i - c_{i-1}) (t - t_i)^2 \cdot D_i, \qquad (2.18)$$

where the D_i are defined as: $D_1 = 1$, if $t_1 < t$ and zero otherwise and $D_2 = 1$, if $t_2 < t$ and zero otherwise. Initialising $t_0 = 0$, (2.18) becomes:

$$\text{Tech} = a_0 + b_0 t + c_0 t^2 + (c_1 - c_0)(t - t_1)^2 \cdot D_1 + (c_2 - c_1)(t - t_2)^2 \cdot D_2$$
$$+ \sum_{i=3}^{n} (c_i - c_{i-1})(t - t_i)^2 \cdot D_i. \tag{2.19}$$

ESTIMATION AND TESTING

The decomposition of production responses into separate effects such as technical change and substitution effects, changes in the number of inputs, scale and scope economies requires estimation of a proper cost function. Intuitively, producers are considered to exhibit cost minimising behaviour subject to production technology that acts as a binding constraint on costs. Assuming the production function is well behaved so that first- and second-order conditions are met, the producer minimises total production cost subject to y^0, the required output level, which acts as a binding constraint on total costs. Hence, the producer's objective function is to minimise:

$$C = \sum_{i=1}^{n} p_i q_i, \tag{2.20}$$

subject to the production function $f(q_1, \ldots, q_n) = y^0$, where p is the price of factor inputs $i = (1, 2, \ldots, n)$ and q denotes input quantity.[9] Respecifying (2.20) as a Lagrangean function yields:

$$L = \sum_{i=1}^{n} p_i q_i + \lambda [y - f(q_1, \ldots, q_n)] \tag{2.21}$$

Minimising (2.21) requires setting the first-order partial derivatives to zero:

$$L_i = p_i - \lambda f_i = 0 \qquad i = 1, 2, \ldots, n$$

and

$$L_\lambda = y^0 - \lambda f_i = 0, \tag{2.22}$$

and the Bordered–Hessian matrix of second-order partial derivatives is negative semi-definite. Given these conditions, the reduced form solutions to the cost-minimisation problem imply the factor demand equations:

$$q_i = q_i^*(p_1, p_2, \ldots, p_n, y^0) \qquad i = 1, 2, \ldots, n$$

and

$$\lambda = \lambda^*(p_1, p_2, \ldots, p_n, y^0), \tag{2.23}$$

where q_i^* denotes the quantity choice of input i that minimises total costs, and λ^* denotes the firm's marginal cost function. Substitution of the solutions to cost minimisation into the original objective function yields the indirect cost function:

$$C^* = \sum_{i=1}^{n} p_i q_i^*(p_1, \ldots, p_n, y^0) + \lambda^*(p_1, \ldots, p_n, y^0) \, [y^0 - \sum_{i=1}^{n} q_i^*(p_1, \ldots, p_n, y^0)], \tag{2.24}$$

where C^* denotes the minimum cost as the exogenous parameters, p_i and y^0, vary. Then, by Shephard's lemma:

$$\frac{\partial C^*}{\partial p_i} = q_i^*(p_1, \ldots, p_n, y^0), \tag{2.25}$$

since $(p_i - \lambda^* f_i) = 0$ and $y^0 - f(q_1^*, \ldots q_n^*) = 0$. This result states that the change in the indirect cost function with respect to a change in the price of the ith factor yields the firm's demand schedule for factor i. Similarly, the change in the unconstrained cost function $C = \sum_{i=1}^{n} p_i q_i$ with respect to a change in the price of factor i yields:

$$\frac{\partial C}{\partial p_i} = q_i. \tag{2.26}$$

Noting that (2.23) implies the unconstrained cost function is linear in input price p_i, it can be seen that when $q_i = q_i^*$, $\frac{\partial C}{\partial p_i} = \frac{\partial C^*}{\partial p_i}$. Hence, the minimal cost is equal to the unconstrained total cost when the optimal input quantity is chosen. However, since C^* always defines the minimum cost path as input prices and output changes, the choice of $q_i \neq q_i^*$ implies that $C^* < C$. Further, since the unconstrained total cost is linear in input prices, C^* must be strictly concave in input prices. In turn, strict concavity of the cost function implies that the second derivative $C_{p_i p_i}^* \leq 0$. Since, by Shephard's lemma

$C_{p_i}^* = q_i^*$, $C_{p_i p_i}^* = \frac{\partial q_i^*}{\partial p_i} \leq 0$ implies downward sloping factor demand schedules. A further implication of a proper cost function is that the demand curves $q_i = q_i^* (p_1, \ldots, p_n, y^0)$ are homogenous of degree zero and the cost function is homogenous of degree one. The implication is that factor demands are affected only by relative price changes rather than absolute changes. Formally, this is demonstrated by minimising the scaled total costs function subject to the factor prices tp_i, that is, $C = \sum_{i=1}^{n} tp_i q_i$ subject to the production constraint $f(q_1, \ldots, q_n) = y^0$, where t denotes an arbitrary scaling factor on input prices. Forming the Lagrangean, the first-order conditions are:

$$L_i = tp_i - \lambda f_i = 0 \qquad i = 1, 2, \ldots, n$$

and

$$L_\lambda = y^0 - f(p_1, \ldots, p_n). \tag{2.27}$$

Solving for factor prices yields:

$$\frac{tp_i}{tp_j} = \frac{p_i}{p_j} = \frac{f_i}{f_j}, \; i \neq j. \tag{2.28}$$

Thus, the relationship between relative factor prices and relative factor production shares is invariant to equi-proportionate changes in prices.

Optimising behaviour is subsumed under the assumption that the producer will attempt to maximise output or minimise costs of production. Cost minimisation is subject to the constraints imposed by the technology available to the producer, which is assumed to reflect decreasing marginal rates of substitution as an input is successively substituted for another. Together, these assumptions define a unique minimum cost for a specified output. Similarly, homogeneity of total cost in input prices is demonstrated by:

$$C^* = [t(p_1, \ldots, p_n), y^0] \equiv \sum_{i=1}^{n} tp_i q_i^*(p_1, \ldots, p_n, y^0)$$

$$\equiv t \sum_{i=1}^{n} p_i q_i^*(p_1, \ldots, p_n, y^0)$$

$$\equiv tC^*(p_1, \ldots, p_n, y^0). \tag{2.29}$$

In addition to these conditions, Young's theorem implies symmetry restrictions on the cross-price effects, that is, $\frac{\partial q_i^*}{\partial p_j} = \frac{\partial q_j^*}{\partial p_i}$. Further propositions for proper cost function estimation are that the cost function is non-decreasing in input prices and that marginal costs are always positive. Collectively, concavity, homogeneity and symmetry requirements together with downward sloping factor demand schedules that are non-decreasing in input prices and display positive marginal costs are referred to as defining the consistency region.

Parameterising the general cost function requires estimation of a flexible functional form such as the Composite cost function or its nested alternatives such as the Translog, Generalised Leontief or Generalised CES-Quadratic functions. For example, consider the following Translog cost function for a two-input two-output production process:

$$\ln C = \alpha_0 + \sum_i \beta_i \ln p_i + \sum_k \beta_k \ln Q_k + \frac{1}{2}\sum_i \sum_j \gamma_{ij} \ln p_i \ln p_j$$

$$+ \frac{1}{2}\sum_i \sum_k \gamma_{ik} \ln p_i \ln Q_k + \frac{1}{2}\sum_k \sum_l \gamma_{kl} \ln Q_k \ln Q_1 \qquad (2.30)$$

where $i, j = L, K$ and $k, l = 1, 2$. Symmetry restrictions imply that $\gamma_{LK} = \gamma_{KL}$ and $\gamma_{12} = \gamma_{21}$. To impose linear homogeneity on the cost function, calculate the percentage change in cost with respect to an equi-proportionate change in input prices, that is, $\dfrac{dC}{C}\bigg|\dfrac{dP}{P}$. Grouping parameters reveal homogeneity restrictions: $\beta_L + \beta_K = 1$, $\beta_{KK} + \beta_{KL} = 0$, $\beta_{LL} + \beta_{LK} = 0$, $\beta_{1K} + \beta_{1L} = 0$, $\beta_{2K} + \beta_{2L} = 0$. The marginal cost with respect to outputs is non-negative when:

$$\frac{\partial C}{\partial Q_i} = (\beta_i + \beta_{ii} \ln Q_i + \beta_{ij} \ln Q_j + \Sigma_n \beta_{in} \ln p_n)\frac{C}{Q_i} \geq 0, \qquad (2.31)$$

for all $i \neq j$ where $i, j = 1, 2$ and $n = K, L$. Non-decreasing input prices imply:

$$\frac{\partial \ln C}{\partial \ln P_n} = \beta_n + \beta_{nn} \ln P_n + \beta_{1n} \ln P_1 + \Sigma_i \beta_{in} \ln Q_i \geq 0 \qquad (2.32)$$

for all $1 \neq n$ where $1, n = K, L$ and $i = 1, 2$. Finally, following Diewert and Wales (1987), the cost function is concave in input prices if the matrix:

$$\Gamma(Q) \equiv \begin{bmatrix} \beta_{LL} - S_L^*(1 - S_L^*) & \beta_{LK} + S_L^* S_K^* \\ \beta_{KL} + S_K^* S_L^* & \beta_{KK} - S_K^*(1 - S_K^*) \end{bmatrix} \qquad (2.33)$$

is negative definite, where * indicate the estimated cost shares.

Estimation of a proper cost function using a flexible functional form yields a rich source of empirical measurements and testable hypotheses such as testing for separability of outputs, marginal costs and estimates of the rate, source and bias of technical change. Other inferences able to be drawn from valid cost function estimations include the interactions of inputs, demand elasticities and elasticity of scale and scope. The elasticity of scale defines the percentage change in output with respect to a percentage change in inputs, that is, $\dfrac{dQ}{Q}\bigg|\dfrac{dq}{q}$. For multi-output, multi-input cost functions, Caves et al. (1981) demonstrate that the elasticity of scale (ε_{Qq}) is the inverse of the elasticity of cost with respect to output:

$$\varepsilon_{CQ} = \sum_{i=1}^{n} \frac{\partial C}{\partial Q_i} = \frac{1}{\varepsilon_{Qq}}. \tag{2.34}$$

For the Translog specification (2.30) the elasticity of cost is defined:

$$\sum_{i=1}^{n} \frac{\partial \ln C}{\partial \ln Q_i} = \beta_i + \beta_{ii} \ln Q_i + \beta_{ij} \ln Q_j + \Sigma_n \beta_{in} \ln p_n. \tag{2.35}$$

Measuring economies of scope is difficult since analysis requires comparing total costs of multiple-output production to the cost of specialised production. Formally, economies of scope exist for outputs Q_1 and Q_2 if $C(Q_1, Q_2) \leq C(Q_1, 0) + C(0, Q_2)$. When there are multiple firms in an industry with some producing a single output and others producing multiple outputs, testing for economies of scope can be assessed by comparing the estimated cost functions of the multi-output firms vis-à-vis the single-output firms. When these data are unavailable, tests of economies of scope require testing for non-jointness. Non-jointness implies the cost of producing several outputs equals the sum of the costs of producing the outputs separately (Evans and Heckman, 1983). Denny and Pinto (1978) show that when the cost function is linear homogenous in input prices, symmetric in the cross-price elasticities and exhibits constant returns to scale, then the Translog cost function is non-joint if $\gamma_{12} = -\beta_1\beta_2$. If this restriction is accepted, the cost function does not exhibit economies of scope. Alternatively, Jorgenson (1986) has suggested use of Cholesky factorisation to test non-jointness. Under the assumptions of competitive markets and a production function which is homogenous of degree one, a necessary and sufficient condition for non-joint production is that the cost of each output is independent of the prices of the inputs used to produce another output. Equating input price ratios to the marginal products in the cost function yields the restrictions on the marginal products. A test on the hypothesis of non-joint production is established by transforming the matrix of second-order restrictions using Cholesky factorisation. When the characteristic values of Hessian matrix are zero production is non-joint.

ESTIMATION RESULTS

An inherent advantage in modelling telecommunications carrier technology via a cost function is the ability to observe and anticipate likely carrier production decisions based on changes in the contemporary economic environment. By estimating the magnitude of economies of scale and scope, policy makers can assess whether incumbent pricing strategy is the result of the employment of highly productive technology or simply predatory. By understanding the contribution of various sources of technical change, management can efficiently allocate scarce investment funds to extract maximal benefit. In addition, accurate measurement of the impact of interactions between technical change, input prices and scale changes on total costs enable investors to better anticipate how carriers will respond to external changes in market conditions. For example, switching from analogue circuit switched technology to digital packet switching is likely to generate new efficiencies, which in turn imply changes in market structure. New sources for scale economies, for example, lead to specialisation, vertical integration and a drive to expand markets. Alternatively, scope economies suggest service bundling, joint production and cross-industry merger. Moreover, given the high fixed costs of changing technology, carriers need to expand output and achieve scale quickly. Knowledge of the magnitude of scale economies provides a means of assessing the future rate of growth in output markets, and the extent to which increased competition will stimulate price reductions in output markets.

Toward that end, available empirical evidence generated by cost function studies in past decades provides a wealth of information about the telecommunications industry. Over that time, successive theoretical developments have permitted a wide variety of models using various flexible functional forms. Such specifications have evolved through time, changing to reflect increasing technical sophistication and radical changes in market structure. Initially, telecommunications industry production was specified as a one-output, two-input technology. Nadiri and Schankerman (1981), for example, estimate a single-output, three-input Translog model and a single-output, four-input model. The models indicate price inelastic demand for labour and capital and unit elastic material demand. Both models suggest increasing returns to scale with a pronounced downward time trend (Kiss and Lefebvre, 1987). Technical change is reported to be capital using, labour saving and material neutral, but to suggest accelerating total cost increases. Smith and Corbo (1979) report inelastic factor demand, substitution between labour and capital and labour and material. Capital and material demands are largely independent while output growth causes declines in the capital to labour ratio. Technical change is capital using, labour saving and material neutral. Denny et al. (1979) report inelastic factor demand, labour–capital and labour–material substitution and economies of scale, accelerating downward cost shifts for Bell Canada. Kiss et al.'s (1983) single output model reveals inelastic factor demand, capital–labour and material–labour substitution, economies of scale and capital saving technical change using a single output Generalised Translog specification. Their two outputs increased the estimates of returns to scale. In order to test the robustness of these findings, Christensen et al. (1983) investigate the robustness of parameter estimates across specifications. In general, estimates are robust to changes in technological change proxies and changes in capacity utilisation, with downward technological shifts, scale economies in the range [1.5, 1.9]. Kiss and Lefebvre (1985) model Bell Canada (1956–81) and Alberta Government Telephones (1952–78) separately and as a two-firm panel set using a single output specification. They consistently find inelastic factor demand, increasing returns to scale and decelerating downward shifts in cost.

The inability of one-output models to assess subadditivity and economies of scope prompted a succession of multiple-output specifications. In their two-output monopoly-competitive output model, Smith and Corbo (1979) found price inelastic demand for three factors of demand and substitution between labour and capital, and labour and materials. The interaction parameters between labour and capital inputs and both outputs are highly significant. The results also show that growth in output strongly increased labour–capital and labour–materials ratios for the monopoly model as well as moderately increasing returns to scale. The competitive output induced downward sloping marginal cost for monopoly output. Denny et al. (1979) add Wide Area Telephone Service output to estimate a three-output, three-input model. Inelastic input price elasticity of demand remains robust to the specification along with scale economies. In addition, the local–toll and local–other output interaction terms indicate cost complementarity. Testing for functional form, Breslaw and Smith (1981) estimate a Cobb–Douglas cost function, implying unitary factor substitution. Their specification supports inelastic factor demand and scale economies. Further, their estimates indicate that increases in output volume reduced the marginal costs of other outputs and report statistically significant cost complementarity between local and message toll services (Kiss and Lefebvre, 1987). Evans and Heckman (1984) introduce a further innovation, adopting the Box–Cox transformation procedure to indicate the adjusted Translog model best represents AT&T production technology. Using this specification, they find the factors are substitutes and report substantial economies of scale. Technical change is neutral, although the second-order output parameters are large and unstable. Re-examining

Evans and Heckman's results, Diewert and Wales (1991a) report their cost function does not exhibit the necessary regularity conditions. Despite earlier concerns of model instability, Fuss and Waverman (1981) estimate a three-output model assuming endogeneity in toll output. Hence, they estimate two demand equations jointly with the cost function and input demand equations. Their results indicate constant returns to scale for local output and increasing returns to scale for the two long-distance outputs. The results also suggest that there are diseconomies of scope between local and toll outputs. Allowing the Box–Cox parameters to vary between outputs, Kiss and Lefebvre re-estimated this model and report constant returns to scale is rejected for local output.

Recognising the increasing competitiveness among US carriers, Bernstein (1988) estimates a dynamic cost-of-adjustment production model, using standard forecast techniques to simulate producer decision-making in relation to future price and output movements. Among his results, Bernstein reports that increasing returns to scale for Bell Canada is due to the limited effect of toll output on variable cost. In a later study, Gentzoglanis and Cairns (1989) estimate increasing returns to scale for Bell Canada and Alberta Government Telephones. However, Fuss (1992) dismisses the results because the cost function violates linear homogeneity. Ngo (1990) and Röller (1990) re-examine economies of scale for Bell Canada and AT&T, respectively. Using an extended version of Evans's and Heckman's data and a CES-Quadratic cost function, Röller also finds economies of scope estimates. Diewert and Wales (1991b) estimate multiple-output cost functions for the US and Japanese telecommunications industries. For the USA, they find a strong degree of labour saving technical progress. Marginal cost for local cost trended up until 1961, changed direction in 1974 and trended up again in 1977. The returns to scale measure initially showed increasing returns in 1947 and trended down in 1977. A similar pattern emerged for Japanese data, although toll marginal cost exhibits a cyclical pattern. Economies of scale initially constant are rising through time to reveal increasing returns in 1987.

In response to long held concerns that available data sets offer insufficient degrees of freedom to estimate stable parameters, Shin and Ying (1992) substantially extend and refine the available data, creating a cross-section, time-series panel data set of 58 local exchange carriers over eight years. In contrast to earlier studies, their results suggest only modest scale economies. Examination of their estimates of Translog parameters suggest technical change is cost increasing, labour saving and capital using. Moreover, output interaction terms indicate cost complementarities between local and toll outputs. Following Shin and Ying, and Krouse et al. (1999) examine the Bell System post-divestiture period using panel data on the divested Bell Operating Companies. Their specification explicitly modelled the effects of changes in the regulatory framework governing US telecommunications. Of particular interest, Krouse et al. find that rate of return regulation does not impose a binding constraint of industry costs. Further, they find that regulatory reform is cost increasing, divestiture is cost saving, but that both effects have tended to increase costs through time. Moreover, divestiture has contributed to labour-saving and capital-using effects. To capture endogeneity in outputs Nadiri and Nandi (1999) estimate both short- and long-run dynamic demand and supply models. The empirical results reveal that productivity growth in US telecommunications is largely explained by mark-up pricing, while the influence of technical change is decreasing. In addition, the post-divestiture period is distinguished by increasing labour productivity, decreasing cost elasticity with respect to toll service, but an increasing cost elasticity of local service. Sung and Gort (2000) further enhance panel data approaches by introducing explicit arguments for changes in input quality. They seek to clarify and investigate Shin's and Ying's finding that telecommunications carriers exhibit constant returns to scale. Their results indicate that returns to scale are a decreasing function of firm size, with small firms exhibiting increasing returns, while larger firms exhaust economies of scale. In addition, they confirm that capital quality improvements are labour saving and capital using.

Finally, Bloch et al. (2001) analyse the Australian telecommunications sector for the period 1926–91. Given the radical changes in technology over the 65-year span, their study incorporates quadratic spline arguments for technical change. Using a Hicks-neutral specification, the results report that technical change is cost increasing with a downward drift through time. Own-price elasticity estimates of demand for labour and capital are inelastic. Returns to scale are constant until the post-war period and then trend upwards revealing scale economies by the 1980s. Overall, despite early controversy surrounding the estimation of inconsistent cost functions, a variety of specifications and studies identify a set of stylised facts. In general, the telecommunications industry exhibits increasing economies of scale, has experienced increasing efficiency as a result of technical change and that technology has increasingly become capital-intensive. Moreover, the twin effects of divestiture and deregulation have created a shift in the sources of productivity growth in regulation. Finally, in spite of early concerns, it has been demonstrated that competition has not had a detrimental impact on efficiency.

CONCLUSION

This chapter seeks to highlight some theoretical and empirical issues relating to the estimation of telecommunications cost functions. The survey is indicative rather than exhaustive, presenting the stylised facts and general methodological approach to analysing telecommunications technology. The theoretical developments provide the ability to flexibly decompose the simultaneous effects on costs of changes in the quantities of inputs, relative prices, scale, scope and technology based on a minimal set of behavioural assumptions. In addition, the recent emergence of the Internet and associated innovations such as the change from circuit switched technology to packet switching means that the cost structure of telecommunications may have changed radically. Clearly, there is a need for future study and innovation in the analysis of the economics and cost structure of telecommunications technology.

NOTES

1. See Kiss and Lefebvre (1987) for a comprehensive survey.
2. Cost function estimation has the advantage of providing direct estimates of marginal cost, output elasticity, factor prices and technological change.
3. Diewert (1971) defines a functional form flexible if any twice-differentiable cost function can be approximated to the second order.
4. L'Hôpital's rule states that $\lim\limits_{x \to a} \dfrac{m(x)}{n(x)} = \lim\limits_{x \to a} \dfrac{m'(x)}{n'(x)}$. Applying this to the Box–Cox transformation yields: $\lim\limits_{\lambda \to 0} \dfrac{x^\lambda - 1}{\lambda} = \dfrac{\dfrac{d(x^\lambda - 1)}{d\lambda}}{1} = x^\lambda \ln x$, since $\lambda = 0$, $x^\lambda = 1$ and the Box–Cox transformation of x is $\ln x$. See Greene (1993).
5. The specification avoids the need to impose strong separability between inputs and outputs. Thus, input ratios and shares need not be independent of output.
6. This is referred to as satisfying the factor reversal test.
7. Diewert (1976) examines several indexes. The Törnqvist and Malmquist are shown appropriate aggregation functions for the Translog.
8. By contrast, the production function under profit maximising conditions and a set of value shares that add to one imply constant returns. For a fuller analysis of the implications of cost function estimation, see Nadiri and Schankerman (1981) or Jorgensen (2000).
9. This analysis is based on Silberberg and Suen (2001).

REFERENCES

Arrow, K.J., Chenery H.B., Minhas, B.S. and Solow, R.M. (1961), 'Capital–labour substitution and economic efficiency', *Review of Economics and Statistics*, **43**, 225–50.

Baltagi, D.H. and Griffin, J.M. (1988), 'A general index of technical change', *Journal of Political Economy*, **96**, 20–41.

Baumol, W.J. (1977), 'On the proper cost tests for natural monopoly in a multi-product industry', *American Economic Review*, **67**, 809–22.

Bernstein, J.I. (1988), 'Multiple outputs, adjustment costs and the structure of production for Bell Canada', *International Journal of Forecasting*, **4**, 207–19.

Bloch, H., Madden, G., Coble-Neal, G. and Savage, S. (2001), 'The cost structure of Australian telecommunications', *Economic Record*, **77** (239), 338–50.

Braunstein, Y.M., Jussawalla, M. and Morris, S. (1999), 'Globalization in telecommunications: Present trends and future options', presented at the Pacific Telecommunication Conference.

Braunstein, Y.M. and Pulley, L.B. (1998), 'Economies of scale and scope in telephony: applying the composite cost function to Bell System data', in *Communication and Trade: Essays in Honor of Merhoo Jussawalla*, Hampton Press, New Jersey.

Breslaw, J.A. and Smith, J.B. (1981), 'Efficiency, equity and regulation: an econometric model of Bell Canada', Discussion Paper No. 81-01, Institute of Applied Economic Research, Concordia University.

Caves, D.W., Laurits, R.C. and Swanson, J.A. (1981), 'Productivity growth, scale economies, and capacity utilization in U.S. railroads, 1955–74', *American Economic Review*, **71** (5).

Charnes, A., Cooper, W.W. and Sueyoshi, T. (1988), 'A goal programming/constrained regression review of the Bell System breakup', *Management Science*, **34**, 1–26.

Christensen, L.R., Cummings D. and Schoech, P.E. (1981), 'Econometric estimation of scale economies in telecommunications', SSRI Workshop Series No. 8124, University of Wisconsin.

Christensen, L.R., Cummings, D. and Schoech, P.E. (1983) 'Econometric estimation of scale economies in telecommunications', in Courville, L., de Fontenay, A. and Dobell, R. (eds), *Econometric Analysis of Telecommunications*, Amsterdam: North-Holland, 27–54.

Christensen, L.R., Jorgenson, D.W. and Lau, L.J. (1971), 'Conjugate, duality and the Transcendental Logarithmic production function', *Econometrica*, **39**, 255–6.

Coelli, T., Rao, D.S.P. and Battese, G.E. (1999*)*, *An Introduction to Efficiency and Productivity Analysis*, Dordrecht: Kluwer Academic.

Davis, B.E., Caccappolo, G.J. and Chaudry, M.A. (1972), 'FORECYT: an econometric forecasting model for corporate planning analysis', Bell System Corporate Planning Organization Report, New York.

Denny, M., Fuss, M. and Everson, C. (1979), 'Productivity, employment and technical change in telecommunications: the case of Bell Canada', Final Report to the Department of Communications, March.

Denny, M. and Pinto, C. (1978), 'An aggregate model with multiproduct technologies', in Fuss, M. and McFadden, D. (eds), *Production Economics: A Dual Approach to Theory and Applications*, Amsterdam: North-Holland.

Diewert, W.E. (1971), 'An application of the Shephard duality theorem: a generalized Leontief production function', *Journal of Political Economy*, **79**, 481–507.

Diewert, W.E. (1983), 'The theory of the output price index and the measurement of real output change', in Diewert, W.E. and Montmarquette, C. (eds), *Price Level Measurement*, Statistics Canada, 1039–113.

Diewert, W.E. (1976), 'Exact and superlative index numbers', *Journal of Econometrics*, **4**, 115–45.

Diewert, W.E. and Wales, T.J. (1987), 'Flexible functional forms and global curvature conditions', *Econometrica*, **55**, 43–68.

Diewert, W.E. and Wales, T.J. (1991a), 'Multiproduct cost function estimation and subadditivity tests: a critique of the Evans and Heckman research on the US Bell System', Discussion Paper, Department of Economics, University of British Columbia.

Diewert, W.E. and Wales, T.J. (1991b), 'On the subadditivity of telecommunications cost functions: some empirical results for the US and Japan', Discussion Paper, Department of Economics, University of British Columbia.

Dixit, A.K. (1976), *The Theory of Equilibrium Growth*, Oxford: Oxford University Press.

Dobell, A.R., Taylor, L.D., Waverman, L., Liu, T.H. and Copeland, M.D.G. (1972), 'Communications in Canada', *Bell Journal of Economics and Management Science*, **3**, 175–219.

Eichorn, W. and Voeller, J. (1983), 'The axiomatic foundations of price indexes and purchasing power parities', in Diewert, W.E. and Montmarquette, C. (eds), *Price Level Measurement*, Statistics Canada, 411–50.

Evans, D.S. and Heckman, J.J. (1983), 'Multiproduct cost function estimates and natural monopoly tests for the Bell System', in Evans, D.S. (ed.), *Breaking Up Bell*, Amsterdam: North-Holland, 253–82.

Evans, D.S. and Heckman, J.J. (1984), 'A test for subadditivity of the cost function with an application to the Bell System', *American Economic Review*, **74**, 615–23.

Fuss, M. (1992), 'Economic evidence on scale, scope and the presence of natural monopoly: What can it tell us about the desirability of competition in public long-distance telephone service?', Working Paper Series 9210, University of Toronto.

Fuss, M. and Waverman, L. (1977), 'Multi-product, multi-input cost functions for a regulated utility: the case of telecommunications in Canada', presented at the NBER Conference on Public Regulation, Washington DC.

Fuss, M. and Waverman, L. (1981), 'The regulation of telecommunications in Canada', Technical Report No. 7, Economic Council of Canada, Ottawa.

Gentzoglanis, A. and Cairns, R.D. (1989), 'Cost comparisons and natural monopoly issues for two Canadian telecommunications carriers: Bell Canada and Alberta Government Telephones', presented to the Bellcore-Bell Canada Conference on Telecommunications Costing in a Dynamic Environment, San Diego.

Gollop, F.M. and Roberts, M.J. (1981), 'The structure of production, technological change, and the rate of growth of total factor productivity in the US Bell System', in Cowing, T. and Stevenson, R. (eds), *Productivity Measurement in Regulated Industries*, New York: Academic Press.

Greene, W.H. (2000), *Econometric Analysis*, 4th edn, New Jersey: Prentice Hall.

Heathfield, D.F. and Wibe, S. (1987), *An Introduction to Cost and Production Functions*, Hampshire: MacMillan Education.

Hicks, J.R. (1932), *The Theory of Wages*, London: Macmillan.

Jorgenson, D.W. (1986), 'Econometric methods for modeling producer behavior', in Griliches, Z. and Intriligator, M.D. (eds), *Handbook of Econometrics*, Volume III, Amsterdam: North-Holland.

Jorgenson, D.W. (2000), *Econometrics: Econometric Modeling of Producer Behavior: Econometrics*, Volume I, Cambridge: MIT Press.

Kiss, F., Karabadajian, S. and Lefebvre, B.J. (1983), 'Economies of scale and scope in Bell Canada', in Courville, L., de Fontenay, A. and Dobell, R. (eds), *Economic Analysis of Telecommunications: Theory and Applications*, Amsterdam: North-Holland.

Kiss, F. and Lefebvre, B.J. (1985), 'Economies of scale in common carriers', presented at the Bellcore Economic Cost Modelling Forum, Atlantic City.

Kiss, F. and Lefebvre, B.J. (1987), 'Econometric models of telecommunications firms', *Revue Economique*, **2**, 307–74.

Krouse, C.G., Danger, K.L., Cabolis, C., Carter, T.D., Riddle, J.M. and Ryan, D.J. (1999), 'The Bell System divestiture/deregulation and the efficiency of the operating companies', *Journal of Law and Economics*, **62**, 61–87.

Nadiri, M.I. (1982), 'Producers theory', in Arrow, K. and Intriligator, M.D. (eds), *Handbook of Mathematical Economics*, Volume II, Amsterdam: North-Holland.

Nadiri, M.I. and Nandi, B. (1999), 'Technical change, markup, divestiture, and productivity growth in the US telecommunications industry', *Review of Economics and Statistics*, **81**, 488–98.

Nadiri, M.I. and Schankerman, M. (1981), 'The structure of production, technological change, and the rate of growth of total factor productivity in the US Bell System', in Cowing, T. and Stevenson, R. (eds), *Productivity Measurement in Regulated Industries*, New York: Academic Press.

Nerlove, M. (1963) 'Returns to scale in electricity supply', in Christ, C. (ed.), *Measurement in Economics: Studies in Mathematical Economics and Econometrics in Memory of Yehuda Greenfeld*, Stanford: Stanford University Press.

Ngo, H. (1990), 'Testing for natural monopoly: The case of Bell Canada', presented at the 1990 Meeting of the Canadian Economics Association, Victoria.

Panzar, J.C. and Willig, R.D. (1975), 'Economies of scale and economies of scope in multi-output production', Discussion Paper 33, Bell Laboratories.

Pulley, L.B. and Braunstein, Y.M. (1992), 'A composite cost function for multi-product firms with an application to economies of scope in banking', *Review of Economics and Statistics*, **74**, 221–30.

Röller, L-H. (1990), 'Proper quadratic cost functions with an application to the Bell System', *Review of Economics and Statistics*, **72**, 202–10.

Shephard, R.W. (1970), *Cost and Production Functions*, Princeton: Princeton University Press.

Shin, R.T. and Ying, J.S. (1992), 'Unnatural monopolies in local telephone', *Rand Journal of Economics*, **23**, 171–83.

Silberberg, E. and Suen, W. (2001), *The Structure of Economics: A Mathematical Analysis*, New York: McGraw-Hill.

Smith, J.B. and Corbo, V. (1979), 'Economies of scale and economies of scope of Bell Canada', Institute of Applied Economic Research, Concordia University, Final Report to the Department of Communications.

Suits, D.B., Mason, A. and Chan, L. (1978), 'Spline functions fitted by standard regression methods', *Review of Economics and Statistics*, **60**, 132–9.

Sung, N. (1998), 'The embodiment hypothesis revisited: evidence from the US local exchange carriers', *Information Economics and Policy*, **10**, 219–36.

Sung, N. and Gort, M. (2000), 'Economies of scale and natural monopoly in the US local telephone industry', *Review of Economics and Statistics*, **82**, 964–7.

TeleGeography (1999), *Global Telecommunications Traffic Statistics and Commentary*, TeleGeography, Washington DC.

Vinod, H.D. (1975), 'Applications of new ridge regression methods to a study of Bell System scale economies', Federal Communications Commission, Docket No. 20003, Bell Exhibit No. 42.

Vinod, H.D. (1976), 'Bell System scale economies and estimation of joint production functions', Federal Communications Commission, Docket No. 20003, Bell Exhibit No. 59.

Zellner, A. (1962), 'An efficient method for estimating seemingly unrelated regressions and tests for aggregation bias', *Journal of the American Statistical Association*, **57**, 348–68.

L96 1u9l

O24

3. Telecommunications productivity

Laurits R. Christensen, Philip E. Schoech and Mark E. Meitzen

INTRODUCTION

This chapter presents a survey of total factor productivity (TFP) methods and studies for the telecommunications industry. TFP is the ratio of total output to total input, where total output consists of all the services provided by the industry and total input includes all resources used to provide those services. TFP is widely recognized as a comprehensive measure of productive efficiency. It is also an important component of price cap regulation. TFP growth can arise from a number of sources, including technical change (shifts in the production function), increasing returns to scale or density and market imperfections. Among these market imperfections are non-marginal cost pricing and factor market imperfections due to regulatory constraints, for example, rate of return regulation. The direct approach to measuring TFP is based on explicit measures of output and input. The measures of output and input are based on economic indexing techniques and, therefore, this approach is referred to as the indexing method. Complementing the indexing method is econometric cost analysis. By estimating the cost function, one can obtain estimates of the sources of TFP growth. In addition to the direct approach to measuring TFP, there have been some studies employing an indirect approach, which focuses on trends in output prices. This chapter is organized as follows. In the next section, we discuss methodological issues in measuring TFP. Next, we survey empirical studies of TFP for the telecommunications industry. We then examine the relationship between TFP growth and output growth, and conclude with a discussion of TFP and its application to price cap regulation.

METHODOLOGICAL ISSUES

While a general discussion of total factor productivity methods can be found in a number of sources, we focus on a few issues that are particularly relevant to the measurement of TFP in the telecommunications industry.[1] These issues include measurement of telecommunications output, measurement of capital input, and methods for aggregating the diverse outputs and inputs into overall indexes of total output and total input.

Measurement of Output

In general, there are two ways to measure outputs. First, one can use physical quantity measures. One industry for which physical measures are used to represent output is the electric utility industry where kilowatt-hours are often used. Second, one can use deflated revenue measures where output is calculated by dividing revenue by an appropriate price index. This is the approach that has typically been used for the telecommunications industry. The choice of method depends upon the quality of data available and the complexity of the outputs being measured. In theory, if one had information on the revenue, price, and quantity of each service provided, then the physical output measure would produce the same result as the deflated revenue method. This is because the physical output measures would be based on the quantities of the different services provided, while the deflated revenue measures would be obtained by dividing the revenue of each service by its price, which equals the quantity provided. Because this detailed information is not generally available, one must choose the method that most closely approximates this ideal. In industries where there are relatively few types of outputs produced and the various outputs do not have significant quality differences, physical output measures generally capture all the relevant dimensions of output. On the other hand, the deflated revenue approach is generally better in those industries where a large number of services are provided, or where there are significant quality differences among the services provided, assuming that price indexes representing those different services are available.

Since the telecommunications industry is noted for the diverse set of services it provides, most telecommunications TFP studies use the deflated revenue approach to measuring output quantities. In some instances, good price indexes are unavailable for some or all outputs, and researchers effectively have no alternative to using physical output measures to represent those outputs. To minimize the possible bias arising from the use of physical measures, it is important to construct physical measures for as detailed service categories as possible. Furthermore, a single physical measure can be used as a proxy for other outputs, if those outputs are generally bundled together in fixed proportions.

Measurement of Capital Input

The second issue concerns the appropriate measure of capital input. In particular, for a given type of asset, for example, switching equipment or cable
and wire, one must determine how to weight assets of different vintages
when constructing an overall measure of capital input. There have generally been three approaches taken in the literature. The first approach is to
assume that an asset maintains a constant level of productive efficiency
over its lifetime. This assumption, commonly known as the 'one-hoss shay'
model of capital, is implicitly employed when the measure of capital input
is based on gross stock measures. The second approach is to assume that
an asset's level of productive efficiency declines at a linear rate as it ages.
This assumption is implicitly employed when the measure of capital input
is based on net stock measures. The third approach is to assume that the
productive efficiency of an asset declines geometrically as it ages. If we
define φ_t to be the relative productivity efficiency of an asset as age t, then

$$\varphi_t = (1-\delta)^{t-1} \tag{3.1}$$

where δ represents the geometric rate of efficiency decline.

   A substantial body of literature has been devoted to determining which
of these three patterns is the most appropriate assumption for capital measurement. Hulten (1990) and Hulten and Wykoff (1996) provide good sum-
maries of these results. Most of the empirical evidence supports the use of
the geometric assumption over the 'one-hoss shay' or straight-line assumptions. The geometric assumption has an additional attractive feature,
namely that one can compute the quantity of capital input in a relatively
straightforward manner. If asset deterioration is geometric, then the quantity of capital available for production at the end of year t is given by the
recursive equation:

$$K_t = (1-\delta)K_{t-1} + I_t \tag{3.2}$$

where K represents the quantity of capital and I the quantity of investment
during the year. Once one has the estimated rate of asset deterioration, a
time-series on the quantity of investment, and a starting value for the quantity of capital, one can compute the quantity of capital in all other years.
The use of the geometric assumption has become widespread in TFP
studies, including those of the telecommunications industry.

Aggregation of Inputs and Outputs

The final methodological issue we address is the appropriate basis for aggregation. Of particular importance for telecommunications total factor productivity studies is the appropriate set of weights to assign the different inputs and outputs when constructing measures of total output and input.

Following the practice initially established by Jorgenson and Griliches (1967), most studies of TFP weight the different outputs and inputs by their respective prices.[2] In the case of outputs, the relevant prices are those received by the firm for their outputs. For inputs, the relevant prices are the prices paid for the utilization of the inputs during the production period. In the case of capital, this is the rental price for those assets that are rented from other economic agents or the annual cost of utilizing an asset the company owns, sometimes known as the implicit rental price.

Jorgenson and Griliches set this conventional computation in an economic framework by considering a constant returns to scale technology (constant returns to density in a network industry such as telecommunications) and a price-taking, profit-maximizing firm with no quasi-fixed inputs.[3] In such a scenario, the prices of outputs represent their respective marginal costs and the prices of inputs are equal to their respective value of marginal products. Furthermore, the implicit rental cost of capital input is equal to property income, that is, total revenue less the payments to other factors of production, and the conventional measure of TFP growth represents the shift in the production function over time, otherwise known as technical change.

Berndt and Fuss (1986) and Hulten (1986) extend the Jorgenson and Griliches results. They consider a model with constant returns to scale technology, quasi-fixed capital, all other inputs and outputs variable, and a price-taking profit-maximizing firm. The profit-maximizing combination of inputs and outputs generates a difference between property income and the implicit rental cost of capital input. In this case, while property income per unit of capital does not equal the price of capital, it equals the value of marginal product. Basing the capital weight on property income when constructing total input leads to a TFP measure that represents technical change. It also represents the change in the long-run cost function due to technical change. Weighting capital by its implicit rental price results in a TFP measure that represents the change in the short-run cost function due to technical change.

In the telecommunications industry, a number of assumptions underlying the Jorgenson–Griliches, Berndt–Fuss and Hulten models are not valid. First, the telecommunications industry exhibits increasing returns to density. Second, telecommunications companies are common carriers, and therefore cannot optimize over output levels. In such a situation output prices will not equal marginal cost, and property income will not equal capital's value of marginal product. Furthermore, property income does not equal the cost of capital input. Consequently, the cost of capital input must be estimated from an implicit rental price formula that uses an *ex ante* opportunity cost or interest rate.

Denny et al. (1981) and Caves and Christensen (1988) provide an analytical framework that decomposes the changes in the conventional TFP index into technical change, exploitation of economies of scale and/or density, capacity utilization, and deviations from marginal cost pricing. In practice, such decomposition requires econometric estimation of the telecommunications production structure, or independent evidence on the cost structure of the industry.

Although the conventional TFP index captures more than the narrow-definition of TFP as technical change only, it provides useful information in the context of price cap-X factors. This can be observed by considering the relationship of telephone industry revenue to telephone industry total cost. An equilibrium condition in competitive markets is that revenue equals economic cost. If we let R represent total revenue, p_i the price of output i, q_i the quantity of output i, C total economic cost, w_j the price of input j and x_j the quantity of input j, we have the following relationship:

$$R = C$$
$$\sum p_i \cdot q_i = \sum w_j \cdot x_j. \tag{3.3}$$

Differentiating (3.3) with respect to time (and denoting the time derivative with a dot over the variable) we get:

$$\sum \dot{p}_i \cdot q_i + \sum p_i \cdot \dot{q}_i = \sum \dot{w}_j \cdot x_j + \sum w_j \cdot \dot{x}_j$$
$$\sum r_i \cdot \left(\frac{\dot{p}_i}{p_i}\right) + \sum r_i \cdot \left(\frac{\dot{q}_i}{q_i}\right) = \sum c_j \cdot \left(\frac{\dot{w}_j}{w_j}\right) + \sum c_j \cdot \left(\frac{\dot{x}_j}{x_j}\right) \tag{3.4}$$

where r_i represents the share of output i in total revenue and c_j represents the share of input j in total cost. Rearranging terms in equation (3.4), we obtain:

$$\sum r_i \cdot \left(\frac{\dot{p}_i}{p_i}\right) = \sum c_j \cdot \left(\frac{\dot{w}_j}{w_j}\right) - \left[\sum r_i \cdot \left(\frac{\dot{q}_i}{q_i}\right) - \sum c_j \cdot \left(\frac{\dot{x}_j}{x_j}\right)\right]. \tag{3.5}$$

Redefining terms in (3.5), we get:

$$\frac{\dot{P}}{P} = \frac{\dot{W}}{W} - \left(\frac{\dot{Q}}{Q} - \frac{\dot{X}}{X}\right) = \frac{\dot{W}}{W} - \frac{\dot{T}}{T}. \qquad (3.6)$$

The left-hand side of the equation is the Divisia index of output prices, the first term on the right-hand side is the Divisia index of input prices and the second term is the growth rate for the conventional measure of TFP. This implies that, in equilibrium, the rate of output price growth will equal the rate of input price growth less the rate of conventionally measured TFP growth. This analytical framework is generally used in setting the parameters of the price cap, and consequently historical studies of TFP can be quite useful.

Finally, the literature contains a few 'indirect' studies of TFP. These studies use a variant of (3.6) to infer the rate of telecommunications TFP growth relative to economy-wide TFP growth. If one can control for historical fluctuations in the ratio of revenue to cost, then differences in telecommunications and economy output price growth can be attributed to differences in input price growth and differences in TFP growth. The indirect studies have generally assumed that input price growth does not differ significantly across sectors, leaving differences in TFP growth as the remaining explanatory factor.

In the following survey of telecommunications TFP studies, most of the studies are based on conventional TFP studies. These studies measure total output by employing deflated revenue measures and consider three factors of production: capital, labor, and materials. In some instances, estimates of marginal cost are used instead of prices when aggregating total output. In others, the Berndt and Fuss model is used to develop estimates of short- and long-run efficiency changes. Finally, some studies are based on a two-factor, that is, capital and labor, model or add research and development as a third factor. These methodological differences will be noted as warranted.

STUDIES OF TELECOMMUNICATIONS TFP

United States Industry Studies

TFP studies of the United States (US) telecommunications industry can generally be divided between those that measure productivity of the industry prior to divestiture of the Bell System in 1984 and those that measure productivity after divestiture of the Bell System. The pre-divestiture studies focus on the Bell System or some measure of the total communications industry. The post-divestiture studies focus on the local exchange carrier (LEC) segment of the industry. Table 3.1 summarizes the results of these studies.

Table 3.1   Summary of US telecommunications TFP studies

Study	Period	TFP Growth (per cent)
Christensen et al. (1981) Bell System	1947–79	3.20
Jorgenson et al. (1987) Telephone Industry	1948–79	2.90
Nadiri and Schankerman (1981) Bell System Four-Factor Model	1947–76	4.10
American Productivity and Quality Center (1988) Communications Industry Two-Factor Model	1948–85	3.90
Crandall and Galst (1991) Total Industry Output Aggregation Based on Price	1960–70 1970–80 1980–88	3.09 3.92 2.65
Crandall and Galst (1991) Total Industry Output Aggregation Based on Marginal Cost	1960–70 1970–80 1980–88	2.70 3.40 0.98
Christensen et al. (1994, 1995a) LEC Industry	1984–93	2.40
Christensen et al. (1995b, 1997) LEC Industry Simplified Study	1984–93 1988–95	2.90 3.00
FCC (1997) LEC Industry	1988–95	3.30

Three of the TFP studies cited in Table 3.1 focused exclusively on the industry prior to divestiture. Christensen et al. (1981) used conventional measures of output and input to estimate Bell System TFP for the period 1947–79. Using similar methods, Jorgenson et al. (1987) measured industry TFP for the 1948–79 period. In their study, they used the broader telephone, telegraph and miscellaneous communications industry, although almost all of the industry was composed of telephone companies. Both studies produced similar results, with the Bell System study showing an average annual TFP growth rate of 3.2 per cent and the industry study showing 2.9 per cent. Nadiri and Schankerman (1981) employed a different methodology to look at Bell System productivity growth over the period 1947–76. While their study relied on many of the same data sources as Christensen et al., their TFP framework includes a fourth input, research and development. Using the 'four-factor' model, they obtained an average annual TFP growth rate of 4.1 per cent.

Three other studies cited in Table 3.1 analysed periods that spanned both the pre-divestiture and post-divestiture periods. The American Productivity and Quality Center (APQC) (1988) conducted a study of communications industry TFP growth over the period 1948–85. This study differed from the conventional approach in that it adopted a 'two-factor' measure of TFP. The framework included two factors of production – capital and labor – and a net output measure.[4] This study also looked at the larger communications sector of the US economy, which includes radio and television broadcasting in addition to telecommunications. Over the 1948–85 period, the APQC model produced an average annual rate of TFP growth of 3.9 per cent. As Domar (1961) has shown, the two-factor framework produces higher estimates of TFP growth than does the conventional TFP methods, with the magnitude of the difference related to the cost share of materials. After accounting for this, the APQC results fall into the same range as the Christensen et al. result.

Another study that spans both periods was by Crandall and Galst (1991). Their study focused on the telephone industry over the 1960–88 period. While they used conventional methods to measure total input, they produced alternative total output indexes. One of the alternatives was to weight the different output categories by their prices. With this total output index, annual TFP growth was 3.09 per cent over the 1960–70 period, 3.92 per cent over the 1970–80 period, and 2.65 per cent over the 1980–88 period. A second total output index was based on a set of estimated marginal costs and produced substantially lower rates of TFP growth, as shown in Table 3.1.[5]

The studies cited above measure TFP growth 'directly', that is, they measure growth in outputs and inputs directly. Spavins and Lande (1990) conducted an indirect study of telephone industry TFP from 1928 to 1988, thus also spanning both periods. They found that over the 1928–88 period, 'real telephone prices' fell at an average rate of 1.7 per cent, while over the shorter 1945–88 period, real telephone prices fell at an average annual rate of 2 per cent. They also found no evidence that the rate of productivity growth either accelerated or decelerated after divestiture. Since TFP growth during these sub-periods was between 0.5 per cent and 1 per cent per year, the results of this indirect study are broadly consistent with the other results cited above.

A number of post-divestiture TFP studies for the US LEC industry were performed in the course of the Federal Communications Commission's (FCC) reviews of its price cap plan for the LECs in the 1990s. Christensen et al. (1994, 1995a) analysed post-divestiture LEC TFP using the methodology developed by Christensen, Christensen and Schoech.[6] Over the 1984–93 period, they found LEC TFP grew at an average annual rate of 2.4 per cent.

Christensen et al. subsequently introduced a simplified method that relied completely on publicly available data. The first of the simplified studies (Christensen et al., 1995b) produced a slightly higher rate of TFP growth than their original study over the 1984–93 period, 2.9 per cent per year.[7] The second simplified study (Christensen et al., 1997) produced results for the 1988–95 period. [8] Average annual TFP growth over that period was 3 per cent.

As part of the price cap performance review for local exchange carriers, the FCC (1997) conducted a TFP study of the Regional Bell Operating Companies.[9] Although the methodology employed by the FCC differed from that used by Christensen et al., the TFP results were similar.[10] Over the 1985–95 period, the FCC study found LEC TFP growth to average 3.3 per cent per year.

Results of Other Studies

A number of studies have estimated Canadian telecommunications industry TFP growth. Those results are summarized in Table 3.2. Denny et al. (1981) estimated TFP growth for Bell Canada over the 1952–76 period. They also decomposed TFP growth into its various sources: nonmarginal cost pricing, nonconstant returns to scale, and technical change. Using prices to weight outputs, they estimated a 3.35 per cent average annual rate of TFP growth over the 1952–76 period. When departures from marginal cost pricing and the unexplained estimation residual were netted out, the estimated rate of residual TFP growth was 2.81 per cent. This latter measure represents efficiency gains from increasing returns to scale and technical change.[11]

Kiss (1983) used conventional methods to estimate the rate of Bell Canada TFP growth over the 1953–80 period. He also decomposed the rate of TFP growth into a scale effect and a technology effect. Over the 1953–80 period the average annual rate of TFP growth was 3.44 per cent. Of this, 2.71 percentage points were explained by the scale effect while 0.83 of a percentage point was explained by the technology effect.

Fuss (1994) estimated TFP growth for Bell Canada and BC Tel over various periods. These TFP measures relied on estimates of marginal cost to aggregate output. Fuss estimated various 'long-run' and 'short-run' rates of TFP growth using the approach laid out in Berndt and Fuss.[12] Applying the long-run model to Bell Canada data, Fuss found that the annual rate of TFP growth ranged from 2 per cent to 3.3 per cent over various sub-periods. Using the short-run model, his estimated rates of TFP growth increased by about 0.1 to 0.2 percentage points. For BC Tel, Fuss did not have sufficient data to estimate results based on the long-run model. Using

Table 3.2   Summary of other telecommunications industry TFP studies

Study	TFP		
	Period	Growth (per cent)	Notes
Denny et al. (1981)	1952–76	3.35	TFP growth
Bell Canada		2.81	Due to scale and tech. change
Kiss (1983)	1953–80	3.44	TFP growth
Bell Canada		2.71	Due to scale
		0.83	Due to tech. change
Fuss (1994)	1953–59	2.00	Long-run model
Bell Canada	1960–69	3.40	Long-run model
Output Aggregation Based	1970–79	3.30	Long-run model
on Marginal Cost	1980–89	2.60	Long-run model
	1980–89	2.70	Short-run model
Fuss (1994)	1953–59	2.20	Long-run model
Bell Canada	1960–69	4.10	Long-run model
Output Aggregation Based	1970–79	3.90	Long-run model
on Price	1980–89	4.60	Long-run model
	1980–89	4.70	Short-run model
Fuss (1994)	1980–89	4.10	Short-run model
BC Tel			
Output Aggregation Based			
on Marginal Cost			
Fuss (1994)	1980–89	5.60	Short-run model
BC Tel			
Output Aggregation Based			
on Price			
Stentor Resource Centre (1996)	1988–95	4.20	Aggregate
Stentor Member Companies		6.00	AGT
		5.70	BC Tel
		3.70	Bell Canada
		3.90	MTS
		2.80	MT&T
		2.40	Island Tel
Oniki et al. (1994)	1958–87	3.40	
NTT	1982–87	5.08	

the short-run model, he found substantially higher rates of TFP growth than those for Bell Canada, ranging from 4.1 per cent to 4.8 per cent per year. Finally, Fuss showed that using prices instead of marginal costs to aggregate output increases the rate of TFP growth. While the magnitude of the increase was relatively small during the 1953–9 period, during the 1980–89 period it increased the reported rate of TFP growth by 1.5 to 2 percentage points.

In their submission of evidence in the Canadian Radio and Telecommunications Commission (CRTC) price cap proceeding, the Stentor Resource Centre (1996) performed TFP studies for a number of the Stentor member companies. These studies were based on conventional TFP methods. For the combined companies, TFP growth averaged 4.2 per cent over the 1988–95 period. The individual company results varied around this average: AGT, 6.0 per cent; BC Tel, 5.7 per cent; Bell Canada, 3.7 per cent; MTS, 3.9 per cent; MT&T, 2.8 per cent and Island Tel, 2.4 per cent. Stentor noted that the 4.2 per cent combined average was consistent with Bell Canada's 30-year average of 4.1 per cent.

Finally, Oniki et al. (1994) estimated TFP for Japan's Nippon Telegraph and Telephone (NTT). For the 1958–87 period, they estimated average annual TFP growth of 3.4 per cent for NTT. However, for the 1982–7 period, TFP growth averaged 5.08 per cent. The authors decomposed NTT's TFP growth into scale effects, technological innovation, and competitive liberalization and privatization of NTT. They conclude that the liberalization polices of the early-1980s that opened Japan's telecommunications market to competition and moved toward the privatization of NTT were major sources of productivity improvement for NTT.

The Effect of Output Growth on TFP Growth

TFP growth can arise from various sources. One primary source of TFP growth is technological change, that is, shifts in the production function that allow a firm to use fewer inputs to produce the same amount of output. A second primary source of TFP growth is the exploitation of economies of density through output growth. Economies of density are present when average cost falls as more output is provided over a network of fixed size. Therefore, when economies of density are present, increases in output reduce the average level of inputs per unit of output. Consequently, increases in output growth lead to increases in TFP growth and, conversely, decreases in output growth lead to decreases in TFP growth.[13]

In addition to the rate of growth in total output, the sources of that output growth can be an important determinant of conventionally measured TFP growth, where prices are used to aggregate output. In industries with economies of density, prices are typically set above marginal cost for the various services provided by the firm, in order to generate revenue sufficient to cover total cost. When the markup of price relative to marginal cost varies over the services provided, growth in high markup services contributes more to TFP growth than growth in low markup services. Conversely, reductions in the growth of high markup services lead to disproportionate reductions in TFP growth.

Caves and Christensen (1988) developed a theoretical framework to analyse the relationship between output growth and TFP growth. Their framework is applicable to firms or industries that provide services over a network, such as the telephone industry. TFP growth in such industries can be related to economies of density, economies of scale, capacity utilization, and technological change. For LECs, network size can be represented by measures such as the number of access lines. When average cost falls as output rises over this given network, economies of density are present. Economies of scale describe the change in average cost when both output and the size of the network increase. If average cost declines when output and network size increase proportionately, economies of scale are present. Capacity utilization describes the impact on cost when capital is not at its optimum level, that is., the industry has too much or too little capital.[14]

Caves and Christensen included the telephone industry among the six industries they analysed. Their analysis of the telephone industry relied on the two major econometric studies of the US telephone industry that had been completed at the time of their study.[15] Both these studies showed a strong relationship between output growth and TFP growth. Though neither study includes measures of network size, Caves and Christensen concluded that the relationship between output growth and TFP growth was largely due to economies of density.

Two limitations of the studies on which Caves and Christensen relied are that neither study addressed the role of network size in TFP growth and both studies focused on the entire Bell System, which included both the Bell Operating Companies (the LECs) and the Long Lines division. Bell Communications Research (1987) provided an econometric cost analysis in 1987 of the Bell Operating Companies that specifically addressed the issue of network size. Using the methods developed by Christensen et al., Bellcore developed quarterly measures of output and input for the Bell Operating Companies, covering the years 1972 to 1982. The econometric models estimated from these data included measures of network size. The estimated models showed substantial economies of density, but constant returns to scale. The Bellcore results imply that a one per cent increase in output, holding network size fixed, leads to approximately a 0.8 per cent increase in TFP.

In two more recent papers, Shin and Ying attempted to focus on local carriers and incorporate measures of network size. While there are some problems in the data used in both of these papers, their results indicate support for large economies of density. The first of these studies (Shin and Ying, 1992) was based on data for 58 local telephone companies over the 1976–83 period. The output measures used in the study were number of local calls and number of toll calls, which do not capture the heterogeneity of services provided by local exchange companies. They characterized a third variable used in the analysis, number of access lines, as an output variable, but this variable defines the network over which services are being provided.

At the sample mean, the cost elasticities of local calls, toll calls, and access lines sum to 0.94, which shows minor economies of scale. However, the sum of the local call and toll call elasticities equals 0.25, which shows considerable economies of density. This would imply that a one per cent increase in local and toll calls would increase TFP by 0.75 per cent. The Ying and Shin (1993) paper reported a similar analysis of 46 local carriers over the 1976 87 period. This paper has the same data limitations, and produces results similar to those of the first paper. Together, the results of these two papers are consistent with those of the other studies.

Two additional recent papers used simpler econometric models in an attempt to directly relate telephone industry TFP growth to industry output growth. Neither study addressed the impact of network size. Kwoka (1993) analysed the former Bell System companies over the period 1948–87. His econometric model related TFP growth to output growth in addition to other structural variables. His model showed that a one percentage point increase in output lead to a 0.535 percentage point increase in TFP. Crandall and Galst (1991) estimated an econometric model that similarly links TFP growth to output growth. They estimated this model for the former Bell System companies, independent LECs and the entire telephone industry for the years 1961–87. They found that a one percentage point increase in output increased TFP growth 0.34 per cent for the former Bell System companies, 0.55 per cent for the independent local exchange carriers and 0.37 per cent for the entire industry.

Finally, we briefly note a number of econometric studies based on the data developed by Christensen et al.. These studies were by Evans and Heckman (1983, 1984 and 1988), Charnes et al. (1988) and Röller (1990a and 1990b). The authors attempted to estimate models with multiple indexes of output, using the pre-divestiture Bell System data. None of the authors attempted to model network size. The results of these models vary widely and the researchers offer conflicting interpretations of the data. This is not surprising, since the indexes of output used are highly collinear and it is not possible to econometrically determine the impact of each index on cost. As noted by Waverman (1989), the collinearity of the variables produces meaningless (negative) estimates of marginal costs for some observations within the samples used for the analysis. This also implies that the estimated cost elasticities are unreliable; hence these models are not of value in determining the relationship between output growth and TFP growth.

In conclusion, recent econometric literature supports the conclusion first reached by Caves and Christensen, that the telephone industry has significant economies of density. This literature suggests that the magnitude of the impact may even be greater than that estimated by Caves and Christensen. This evidence also shows that economies of density exist for the LECs. Using the more conservative Caves and Christensen results, a one percentage point decrease in output will lead to a reduction in TFP growth of between 0.3 and 0.5 percentage points.

THE ROLE OF TFP IN PRICE CAP REGULATION

Conceptual Basis

Many price cap plans in the USA and Canada base the price cap index (or price ceiling) on a general measure of output price inflation, such as the Gross Domestic Product Price Index (the GDPPI). These formulas are generally known as 'GDPPI-X' formulas, since they relate the percentage change in the price cap index to the percentage change in the GDPPI offset by some predetermined amount represented by an X factor. TFP growth is an important component of this X factor.

A useful conceptual framework for setting the X factor can be developed by using (3.6) above. Equation (3.6) establishes a relationship between telecommunications output price growth, telecommunications input price growth and TFP. Because a similar relationship exists between output prices, input prices and TFP in the economy, one can establish a relationship between telecommunications output price growth and the rate of economy-wide output price growth:

$$\left(\frac{\dot{P}}{P}\right)_{tel} = \left(\frac{\dot{P}}{P}\right)_{econ} - \left[\left(\frac{\dot{T}}{T}_{tel} - \frac{\dot{T}}{T}_{econ}\right) + \left(\frac{\dot{W}}{W}_{econ} - \frac{\dot{W}}{W}_{tel}\right)\right] \qquad (3.7)$$

Equation (3.7) is the framework for a price cap formula. The left-hand side of (3.7) represents the growth in telecommunications output prices that are subject to the price cap. The first term on the right-hand side of (3.7) represents a measure of economy-wide output price growth, such as the GDPPI used in the FCC's price cap plan. The second term in brackets on the right-hand side provides the basis for the X factor. That is:

$$X = \left[\left(\frac{\dot{T}}{T}_{tel} - \frac{\dot{T}}{T}_{econ} \right) + \left(\frac{\dot{W}}{W}_{econ} - \frac{\dot{W}}{W}_{tel} \right) \right] \tag{3.8}$$

The first term in the brackets represents the difference between telecommunications TFP growth and economy-wide TFP growth. It is generally known as the TFP differential. The second term represents the difference between economy-wide input price growth and telecommunications input price growth, and is known as the input price differential.

Empirical Implementation

In using (3.8) to establish an X factor, it is necessary to develop expectations regarding the TFP and input price differentials. Generally, these expectations are formed by examining historical evidence on TFP series and input price differentials, and determining whether adjustments are necessary to provide appropriate forward-looking estimates. We next turn to these implementation issues.

Given the historical evidence on TFP, the question arises in price cap proceedings whether the efficiency incentives of greater industry competition and the implementation of price cap regulation will have an impact on TFP growth. The assessment of these impacts is found in the form of a 'consumer productivity dividend' that is a part of many price cap plans. In practice, the consumer productivity dividend is typically determined by the regulatory process, without any empirical basis to substantiate the value chosen.

Competition may produce conflicting effects on TFP growth. In the previous section, we indicated that output growth has an impact on TFP growth (and presumably the X-factor). If, for example, competition reduces incumbent LEC output growth, it can be expected that TFP growth will diminish. This effect is magnified if high-margin services are the target of competition. However, the results of Oniki et al. in the study of NTT indicate that the introduction of competition and liberalization may enhance TFP growth. The question is whether these enhancements are sufficient to offset any losses due to reduced output growth.

The fact that many, if not all, price cap plans include a non-zero consumer productivity dividend is an assessment by regulators that they expect the efficiency incentives of competition and incentive regulation to outweigh any negative impacts on TFP growth.

As (3.8) shows, the input price differential is a conceptual component of the price cap formula. However, whether the differential should have a non-zero value in the formula is an empirical matter that hinges upon whether a differential is expected over the proposed price cap period.

Christensen (1995) conducted an analysis of input prices in the USA. He found that while there were large year-to-year variations in the input price differential, over the 1948–92 period, input prices for US telephone companies grew at an average annual rate of 4.7 per cent and input prices for the US economy grew at an average annual rate of 4.75 per cent. Christensen also conducted statistical tests of the annual input price differentials over various time periods, and found that there was no statistically significant difference. Christensen et al. (1996) found average annual input price growth of 2.9 per cent for the US LEC industry over the 1984–94 period (a differential of −1.3 per cent), and 4.1 per cent over the 1989–94 period (a positive differential of 0.4 per cent). Statistical tests found that the input price differential was not statistically different from zero in either case.

While the FCC (1997) also attempted to measure the input price differential over the post-divestiture period, the analysis was flawed by the fact that property income was used to measure the price of capital. As mentioned above, this will generally not be appropriate for the telecommunications industry. One method of determining the reasonableness of this assumption is to compute the effective rate of return implicit in the property income. In calculations we have made using the FCC data, we have determined that the effective return implicit in the FCC calculation fell dramatically over the period of analysis, indicating that the FCC approach significantly understated the rate of input price growth. Nevertheless, in determining the price cap X factor for LECs, the FCC used its calculations to infer a substantial input price differential.[16]

In the Canadian price cap decision, the CRTC (1997) established an input price differential of 0.3 per cent. This was based on the input price differential for Bell Canada over the 1962–95 period.

CONCLUSION

TFP growth for the telecommunications industry has historically outpaced TFP performance for the overall economy in the USA, as well as other countries. TFP performance of the telecommunications industry can be attributed to economies of density, the relatively high growth of high-margin services, industry technical change and regulatory reform.

In recent years TFP has taken on importance as a key component of price cap regulation. While the historical record typically indicates a remarkable stability in telecommunications TFP performance relative to the overall economy, particularly in the USA, this stability has not created a consensus about how the results should be applied to price cap regulation. The crux of the debate is how the factors mentioned above will affect expected TFP growth.

NOTES

1. For example, see Jorgenson et al. (1987) and Caves et al. (1980).
2. This is equivalent to weighting the growth rate for each output by its share of total revenue and the growth rate of each input by its cost share.
3. See Caves et al. (1985) for a discussion of scale and density issues in network industries.
4. Net output is defined to be total output net of materials input.
5. Crandall and Galst also calculate a third total factor productivity index, based on a deflated revenue output measure (with 1977 as the base year). This index suffers the methodological problems shared by all fixed weight indexes, and therefore we do not report those results herein.
6. These studies included data from nine local exchange carriers: Ameritech, Bell Atlantic, BellSouth, GTE, NYNEX, Pacific Telesis, Southern New England, Southwestern Bell and US West.
7. The results of the two studies were closer over the 1988–93 period with the simplified study estimated average annual TFP growth of 3 per cent, compared to 2.8 per cent for the first study. One source of the large difference in the earlier years was that the FCC changed its account rules beginning in 1988, necessitating adjustments to pre-1988 data for comparability with post-1988 data.
8. Two additional companies – Lincoln and Sprint – were added to the dataset for the simplified studies.
9. Over the time period of this study, the Regional Bell Operating Companies consisted of Ameritech, Bell Atlantic, BellSouth, NYNEX, Pacific Telesis, Southwestern Bell and US West.
10. In terms of TFP measurement, the largest differences arose in the measurement of output. The FCC relied on physical output measures.
11. The authors also demonstrate that their results are sensitive to the measures of output used. An alternative using physical measures of output results in TFP growth of 1.38 per cent (compared to 3.35 per cent) for the 1952–76 period.
12. One should note that using estimated marginal costs that differ from prices is inconsistent with the assumptions of the Berndt–Fuss model.
13. Economies of scale and capacity utilization are other potential sources of TFP growth. Empirical studies have not found either to have a significant impact on TFP growth in the telephone industry.
14. Using the Caves–Christensen model as a point of departure the rate of TFP growth can be decomposed in the following manner:

$$tfp = \sum m_i y_i - \sum s_j x_j$$
$$= \sum (m_i - \varepsilon_i) y_i - \varepsilon_n n + v$$

where tfp is the rate of TFP growth, m_i is the share of output i in total revenue, y_i is the growth rate in output i, s_j is the share of input j in total cost, x_j is the growth rate in input j, ε_i is the cost elasticity of output i, ε_n is the cost elasticity of network size n and v is the rate of technological change. Economies of density are present when the sum of the cost elasticities of output (the m_i) is less than one; economies of scale are present when the sum of the cost elasticities of output and the network elasticity $\sum \varepsilon_i - \varepsilon_n$ is less than one.

15. Christensen et al. (1983) and Nadiri and Schankerman (1981).
16. The FCC did not specify the input price differential in establishing the X factor for the LECs in their May 1997 Decision. Rather, the input price differential contained in the X factor must be inferred from the evidence placed on the record. Based on that evidence, it appears that the FCC relied on a range of 1.9 per cent to 2.8 per cent for the input price differential.

REFERENCES

American Productivity and Quality Center (1988), *Multiple Input Productivity Indices*, cited in FCC Docket 87-313, *Second Notice of Proposed Rulemaking*, 208.

Bell Communications Research (1987), 'Econometric Estimation of the Marginal Operating Cost of Interstate Access', Special Report SR-FAD-000552.

Berndt, E.R. and Fuss, M.A. (1986), 'Productivity measurement with adjustments for variations in capacity utilization and other forms of temporary equilibrium', *Journal of Econometrics*, **33**, 7–29.

Canadian Radio and Telecommunications Commission (1997), *Price Cap Regulation and Related Issues*, Telecom Decision CRTC 97-9.

Caves, D.W. and Christensen, L.R. (1988), 'The importance of scale, capacity utilization, and density in explaining interindustry differences in productivity growth', *The Logistics and Transportation Review*, **24**, 3–32.

Caves, D.W., Christensen, L.R. and Swanson, J.A. (1980), 'Productivity in U.S. railroads', *Bell Journal of Economics*, **11**, 166–81.

Caves, D.W., Christensen, L.R., Tretheway, M.W. and Windle, R.W. (1985), 'Network effects and the measurement of returns to scale and density for U.S. railroads', in Doughety, A.F. (ed.), *Analytical Studies in Transport Economics*, Cambridge, Cambridge University Press, 97–120.

Charnes, A., Cooper, W.W. and Sueyoshi, T. (1988), 'A goal programming/constrained regression review of the Bell System breakup', *Management Science*, **34**, 1–26.

Christensen, L.R. (1995), 'An input price adjustment would be an inappropriate addition to the LEC price cap formula', Christensen Associates.

Christensen, L.R., Christensen, D.C. and Schoech, P.E. (1981), 'Total factor productivity in the Bell System', Christensen Associates.

Christensen, L.R. Christensen, D.C. and Schoech, P.E. (1983), 'Econometric estimation of scale economies in telecommunications', in Courville, L., de Fontenay, A. and Dobell, R. (eds), *Econometric Analysis of Telecommunications*, Amsterdam: North-Holland, 27–54.

Christensen, L.R., Schoech, P.E. and Meitzen, M.E. (1994), 'Productivity of the local operating telephone companies subject to price cap regulation', Christensen Associates.

Christensen, L.R., Schoech, P.E. and Meitzen, M.E. (1995a), 'Productivity of the local operating telephone companies subject to price cap regulation, 1993 Update', Christensen Associates.

Christensen, L.R., Schoech, P.E. and Meitzen, M.E. (1995b), 'Total factor productivity methods for local exchange carrier price cap plans', Christensen
Associates.

Christensen, L.R., Schoech, P.E. and Meitzen, M.E. (1996), 'Total factor productivity methods for local exchange carrier price cap plans: reply comments',
Christensen Associates.

Christensen, L.R., Schoech, P.E. and Meitzen, M.E. (1997), 'Updated results for
the simplified TFPRP model and response to productivity questions in the FCC's
access reform proceeding', Christensen Associates.

Crandall, R.W. and Galst, J. (1991), 'Productivity growth in the U.S. telecommunications sector: the impact of the AT&T divestiture', Washington DC, The
Brookings Institution.

Denny, M., Fuss, M. and Waverman, L. (1981), 'The measurement and interpretation of total factor productivity in regulation industries, with an application to
Canadian telecommunications', in Cowing, T. and Stevenson, R. (eds),
Productivity Measurement in Regulated Industries, New York: Academic Press,
179–218.

Domar, E. (1961), 'On the measurement of technical change', *Economic Journal*,
71, 709–29.

Evans, D. and Heckman, J. (1983), 'Multiproduct cost function estimates and
natural monopoly tests for the Bell System', in Evans, D.S. (ed.), *Breaking Up
Bell*, Amsterdam: North-Holland, 253–82.

Evans, D. and Heckman, J. (1984), 'A test for subadditivity of the cost function with
an application to the Bell System', *American Economic Review*, **74**, 615–23.

Evans, D. and Heckman, J. (1988), 'Natural monopoly and the Bell System:
response to Charnes, Cooper, and Sueyoshi', *Management Science*, **34**, 27–38.

Federal Communications Commission (1997), *Fourth Report and Order in CC
Docket No. 94-1 and Second Report and Order in CC Docket No. 96-262 (Fourth
Report and Order)*, Adopted 7 May, 1997.

Fuss, M.A. (1994), 'Productivity growth in Canadian telecommunications',
Canadian Journal of Economics, **27**, 371–92.

Hulten, C.R. (1986), 'Productivity change, capacity utilization, and the sources of
efficiency growth', *Journal of Econometrics*, **33**, 31–50.

Hulten, C.R. (1990), 'The measurement of capital', in Berndt, E.R. and Triplett,
J.E. (eds), *Fifty Years of Economic Measurement*, Chicago: University of Chicago
Press, 119–52.

Hulten, C.R. and Wykoff, F.C. (1996), 'Issues in the measurement of economic
depreciation: introductory remarks', *Economic Inquiry*, **34**, 10–23.

Jorgenson, D.W., Gollop, F.M. and Fraumeni, B.M. (1987), *Productivity and U.S.
Economic Growth*, Cambridge, MA: Harvard University Press.

Jorgenson, D.W. and Griliches, Z. (1967), 'The explanation of productivity change',
Review of Economic Studies, **34**, 249–80.

Kiss, F. (1983), 'Productivity gains in Bell Canada', in Courville, L., deFontenay,
A., and Dobell, R. (eds.), *Economic Analysis of Telecommunications*, Amsterdam: North-Holland, 85–114.

Kwoka, J. (1993), 'The effects of divestiture, privatization, and competition on productivity in US and UK telecommunications', *Review of Industrial Organization*,
8, 47–62.

Nadiri, M.I. and Schankerman, M.A. (1981), 'The structure of production, technological change, and the rate of growth of total factor productivity in the US

Bell System', in T. Cowing and R. Stevenson (eds), *Productivity Measurement in Regulated Industries*, New York: Academic Press, 219–47.

Oniki, H., Oum, T.H., Stevenson, R., and Zhang, Y. (1994), 'The productivity effects of the liberalization of Japanese telecommunications policy', *Journal of Productivity Analysis*, **5**, 63–79.

Röller, L-H. (1990a), 'Proper quadratic cost functions with an application to the Bell System', *Review of Economics and Statistics*, **72**, 202–10.

Röller, L-H. (1990b), 'Modelling cost structure: the Bell System revisited', *Applied Economics*, **22**, 1661–74.

Shin, R. and Ying, J. (1992), 'Unnatural monopolies in local telephone', *Rand Journal of Economics*, **23**, 171–83.

Spavins, T.C. and Lande, J.M. (1990), 'Total telephone productivity in the pre and post-divestiture period', FCC CC Docket 87-313, *Supplemental Notice of Proposed Rulemaking*, Appendix D.

Stentor Resource Centre, Inc. (1996), 'Price cap regulation and related issues', CRTC, 96–8.

Waverman, L. (1989), 'US interexchange competition', in Crandall, R.W. and Flamm, K. (eds), *Changing the Rules: Technological Change, International Competition, and Regulation in Communications*, Washington DC: Brookings, 62–113.

Ying, J. and Shin, R. (1993), 'Costly gains to breaking up: LECs and the Baby Bells', *Review of Economics and Statistics*, **75**, 357–61.

4. Competition in local and long-distance telecommunications markets

T. Randolph Beard and George S. Ford

INTRODUCTION

An encyclopedic analysis of competition in the local and long-distance telecommunications markets can take two forms. An empirical review of current and past market conditions can be provided with a focus on the existing degree of competition and the evolution, or demise, of competitive forces in various segments of the industry. This empirical analysis must be constantly updated, as the forces of technology and regulation are always exerting influence on the industry, invoking change at a rapid and some-times unnatural pace. As a product of regulatory proceedings, such empir-ical reviews are widely available and continually produced. An alternative approach is to present economic tools that can be used to evaluate the extent and effects of competition in telecommunications markets. To fully understand competition in telecommunications markets, economists must incorporate the institutional details of the telecommunications industry into their formal competitive analyses, whether theoretical or empirical. The standard models of competition, though frequently applied to the telecommunications industry, are often unsatisfactory due to the use of simplifying assumptions that fail to capture the salient features of the industry and its participants. The shortcomings of tractable models, to a large extent, cannot entirely be overcome. Nevertheless, the better the model represents the actual structure of an industry, the more useful that model will be.

This chapter takes the latter approach and its purpose is to provide a review of conceptual and empirical analyses of competition that have a high degree of relevance for the telecommunications industry. By no means can a single chapter, nor a single theoretical framework, encompass the myriad institutional details that influence the nature and extent of compe-tition in the telecommunications industry. A few of the more common threads of competition analyses of the industry, however, are evaluated herein. Some of the tools considered in this chapter are specific applications of general ideas to the telecommunications industry, while others are more general treatments with obvious implications for communications. A number of themes important to the analyses of competition in the telecommunications industry are considered in this chapter. For example, our expectation is that telecommunications industry structure always will be somewhat concentrated, due to the presence of sunk costs in network deployment and customer acquisition. In the next section, a formal model of industry concentration that has direct relevance to a number of communications markets is summarized. Cross-subsidies, alleged and real, are prevalent in regulated communications markets. The following section shows, in contrast to the expectation of traditional economic models, that competition may increase rather than decrease the amount of cross-subsidy in a market. In the regulatory and academic arenas, the relationship of cost changes and price changes serves as an important indicator of the degree of competition. The flow through of costs in competitive markets, and some tools used for the analysis of flow through are then discussed. Competition that occurs only over parts of a geographic market is also a common feature of competition in the communications industries. This issue, termed fragmented duopoly, is then considered formally. Next, as the communications industry responds to deregulation, mergers abound. The simulation of merger effects is a new tool for antitrust authorities. However, simulations can be unsatisfying, in that specific forms of competition must be assumed. Starting from the concepts of simulations analysis, an empirical model for evaluating competition in the long-distance industry is presented. Conclusions are then provided.

COMPETITION, SUNK COSTS AND MARKET STRUCTURE

Perhaps the most important role of competition analysis, at least in practical or measurable terms, is its influence over regulatory and competition policy. In the intensely regulated telecommunications industry, changes in industry structure, conduct and performance are governed as much by the visible hand of policy as by the invisible hand of market forces. Recently, policymakers around the world have established competition and deregulation as the foundation for public policy towards the telecommunications industry. For the most part, regulatory agencies have not articulated an explicit model of competition as the conceptual foundation for their pro-competition and deregulation goals. In general, extant analysis appears to invoke a view of competition that resembles Bain's structure–performance paradigm, stressing the reduction or elimination of entry barriers that prevent both the fragmentation of market structure and increases in the number of competitors (Bain, 1956). Presumably, as entry barriers are eliminated, market concentration will decline, eventually producing sufficient fragmentation that competitive rivalry will obviate the need for regulation.[1]

Suppose, however, that the process of competitive entry is limited so that market concentration reaches a lower bound, that is, an equilibrium level of industry concentration. There is some evidence suggesting that such a lower bound may, in fact, exist in local and long-distance telecommunications markets, notwithstanding efforts to reduce barriers to entry. If this equilibrium level of concentration is sufficiently high, efforts to deregulate may be derailed by concerns regarding the exercise of market power. In this section, a general conceptual framework illustrating this notion of equilibrium industry structure is presented. From the policy perspective, this analysis is useful because it provides a general framework with which alternative policies can be evaluated and tempers expectations regarding the extent of competitive entry in some telecommunications markets. Further, the analysis demonstrates the shortcomings of applying overly simplistic models of competition to industries in which sunk costs are important.

Following Sutton (1991) and Duvall and Ford (2001), a two-stage game is considered in which each of a number of potential firms decides whether or not to enter the market in the first stage of the game. Further, assume that entry requires setup costs that are sunk (labeled as κ). Although the precise extent of how sunk an investment is cannot be determined *ex ante*, it is likely that a non-trivial proportion of the investment in network switches, transmission facilities, marketing, and even the lobbying of regulatory and legislative bodies will be sunk, since it is difficult or impossible to re-deploy such assets to purposes other than those initially intended. At the second stage of the game, those firms that have entered engage in price competition.

Let the demand curve be $Q = S/p$ where Q measures the quantity demanded of a particular communications service which for present purposes is assumed to be homogeneous; p measures the unit price of the product or service; and S measures total consumer expenditure on a product or service during some relevant market period and is independent of market price.[2] S also provides a measure of market size, and it is assumed the market is equally divided among actual entrants. Since this market demand function has a constant, unit own-price elasticity (the demand curve is iso-elastic), it can be shown that the profit-maximizing monopoly price approaches infinity for any marginal cost. For analytical convenience, therefore, it is assumed that sales fall to zero above some cutoff price p_m. Thus, p_m corresponds to the profit-maximizing monopoly price.[3]

Suppose N facilities-based carriers decide to enter the market at Stage 1 of the game. The profit function of a representative Firm i in Stage 2 of the game is given by $\pi_i = (p(Q) - c)q_i$, where q_i is Firm i's level of output and p is market price, which is a function of total market output $\{p = p(Q)\}$, and c is marginal cost, which for convenience is assumed constant across all output levels and firms. Defining the common conjectural variation term, dQ/dq_i, as ϕ (where $\phi \geq 0$), it is straightforward to show that the symmetric conjectural variation equilibrium price is

$$p = c \left\{ \frac{N}{N - \phi} \right\} \tag{4.1}$$

unless p exceeds p_m, that is, the price at which sales become zero, in which case $p = p_m$ (the monopoly price).[4] Consistent with the typical expectation with respect to increases in the number of competing firms, for any given $\phi > 0$, a greater number of firms results in a lower equilibrium price. In the limit, price approaches marginal cost as the number of firms increases. The conjectural variation term for the Cournot model equals one ($\phi = 1$) so that $p = c\{N/(N - 1)\}$ in that case.[5] For Bertrand competition, $\phi = 0$ so that $p = c$ for any number of firms exceeding one.[6] In this general specification of industry price, less-than-Cournot competition is indicated by values of ϕ in excess of one.[7] Generally, ϕ can be viewed as a measure of the weakness of price competition, with higher values of ϕ indicating less intense price competition. At equilibrium market price p, equilibrium output per firm is $q_i = S/Np$. Firm i's profit, therefore, is $\pi_i = \phi S/N^2$. Assuming S or market size is constant, profits realized are clearly dependent on the number of competitors, N, that enter the market and the intensity of price competition (ϕ). For a fixed level of the intensity of price competition (4.1) shows that, as the number of firms increases, the equilibrium level of profit approaches zero (as is commonly expected).

Turning now to the first stage of the game, the entrant's strategy takes one of two forms: (a) do not enter; or (b) enter, incurring sunk costs κ, and set output at the second stage of the game as a function of the number of firms that enter the market in the first stage. The entrant's payoff is either zero (if the firm chooses not to enter), or else it is equal to the profit earned at the second stage of the game less κ. The net profit of firm i is $\{\phi S/(M + 1)^2\} - \kappa$, where M is the number of other firms choosing to enter. Entry occurs if the net profit is positive, and continues in Stage 1 of the game until profits just equal the sunk cost of entry. The number of firms in equilibrium is the integer part of

$$N^* = \sqrt{\phi S/\kappa} \tag{4.2}$$

where N^*, or $1/N^*$, can be taken as the equilibrium level of concentration. Because we have assumed all firms are identical, $1/N^*$ also is equal to the Herfindal–Hirshmann index.[8] Note that the equilibrium number of firms N^* is expressed as a positive function of market size (S) and the weakness of price competition (ϕ), but a negative function of the level of sunk entry costs (κ). In the Cournot case, $\phi = 1$ and $N^* = \sqrt{S/\kappa}$. Alternatively, Bertrand competitors will force prices down to marginal cost so that each firm, after the first that enters, realizes a loss equal to the investment sunk in setup costs, κ. Bertrand price competition implies, therefore, that only one firm enters the market in the first stage of the game and sets a profit-maximizing monopoly price in the second stage, so long as setup costs are greater than zero.

In the context of competition policy, where N is typically the target variable, (4.2) is important for (at least) three reasons. First, (4.2) provides a general expression for the factors influencing the equilibrium level of industry concentration. For example, (4.2) shows that the number of firms in equilibrium is inversely related to the level of setup cost. Reducing sunk costs, therefore, will lower the equilibrium level of concentration. Recent efforts to unbundle the local exchange network are examples of competition policies that reduce the sunk cost of entry. The impact of unbundling requirements, by reducing sunk costs, is to lower equilibrium industry concentration. Regulation also can increase sunk entry costs and, as a consequence, increase equilibrium industry concentration. In the US cable television industry, the level playing field law requires the entrant to incur sunk investments identical to those of the incumbent. The effect of this law is to impede competitive entry by raising sunk costs.[9]

In addition, regulation and competition policy can influence the potential success of entrants by affecting the size of the market (S). Limited access to subsidies, for example, creates asymmetry in the addressable markets available to entrants and incumbents.[10] Similarly, in the US domestic local exchange market, the Federal Communications Commission (FCC) does not require ILECs to provide unbundled local switching for small business customers with more than three access lines in highly dense markets.[11] The lack of access to unbundled switching limits the ability of potential entrants to serve this particular (and related) local exchange markets, thus increasing concentration. Second, this two-stage game demonstrates, in a simple way, the important concept of equilibrium levels of concentration. In many cases, competition policy ignores the fact that sunk costs limit the number of potential entrants and, as a consequence, precludes the large-numbers competition desired by policymakers. Competition policy can be improved, perhaps, by integrating into policy analysis the economic concept of equilibrium concentration. Further, if deregulation is tied to large-numbers competition, then deregulation may never occur in industries or markets with sunk costs, intense price competition, or a relatively small market size. Third, the paradox between the conclusions of the two-stage game and more traditional views of competition is as apparent as it is important. The typical view of competition has price competition increasing with declines in industry concentration. In other words, the more firms in a market, the more competitive is that market. This more traditional view of the relationship between concentration and price competition is the core of competition analysis for both regulatory and antitrust agencies. The two-stage game, alternatively, shows that high concentration can be the result of intense price competition. Thus, perhaps the most important insight from the two-stage game is that it exposes the limitations of applying traditional competition analysis to the communications industries, or to any markets for which sunk costs are an important element of the cost structure of firms.

COMPETITION AND CROSS-SUBSIDIES IN PRICING

An unintended benefit of the Telecommunications Act of 1996 is the impetus it has provided to economists for research into novel aspects of economic behavior. The related issues of flow through and cross-subsidization in pricing provide an excellent example of this phenomenon. As is discussed later in this chapter, an important controversy has arisen with respect to the effects of (often mandated) reductions in interstate, switched access charges on long-distance prices. This problem illustrates a more general issue of great policy significance: with switching costs, multi-product sellers and service bundling, what patterns of observed prices can be used as evidence in determining the degree of competition in telecommunications markets? This section provides a partial answer to this question by reviewing recent findings on the possibility of cross-subsidization in pricing by multi-product retailers, and the role of competition in this result. The discussion will reinforce later conclusions on flow through. As a general proposition, competition in telecommunications will involve relatively few firms selling bundled services with price systems that substantially deviate from those arising in simple textbook models.

The point of departure is the well-known analysis of Bliss (1988), which offers a useful characterization of retail shop competition. Unlike other firms, retail shops largely resell products produced by others, their value arising from the retailing function itself. By gathering products/services together, retailers allow buyers to save on transaction costs by concentrating their products at a single location. This description of retail competition has several important implications. First, competition between such firms will focus on the welfare implications of the prices of the offered bundle. In other words, a customer patronizing such a seller is not generally interested per se in the price of any single offered good or service; it is the welfare consequence of the overall offering that matters. This occurs because, when visiting a seller is expensive, buyers will not patronize multiple sellers in response to small differences in the price of individual goods or services. As a result, sellers have some freedom in setting individual prices. Second, as argued by Bliss (1988), the pricing problem of the retailer can be conceptualized as 'set prices to maximize profits subject to a competitively determined minimum welfare constraint'. In other words, for some representative or marginal customer, the prices of goods sold are selected to maximize seller profit subject to the constraint that the selected prices $\{p_1, p_2, \ldots, p_n\}$ provide the customer with a level of welfare or utility sufficient to attract their trade. This level of utility is determined by the consumer's alternatives, that is, by the degree of competition. Put another way, a purely competitive retail firm is merely a monopoly operating subject to a suitably severe welfare constraint.

If one accepts the conceptualization provided by Bliss (1988), then retail prices are immediately seen to be some form of Ramsey prices. Prices are marked up proportionately to the inverse price elasticities of in-store demand for the goods. Curiously, this result is identical to that obtained by the celebrated 'Weak Invisible Hand' theorem of Baumol et al. (1977) for a perfectly contestable market, although the mechanism used to obtain this result is quite different. The implications of Ramsey pricing for policy analysis are critical. First, a one-to-one flow through of cost changes to prices is ordinarily impossible because any such change would violate the Ramsey condition applying to any optimal prices. Second, the producer's surpluses attributed to different services will probably vary dramatically due to differences in the elasticities of demand for the services.

Although the retail competition view of Bliss (1988) sheds considerable light on competition among multi-product sellers when consumers incur costs that arise from seller contacts, Bliss's model does not explain a set of phenomena of relevance to telecommunications and many other industries. In particular, as with the Weak Invisible Hand theorem, cross-subsidies never arise in Bliss's analysis unless one postulates demand complementarities of the conventional sort. Bliss (1988, p. 383) states, 'Most examples [of loss leading, that is, subsidized pricing] are probably spurious. In characterizing prices in a perfectly contestable market, Spulber (1989, p. 142) notes that, 'at a sustainable market equilibrium, prices are greater than or equal to marginal costs'. Thus, neither retail competition nor contestability theory alone provides any ready explanation of subsidized pricing. Of course, subsidized pricing in environments of market power has several conventional explanations familiar in the literature of industrial economics. These explanations include demand complementarities, network externalities, asymmetric information, and predatory pricing. In each case, as detailed by Beard (2001), prices below marginal costs may arise although some degree of market power is required.

In the absence of market power, can cross-subsidization occur? This question is important from a policy perspective because, if the answer is 'no', then evidence of cross-subsidization is prima facie evidence of market power. If the answer is 'yes', then a plausible explanation must be sought for the subsidies in some new theory, or else the evidence itself must be discredited. Attention here is restricted to the affirmative case. Beard (2001) provides a generalization of the Bliss (1988) retail competition model that explains how cross-subsidization can arise, provides a prediction of what goods will be subsidized and explains the role of competition in this result. This generalization merely postulates that consumers are differentiated, and involves no assumptions of complementarity at the level of individual consumer demands, asymmetric information or externalities. Zero long-run profit and free entry conditions are found to be compatible with cross-subsidies. The Beard (2001) analysis works by assuming that, for any given retailer, there is more than one type of customer. Each customer-type/retailer pairing involves a different minimum welfare constraint that must be satisfied to obtain a commitment from that customer type. These constraints are created by competitive conditions in the market, and reflect a variety of factors, for example, buyer location, that are not specified. However, if the types are generated by variations in a continuous variable, then with probability equal to one only a single constraint is binding at any optimum. This fact generates the results obtained.

The reader interested in the technical details of the argument may consult Beard (2001). However, a simple example will illustrate the mechanism. Suppose a seller has the potential customers Type-0 and Type-1, and sells two goods, A and B. Demands are $X_A^0 = 1 - p_1$, $X_B^0 = 1 - p_2$, $X_A^1 = 1 - p_2$, $X_B^1 = 2 - p_2$, where X_K^i is consumer type i's demand for good $K \in \{A, B\}$. Let marginal costs be $MC_A = MC_B = 0$ and let V_0 be the market determined minimum utility required to attract Type-0, and similarly for V_1. Actual utilities enjoyed by the consumer are taken to equal their consumer surpluses, that is, income effects are ignored. When the welfare constraint needed to attract the Type-0 seller is binding, profit maximizing prices, $p_A^* = 1 - 2\sqrt{2V_0/5}$ and $p_B^* = 1 - \sqrt{2V_0/5}$, are obtained so $p_A^* < 0$, that is, good A is subsidized, whenever $V_0 > 5/8$. Direct calculation shows that the seller obtains higher profits by attracting both types of customer, and that a subsidy of product A in the economic sense, price below incremental costs, occurs at profit-maximizing prices.

It can also be shown that the goods that receive subsidies have a uniform characteristic. The good in relatively greater demand by the customer who spends less receives a subsidy. If, for two goods, A and B, the Type-0 buyer spends less than Type-1 at all prices, and the Type-0 buyer represents the binding constraint, then if $(X_A^0/X_B^0)<(X_A^1/X_B^1)$, good A is a candidate for a subsidy. Most importantly, this analysis establishes a curious and initially unintuitive conclusion that subsidies occur only when competition is sufficiently intense. The example given above illustrates why this is so: if V_0 is 'small' (but still binding), $p_A^*>0$. Put another way, a firm facing weak welfare constraints, a circumstance interpreted here to imply little vigorous competition, prices all goods above marginal costs.

This analysis is immediately relevant to several issues in policy debates over competitiveness in telecommunications markets. In reference to the retail inter-exchange market, most large sellers offer multiple services and indeed, many have business plans that contemplate bundled sales of long-distance, local service, Internet access, cellular service and so on. Also, telecommunications consumers vary widely in their usages of different types of service, and their expenditures. Although the options of sellers to offer multiple service packages complicate any analysis, the results of this section suggest that cross-subsidies in pricing, rather than proving market power, may actually reflect the intensity of competition. Thus, credible evaluations of telecommunications competition, particularly as the contents of service bundles expand, should presumably focus on returns or demand elasticity, rather than on subsidies in pricing.

FLOW THROUGH OF COST CHANGES

Flow through refers to the effect of a change in incremental costs on equilibrium prices, and whether cost reductions, particularly reductions resulting from regulatory action, flow through to consumers in the form of lower prices. Flow through is an important political issue in the telecommunications industry. Efforts to ease regulation and deregulate prices typically require at least some evidence that the target market is competitive, and the relationship between cost and price changes (flow through) is a common indicator of competitiveness.

Despite the political and regulatory importance of the concept of flow through of cost changes, relatively little formal work on this topic exists. It appears, in fact, that most analysts fail to question the nature of the link, if any, between cost changes and price changes. Rather, it is often the case that the relationship is simply assumed, without any supporting theoretical analysis. However, the linkage between price and cost changes in the telecommunications industry (and those like it) is complex, so that any narrow focus on the prices of a few services, or even a price index, is perhaps misplaced.

In addition to evaluating the price–cost relationship within the context of more traditional economic models, the purpose of this section is to present a theoretically more useful and rigorous definition of flow through. This alternate view of flow through illustrates some of the dangers inherent in applying the more familiar definition of the concept. In general, the analysis draws from the proposition that flow through implies that a cost change does not generate a windfall gain or loss for sellers. With the multi-product pricing common in telecommunications markets, this windfall approach to flow through implies almost nothing about the price of an individual service within some bundle of services. Rather, an incremental cost change will result in a pattern of price changes that reflect both the levels of initial prices and relative demand elasticities. One implication of this conclusion is that individual services may see price changes in excess of, or well below, the corresponding cost change, even if markets are fully competitive.

Sumner (1981) offered an early empirical analysis of flow through. Sumner uses the condition of marginal revenue equals marginal cost to derive: $P = c[\varepsilon/(\varepsilon - 1)]$, where P is price, ε is (minus) the own-price elasticity of demand, and c is incremental cost. With a constant elasticity of demand, $\Delta P = [\varepsilon/(\varepsilon - 1)]\Delta c$, so a regression of price on incremental cost (or changes in these variables) provides an estimate of ε. Because a firm's own-price elasticity of demand (ε) is larger the more competitive the market in which the firm operates, an empirical estimate of a one-to-one relationship between changes in price and cost is interpreted as evidence of a competitive market.

Bulow and Pfleiderer (1983) illustrate a flaw in Sumner's approach by noting that, in general, the curvature of the demand curve could produce the result that price could change more than, less than, or by an equal amount to a change in incremental cost, irrespective of industry structure or the toughness of price competition. This range of potential outcomes results because, for a given increase in marginal costs, the monopolist will contract output so that marginal revenue rises by the same amount as marginal cost. The contraction in output depends critically on the shape, that is, slope, of the marginal revenue curve. Thus, because the slope of marginal revenue logically bears no simple relationship to the slope of demand, a wide variety of potential relationships between price and cost changes are possible, even with monopoly. Extending the analysis of Bulow and Pfleiderer (1983) to the Cournot case with n firms, it can be shown that

$$\frac{\Delta P^*}{\Delta c} = \frac{P'}{P'' \cdot Q/n + P'(1 + 1/n)}, \tag{4.3}$$

where P^* is the equilibrium price, Q is quantity demanded for the industry, and (4.3) is identical to the analysis of Bulow and Pfleiderer (1983) for $n = 1$. Equation (4.3) implies that the effect of a change in marginal cost on equilibrium price is given by the ratio of the slope of the industry demand curve and the slope of the representative firm's marginal revenue curve. For linear demand, (4.3) simplifies to $n/(n+1)$, which is always less than unity and where near-perfect flow through occurs only when n is arbitrarily large, that is, perfect competition. Thus, the conventional analysis of flow through relies heavily on particular assumptions about the curvature of demand.

The inadequacy of the conventional analysis of flow through is not limited to its sensitivity to the curvature of demand. A number of other economic factors influence the observed relationship between price and cost changes. Almost all industries for which flow through might be an issue sell multiple products or services. Long-distance carriers sell, for example, interstate and intrastate toll, operator service and calling card services. The services often are sold under a variety of complex, nonlinear tariffs. Similarly, electric utilities may sell peak and off-peak services, or services subject to interruption, and so forth. Thus, there is generally no single price one may use to evaluate flow through. Price indices must be used, and this substitution for price involves numerous theoretical and practical problems (some of which are discussed later). Also, the conventional analysis assumes that there is a centralized market in which the good(s) and service(s) are sold. In this market, there is no need for relationships between customers and sellers, as the market acts as an intermediary. However, most industries for which flow through may be an important issue involve competition for customers.[12] In particular, providing telecommunications or electrical services to a customer involves signing up and connecting with that customer. This fact, combined with the multi-product nature of these relationships, leads to two important conclusions. First, if switching costs are positive, then there exist degrees of freedom in the pricing of multiple services, a point emphasized in the previous section. Because one cannot generally buy electricity from one firm in the afternoon, and use another supplier for the evening, an increase in the price of one service will not drive buyers away so long as the price reduction of another service compensates the buyer. Second, consumers will select the seller who offers them the highest welfare, which depends on all seller prices together, rather than any particular price or subset of prices. Customers switch suppliers only when a competing offer provides adequate additional benefit to outweigh the cost of switching.

Considering the complex relationship between buyers and sellers of telecommunications services exactly what flow through means is unclear. A more realistic analysis of telecommunications markets indicates that flow through does not imply a simple relationship between a price (or a price index) and incremental cost (or an index of incremental cost). Rather, flow through refers to the condition that the sellers of telecommunications services earn no windfall from a cost reduction. Competition, if it is sufficiently strong, will impose the condition that the price changes resulting from a cost reduction must, in equilibrium, dissipate all cost savings. If they did not do so, then another seller could, in principle, offer a set of prices that produced higher welfare while still increasing its profits.[13]

To illustrate, return to the conceptualization of the last section, due to Bliss (1988). Following Beard (1999), assuming only one type of customer and two goods, a consumer who subscribes to the services of firm i has welfare W of

$$W = \int_{P_1}^{\infty} q_1(s)ds + \int_{P_2}^{\infty} q_2(s)ds \qquad (4.4)$$

where P_i is the price of service i. For expositional convenience, assume uniform pricing, identical consumers and zero income effects. Firm i's profit is:

$$\pi_i = (P_1 - c)q_1(P_1) + (P_2 - c)q_2(P_2) - F_i \qquad (4.5)$$

where c is a common incremental cost, for example, access charges, and F is fixed costs. The firm's objective can be restated as:

$$\max_{P_1 P_2} W \quad \text{s.t. } \pi_i = \pi^0 \qquad (4.6)$$

where π^0 is determined exogenously. Optimal prices are the familiar Ramsey prices for constrained, welfare maximizing prices. For any given cost change Δc, the change in profit, that is, windfall, is zero ($\Delta \pi = 0$) if

$$q_1(\Delta P_1 - \Delta c) + q_2(\Delta P_2 - \Delta c) = -q_1 \varepsilon_1 L_1 \Delta P_1 - q_2 \varepsilon_2 L_2 \Delta P_2 \qquad (4.7)$$

where ε_j is the own-price elasticity of demand for service j, and $L_j = (P_j - c)/P_j$ (the Lerner index). The right-hand side of (4.5) is strictly positive whenever there exists above-marginal-cost pricing.[14] Thus, zero windfall gain from a cost reduction implies that $q_1(\Delta P_1 - \Delta c) + q_2(\Delta P_2 - \Delta c)$ is positive. This result is unsurprising; if the price changes exactly equaled the cost change, then profits rise with a cost reduction because sales increase as price falls. The common expectation of flow through, in other words, is not compatible with the zero-windfall requirement. If one begins with a set of Ramsey prices, price changes that result in a new set of Ramsey prices for a common incremental cost change must satisfy:

$$\Delta P_1 = \left(\frac{P_1^2}{P_2^2}\right)\left(\frac{\varepsilon_2}{\varepsilon_1}\right)\Delta P_2. \tag{4.8}$$

The implications are immediate as efficient changes in prices must satisfy (4.8) and therefore price changes will not all equal the change in marginal cost. Further, some services may exhibit much smaller price changes, depending on the relative elasticities and the level of prices.

A simple example will clarify these results. Suppose that a firm sells two services, Service 1 and Service 2, with prices of USD 0.10 and USD0.15 per unit. A reduction in common incremental cost of USD0.01 occurs. Ordinarily, one would say that prices should fall to USD0.09 and USD0.14 per unit, respectively. Yet, this conclusion is incorrect. Suppose initially that marginal costs are USD0.05 and USD0.06 per unit. At optimal prices, $(0.5)\varepsilon_1 = (0.6)\varepsilon_2$, so $\varepsilon_1 = (1.2)\varepsilon_2$. Suppose $\varepsilon_1 = -1.2$, then $\varepsilon_2 = -1$. The resulting price changes ΔP_1, ΔP_2 must satisfy $\Delta P_1 = (0.367)\Delta P_2$, so Service 1 will exhibit a smaller (in an absolute sense) price reduction than Service 2. Depending on costs and the profit constraint, one could well obtain the result that $|\Delta P_1| < |\Delta c| < |\Delta P_2|$. Therefore, an analysis that focused solely on the relationship between price and marginal cost for Service 1 would, in this case, conclude that pass through did not occur when, in fact, the price changes dissipated all profits from the cost change.

This analysis highlights the complexity of even defining pass through in a model with a somewhat more realistic conceptualization of seller–buyer relationships. However, the real world exhibits far greater complexity than this model. In particular, ordinarily there are: (a) non-constant marginal costs; (b) interdependent demands; (c) income effects from some price changes; (d) consumer differentiation and search; and (e) entry and exit. All of these modifications are quite complex. Yet, the basic lesson of this analysis would not seem to be dependant on the simple formulation adopted here. In particular, the main conclusion is merely that, with competition for customers and multiple products, price systems will be of the Ramsey type and, thus, price changes will be more complex than changes in common incremental costs.

Given this conceptualization, a reduction in per unit costs will lead to price reductions that, in an average sense, exceed the cost reduction. Symmetrically, a cost increase will lead to greater price increases. However, it is patently false that either (a) any given price change will equal the (common) incremental cost change, or (b) any subset of price changes, on price index change, will equal any predetermined value. These conclusions imply that there is considerable danger in applying the naive methodologies previously used to address this topic for public policymaking.

FRAGMENTED DUOPOLY

A singular feature of facilities-based competition in local telephone service, cable TV, electric power and other network industries is that providers, if they compete at all, typically do so only over fractions of their networks. Because customers can use a provider only if the provider's physical distribution plant passes the customer's home, only those potential subscribers passed by two or more sellers have any meaningful choice of service. Additionally, regulation and/or the threat of litigation by potential entrants ordinarily precludes price discrimination by sellers, so that the price paid by customers of a provider are generally the same whether they have a competitive choice or not. Despite this, one ordinarily expects the degree of system overlap, or overbuild, to affect prices and subscription when pricing flexibility is present.

Any facilities-based competition in local telecommunications, as in cable TV, electric power and broadband video services, will involve overbuilds of one sort or another. This feature of potential competition is, unfortunately, wholly absent from most traditional antitrust analyses familiar in other industries. Any credible evaluations of the competitive effects of entry, mergers and other phenomena in facilities-based local telecommunications must take this special feature into account. Fortunately, an economic model exists which makes possible just such an accounting.

Fragmented duopoly (or, more generally, fragmented oligopoly) arises when two (or more) firms compete only over certain geographic areas, but maintain monopoly control of others. When prices are not allowed to vary across market segments, the existence of an overlap can be expected to affect prices generally. Introduced by Basu and Bell (1991) to analyse informal credit markets in India, the theory of fragmental duopoly is immediately useful as a model of product market competition in telecommunications, cable TV services and similar network industries.

Beard and Ford (2001) provide a detailed technical description of a fragmented duopoly model with differentiated goods and apply the model to analyse the pricing and sales consequences of overbuilds in cable television markets in the USA in the early 1990s. Although, to our knowledge, this type of model has not been applied to local telecommunications service competition, presumably because there is very little overlapping of local plant, the lessons drawn from the cable TV experience are relevant to the problem of competition in local wire-line services generally. For this reason, an outline of the Beard and Ford (2001) analysis is given here, and its findings are related to telecommunications.[15]

Imagine two providers, A and B, which sell similar, though not necessarily identical, products. Provider A sells in one area with N_A potential subscribers, while B sells in another with N_B potential buyers. A number of buyers, N_C, can buy from either A or B, and represent the overlapping portion of the firms' geographic markets. Thus, A is the sole provider of service to N_A households, competes with B for the patronage of N_C households, and so on. Let $X_A(P_A, P_B)$ represent the probability that a given household buys service from A when A's service costs P_A and B's costs P_B. Unavailability of B's service is equivalent to P_B being some very high value, so one can suggestively write the probability that a captive customer of A buys A's service as $X_A(P_A, \infty)$. Assume that $0 \le X_A(P_A, P_B) \le 1$, X_A decreases as P_A rises, and X_A does not decrease as P_B rises. Thus, we assume away network externalities at this point. A similar set of assumptions governs $X_B(P_B, P_A)$.

This analysis is initially concerned with determining equilibrium prices, P_A^*, P_B^* such as might be chosen freely by A and B in the absence of collusion or regulation, and in discovering how these prices are related to N_A, N_B and N_C. Beard and Ford (2001) offer an extended theoretical analysis of a problem closely related to that described above, and come to three primary conclusions. First, P_A^* rises as N_A rises, that is, A's prices are higher when A's monopoly market is bigger. Second, P_A^* falls as N_C rises. Finally, and interestingly, P_A^* rises as N_B rises. This last effect occurs because the larger B's monopolized segment, the lower the incentive for B to compete vigorously in the contested segment. Since prices are strategic complements, any increase in B's price causes A to respond in kind.

The discussion above highlights an aspect of fragmented market competition that is likely to be important in local telecommunications markets. The complete duplication of local loop plant is unlikely, leading to fragmented competition. Further, cable TV networks upgraded to provide local telephone services create fragmented competition, since such networks are not geographically identical to local switched networks currently providing services. Even wireless networks, particularly line-of-sight technologies, are limited in coverage, creating the potential for fragmented competition among different technologies. Thus we expect that fragmented duopoly competition will characterize future facilities-based local competition.

An examination of the model description above illustrates the conceptual problems regulatory or antitrust officials are likely to experience in applying conventional models of competition in an uncritical fashion to fragmented markets. This application is not a practical difficulty, however, unless it can be argued that fragmented competition is a significant phenomenon that should affect public policy to some extent. The empirical evidence presented by Beard and Ford (2001), for overbuilt cable markets, suggests that the effects of fragmented competition are of great significance.

Utilizing a unique data set that allows determination of the extent of actual overbuilds (rather than their mere presence), Beard and Ford (2001) estimated a three-equation model that examined prices, penetration of services, and the extent of overbuilding in cable services. They found two significant results relevant to the present inquiry. First, and unsurprisingly, the extent to which a firm's service area is monopolized (not overbuilt) significantly and positively increases prices. Substantial price increases result when the extent of overbuilding is less. Thus, overbuilding is effective as stimulation to competition. Additionally, though, the empirical analysis establishes the result that one firm's prices are higher when the other firm has a larger monopoly component in its market area. The magnitude of this effect is virtually identical to the first effect described above. This finding has an important implication. Given two competing systems with at least some degree of overbuilt plant, if one system expands into a previously not served area, then the result is higher prices charged by both systems.

While one may object that, with respect to local telephone services, there are almost no significant areas in the USA that are unserviced, it is unlikely that competition will take the form of duplication of conventional loop plant. Rather, cable systems, or other networks, may provide telephony in some cases. Given this, the rise of build-outs by overlap competitors is clearly more relevant than is first apparent. An additional complexity needs to be highlighted here. When fragmented competition arises between different sorts of networks, for example, a local wireless telephone network and a cable network, to provide a competing service, for example, local voice messages or switched access, the incremental, per customer costs of one provider are likely to be far lower than those of the other, at least in some cases. The results of such cost differences may well alter the competitive consequences of overbuilding because, when one firm has much lower incremental costs, its optimal price is lower to begin with, affecting demand elasticity.

On balance, the analysis discussed in this section illustrates yet another unique aspect of telecommunications competition. There is a clear danger attendant on routine application of standard models. For example, one might easily ignore the effect on one firm's price of the expansion of the firm's rival into some other monopoly market area. Additionally, build-out requirements must be examined carefully else the public's goal of competition be short-circuited.

MERGER SIMULATIONS

In recent years, antitrust authorities in the USA, such as the Department of Justice Antitrust Division (DoJ), have made increasing use of relatively sophisticated simulations to evaluate the effects of proposed mergers on competition and prices. This trend is understandable given the great promise exhibited by a number of such models. Thus, rather than relying solely on general arguments and indices, such as the Hirschmann–Herfindal index used in the Merger Guidelines, in formulating their public position, regulators can evaluate evidence that, at least on the surface, offers an alluring appearance of both precision and wide applicability.

Merger simulation, as practiced by the DoJ and exemplified in recent cases such as *US v. Interstate Bakeries*, ordinarily requires that conventional econometric analysis to obtain demand elasticity estimates be performed first. The results of such a study are then used as inputs to the simulation process itself. Simulation requires additional assumptions on the nature of competitive interactions but, having made these assumptions, the analyst is able to predict post-merger prices and market shares. As with most simulations, the resulting estimates are not statistical estimates, and testing of hypotheses is replaced by sensitivity analysis. On the other hand, simulations offer a substantial advantage in that markets need not be as carefully defined as is often otherwise the case.

Broadly speaking, simulations adopt one of two possible competitive market specifications, the choice depending on the nature of the relevant products. When goods are not differentiated, Cournot models are used and firms then differ only with regard to their capacities or costs. Alternatively, when product differentiation is important, Bertrand models, which assume firms select prices as competitive instruments, are favored. Bertrand analyses often utilize the residual demand curve format, though such an approach has drawn substantial criticism (Froeb and Werden, 1991).

A primary advantage of simulations is their ability to encompass efficiency gains resulting from mergers. In most models of competition, mergers result in decreased product market competition, but from the economic point of view any such finding is but a part of the efficiency puzzle. Because firms produce under varying cost conditions, mergers can be motivated by, and can result in, real cost savings. For example, the merger of a higher cost firm with a lower cost rival may well reduce production costs, creating a countervailing effect to the diminution of competition which such a merger might imply.

Simulations therefore offer many advantages unobtainable from the purely structural approach traditionally associated with the merger guidelines. These advantages, of course, come at a price. Simulations make several powerful assumptions that, if accepted, allow analysts to perform welfare calculations of interest. Primary among these assumptions, in our view, is the model of product market competition adopted. Both the Cournot and Bertrand simulation approaches adopt models of product market competition that are highly restrictive and, given certain aspects of telecommunications competition, unrealistic. The remainder of this section outlines our concerns in this regard, and illustrates an alternative approach that highlights these issues.

This analysis begins by considering the typical Bertrand price competition model for differentiated products. Suppose there are n firms, $i = 1, 2 \ldots n$, and let $\pi_i(p)$ be the profits of Firm i when prices are p. A Bertrand price equilibrium p^* consists of a set of prices, one for each firm, such as $p_i^* \in \arg \max \pi_i(p_i, p_{-i}^*)$, where p_{-i}^* are the (equilibrium) prices of the other firms. Thus, p^* is a Nash equilibrium in prices. In general, firm costs and demands may differ, and Stage 1 of the simulation exercise is directed at 'filling in the blanks' on costs and demand characteristics. To make the model tractable, one can assume constant demand elasticities, constant average costs, and so on, and such an approach is common.

Two immediate difficulties are apparent in this approach. First, real firms do not engage in a sequence of one-shot competitive interactions, but instead presumably formulate longer-run strategies that game theory identifies as potential equilibria in repeated games. While a sequence of one-shot Nash equilibria is ordinarily a perfect equilibrium for the repeated game many other equilibria involving richer sets of behaviors are permissible. As a consequence, competition in the industry may be quite different from the simple one-shot Bertrand model. After all, mergers themselves are a form of behavior (an action) not ever contemplated in the Bertrand pricing game used to analyse the effects of the merger.

Related to the issue of the nature of competition in a dynamic environment is the related problem of customer–seller attachments and market share. In typical Bertrand models, market shares differ at time t because of differences in costs and demand/goods characteristics at time t. However, the history of long-distance communications in the USA, for example, is one of the steady erosion of AT&T's once 100 per cent market share. This fact should presumably be accommodated somehow in the model.

It seems fair to say that switching among carriers is common, but is neither frictionless nor costless for consumers. This consumer inertia imposes an evolutionary character on competition, and almost certainly affects the pricing policies of firms. Thus, in Markov fashion, structural characteristics at t, inherited from period $t-1$, affect decisions in period t. Of course, the demand curves of a Bertrand price competition model can be modified to reflect these facts. Often, though, considerations of tractability may limit the extent to which this is feasible.

As is well known, conventional random utility models can lead to useful and simple statements of the probabilities that a customer chooses a given vendor. However, such an approach does not account for the circumstance that almost all customers are already signed up with one or another vendor, and thus face the prospect of switching or staying put.

The following approach is offered as an alternative to, or perhaps as a complement to, simulation analysis. Suppose there is a larger number of customers and M providers $i, j = 1, 2, \ldots , M$. At time $t-1$, customers are signed up with providers with the numbers of accounts equal to $N_1, N_2 \ldots N_N$, where $N_1 + N_2 + \ldots + N_N = N$. The market share of seller i in period t, MS_i^t, is equal to

$$MS_i^t = \pi(i, i)MS_i^{t-1} + \sum_{j \neq i} \pi(i, j)MS_j^{t-1} \tag{4.9}$$

where $\pi(i, j)$ is the probability that a customer using vendor i in period $t-1$ uses vendor j in period t, so that $\pi(i, j)$, $i \neq j$, is a switching probability and $\pi(i, i)$ is a staying probability. Thus, any firm's market share at time t equals that part of its market share it retains from the previous period, plus that part it captures from rivals.

Under conventional extreme value distribution assumptions, logistic expressions are written for $\pi(i, j)$:

$$\pi(i, i) = \exp(d_i) / \left(\exp(d_i) + \sum_{j \neq i} \exp(d_j - T) \right)$$

$$\pi(i, j) = \exp(d_j - T) / \left(\exp(d_i) + \sum_{k \neq i} \exp(d_k - T) \right) \tag{4.10}$$

where $d_i = \alpha_i - \beta p_i$, α_i and β are constants, p_i is the price of service from seller i (usually measured as average revenue per minute, or ARPM), and $T > 0$ is the implicit average cost of switching suppliers. This formulation allows for inertia in customer switching, but is otherwise not terribly different from logistic models familiar in the simulation literature (Werden and Froeb, 1994 and 1996). One can use the model outlined above to implement a conventional Bertrand price equilibrium to evaluate a merger. To do this, one uses the equations that describe equilibrium prices. In particular, an optimal price satisfies:

$$-\beta \left(N_i - \sum_j N_j \pi(i, j) \pi(i, i) \right) (p_i - c_i) + N_i = 0. \tag{4.11}$$

When T=0 (switching is costless), this reduces to the very conventional expression

$$(p_i - c_i) = \beta(1 - MS_i^t)^{-1}. \tag{4.12}$$

However, the argument here is that such an approach is, in general, not appropriate because it is based on a very strong, one-shot equilibrium story that is not credible in telecommunications (and many other) markets. Additionally, the more complex condition above is virtually intractable from the econometric perspective.

As a result of these considerations, the following theoretical approach is proposed. First, we assume that a customer probabilistically switches from seller i to j based solely on a comparison of the prices p_i^{t-1} and p_j^t, that is, based on a pair-wise comparison of the current vendor's last period price and the target vendor's current price. Given this substantial theoretical and econometric simplification, the market share of seller i in period t depends on p_i^t, all prices p_j^{t-1} from last period, market shares from $t-1$, and idiosyncratic or firm specific effects. This formulation captures, if in an approximate way, the role of consumer inertia and past market shares in competition.

There is another important and critical aspect that any analysis of mergers and competition in telecommunications should include. Because the interactions between sellers are ongoing, and because consumer inertia implies that history matters, any model used should not be based on a restrictive assumption about the nature of competition. Rather, the approach offered here, being econometric, allows the data to determine the nature of competition, at least to the extent allowed by the specification used.

Therefore a two-equation system is arrived at that determines prices p_t and market share MS_t simultaneously. The general form of such a model is:

$$MS_i^t = (\alpha_i + \beta(p_j^t - p_i^{t-1})) MS_i^{t-1} + \sum_{j \neq i} (\gamma + \beta(p_i^t - p_j^{t-1})) MS_j^{t-1}$$

$$p_i^t = \delta_i MS_i^t + \omega' X_i^t \tag{4.13}$$

where α, β, γ, δ, and ω are parameters to be estimated, and X_i^t is a potentially lengthy vector of exogenous variables relating to costs, lagged concentration indices, access charges, and so on. The template given above is a very simplified one that captures the main points; more sophisticated formulations are possible.

To illustrate this approach, the model above is estimated using data collected from end-users by Paragren to compute the ARPM, that is, price, for four inter-exchange carriers including AT&T, MCI-WorldCom, Sprint and an aggregated class of all other carriers.[16] Data is monthly for the period January 1998 through September 1999. The results are briefly described to illustrate the potential fruitfulness of the approach. The description is for demonstration purposes, and full details are not provided here because they are unnecessary to the methodological point (for full results, see Beard et al., 2001).

Turning to the market share equation, Sprint's customers are found more likely to disappear than those of AT&T, MCI-WorldCom, or other carriers (α_i for Sprint is only 0.92, while other values are around 0.98). Prices, however, while deterring customers, are less significant than one might imagine. Thus, a generally high level of short-term customer attachment to providers is observed, although this varies among carriers. Of more interest is the price equation. The effects of each firm's market share on their average prices vary quite considerably, and in an interesting manner. Notably, this variation can be interpreted as representing differences in pricing strategies among firms, with some exploiting captive buyers and others aggressively seeking new customers. For example, results suggest that AT&T raises its price about three-tenths of a cent per minute for each additional point of market share, while MCI-WorldCom exhibits no statistically significant feedback.

The potential usefulness of these results for merger policy is apparent. Since the approach described here does not assume all firms behave in a particular manner, nor that one firm behaves in the same manner as another, a more realistic assessment of dynamic competition in the telecommunications industry is at least possible. For example, in the case of a merger between Firm 1 and Firm 2, the managers of the acquiring firm might generally be thought to be in charge post-merger, so the new, combined entity would have a market share of $MS = MS_1 + MS_2$ but a market share price formula coefficient equal to that of the acquirer. This leads to some conceivably complex dynamics. In the illustration considered here, a merger between MCI-WorldCom and Sprint raises prices only trivially despite the effects on concentration because MCI-WorldCom does not increase its prices in response to increased market share. Concentration increases, but the distribution of market share among firms actually changes in a favorable manner, since Sprint is seen to price up to its market share more than MCI-WorldCom. The approach outlined here goes some distance towards answering the concerns about simulations described above. In particular, the use of a simulation model that assumes a particular form of product market competition raises concerns that model conclusions may be quite inaccurate. Of course, the alternative methodology outlined here exhibits problems as well, but should this latter approach produce results consistent with an accepted simulation approach, much greater reliance might be placed on those conclusions.

CONCLUSION

Coherent public policy for telecommunications markets requires a sensible and realistic model(s) of competition. Unfortunately, models familiar from most textbooks are not adequate for this purpose. Telecommunications markets exhibit a pattern of structural characteristics that are simply not included in most conventional analyses. These characteristics include high sunk costs, customer switching costs, competition for customers, fragmented geographic competition, and, most importantly, extreme regulatory interventions that take place in a politically-charged atmosphere. There is plenty of money at stake and, as a consequence, politics often replaces economic reasoning. Economics, however, absolutely is required if the potentially huge benefits of sensible telecommunications policy are to be enjoyed by the public. Despite the fact that telecommunications deregulation presents us with some new and difficult economic problems, the situation is by no means impossible. This essay attempts to identify and examine several stylized facts about telecommunications markets that can be and should be integrated into current policy thinking. It should be recognized that concentration levels in some telecommunications markets are subject to lower bounds arising from sunk costs considerations. Any public policy that fails to recognize this fact is likely to fail to deliver potential benefits. Worse, policies can actually increase the extent of sunk costs, thereby reducing competition. Finally, while most authorities believe that weak price competition is a natural by-product of a concentrated market structure, high levels of concentration are more likely when price competition is intense, since the resulting lower margins reduce the ability of markets to support entrants who incur sunk costs.

Once entry has occurred, the nature of the resulting competition also violates naïve textbook treatments. First, since providers are multi-product firms, customers must sign up or pre-subscribe for service, and switching is costly, while competition takes the form of a sort of Ramsey pricing. As a consequence, in some cases cross-subsidies can arise in the pricing structure, even without market power. Additionally, flow through of incremental cost changes will not generally occur as a simple, dollar-for-dollar reduction in specific goods prices. Entry is also likely to be geographically fragmented, with facilities-based providers competing only over areas where physical distribution plants overlap. The extent of overlapping and monopoly market areas can be expected to materially affect prices.

Mergers in telecommunications markets undoubtedly will continue to be a topic of significant public policy concern, and techniques for evaluating the probable effects of such mergers will be of great interest. Recent developments in simulation analyses hold much promise in this regard. However, it may well be misleading to apply simulation models that ignore customer–seller attachments, or to assume that one-shot competition characterizes firm interactions. A simple alternative method of analysis is offered that at least partially addresses these issues.

The stakes involved in regulatory and legal authorities getting it right in telecommunications policy are enormous, and are unlikely to diminish over time. Recognition of the complications described in this chapter may assist us in realizing the benefits that good competition policies can produce.

NOTES

1. In other words, given existing barriers to entry, growth in the size of the market increases the profitability of incumbents, which induces the entry of firms that find it profitable to overcome the entry barriers. The resulting entry decreases the level of market concentration.
2. This specification of demand is discussed in Sutton (1991, p. 32).
3. For the iso-elastic demand curve, sales are positive regardless of price so that the monopoly price is undefined.
4. The more traditional manner in which to describe the conjectural variation term is $dQ/dq_i = 1 + \lambda$ (Waterson, 1984, p. 18). For convenience, the term $1 + \lambda$ is written here as ϕ.
5. In the Cournot model, rival firms choose the quantity they wish to offer for sale. Each firm maximizes profit on the assumption that the quantity produced by its rivals is not affected by its own output decisions. In other words, the conjectural valuation of the Cournot firm is equal to one. The Cournot equilibrium asserts that prices and quantities approach competitive levels as the number of firms supplying the market increases.
6. Several characteristics of communications markets make the distinction between Cournot- and Bertrand-type competition less fundamental. For instance, the decision to enter a market may require investing in plant of a given capacity. This capacity decision may be viewed as a commitment to produce a level of output equal to the output capacity of the plant. Indeed, recent formulations of oligopoly models show that when firms must first choose capacity plant size, the equilibrium of Bertrand competition in prices is identical to that of the simple Cournot model (Kreps and Scheinkman, 1983). Note that the Cournot outcome of the two-stage capacity game of oligopoly pricing is not robust when excess capacity exists. Given the lumpiness and long life of telecommunications facilities as well as the nontrivial potential for partial network failure, the total capacity of existing network may well exceed its utilization in the short term. The telecommunications carrier, however, will not likely view its maximum network capacity as the relevant index of its contribution to satisfying market demand.
7. The maximum of ϕ is equal to the value producing price p_m, the monopoly price.
8. For a thorough theoretical analysis of equilibrium market structures see Baumol and Fischer (1978).

9. Hazlett and Ford (2001) provide a conceptual and empirical analysis of the effects of the level playing field laws in cable TV.
10. See Baumol et al. (1988, p. 362).
11. FCC, *Third Report and Order and Fourth Further Notice of Proposed Rulemaking,* CC Docket No. 96-98 (FCC 99-238), 15 September 1999.
12. Customers typically subscribe to a portfolio of calling services when subsubscribing to a particular telecommunications carrier. Each long-distance minute is not purchased in a centralized market on a real-time basis, which is an implicit assumption in the more traditional analyses of flow through.
13. Of course, the existence of switching costs implies that not all of any cost reduction must be immediately capitalized into increased consumer welfare, but this fact is a market imperfection that merely reduces the probability of finding pass through, rather than a circumstance that requires a redefinition of what pass through means.
14. By assumption, this is required.
15. Alternative technologies that compete with local wire-line services, such as mobile and fixed wireless services, may provide further examples relevant for telecommunications.
16. *Paragren Teletrend*, Reston, VA: Paragren Technologies (1999).

REFERENCES

Bain, J.S. (1956), *Barriers to New Competition*, Cambridge, MA: Harvard University Press.
Basu, K. and Bell, C. (1991), 'Fragmented duopoly: theory and applications to backward agriculture', *Journal of Development Economics*, **36**, 145–65.
Baumol, W.J. and Fischer, D. (1978), 'Cost-minimizing number of firms and determination of industry structure', *Quarterly Journal of Economics*, **92**, 439–67.
Baumol, W., Bailey, E. and Willig, R. (1977), 'Weak invisible hand theorems: On the sustainability of multiproduct natural monopoly', *American Economic Review*, **67**, 350–65.
Baumol, W.J., Panzar, J.C. and Willig, R.D. (1988), *Contestable Markets and the Theory of Industrial Structure*, San Diego, CA: Harcourt Brace Jovanovich.
Beard, T.R. (1999), *Pass-Through of Cost Changes: Theoretical Issues and Policy*, Auburn Policy Research Center Report.
Beard, T.R. (2001), *Competitive Cross Subsidies*, mimeo, Auburn University.
Beard. T.R. and Ford, G.S. (2001), *An Empirical Investigation of Overlap Competition in the Cable Television Industry*, mimeo, Auburn University.
Bliss, C. (1988), 'A theory of retail pricing', *Journal of Industrial Economics*, **36**, 375–91.
Bulow, J.I. and Pfleiderer, P. (1983), 'A note on the effect of cost changes on prices', *Journal of Political Economy*, 91, 182–5.
Duvall, J.B. and Ford, G.S. (2001), 'Changing industry structure: the economics of entry and price competition', *Telecommunications and Space Journal*, **7**, 19–48.
Froeb, L.M. and Werden, G.J. (1991), 'Residual demand estimation for market delineation: complications and limitations', *Review of Industrial Organization*, **6**, 33–48.
Hazlett, T.W. and Ford, G.S. (2001), 'The fallacy of regulatory symmetry: an economic analysis of the "level playing field" in cable TV franchising statutes', *Business and Politics*, **3**, 21–46.
Kreps. D. and Scheinkman, J. (1983), 'Quantity precommitment and Bertrand

competition yield Cournot outcomes', *Bell Journal of Economics and Management Science*, **14**, 326–37.

Spulber, D.F. (1989), *Regulation and Markets*, Cambridge, MA: MIT Press.

Sumner, D.A. (1981), 'Measurement of monopoly behavior: an application to the cigarette industry', *Journal of Political Economy*, **89**, 1010–1019.

Sutton, J. (1991), *Sunk Costs and Market Structure: Price Competition, Advertising, and the Evolution of Concentration*, Cambridge, MA: MIT Press.

Waterson, M. (1984), *Economic Theory of the Industry*, Cambridge, MA: Cambridge University Press.

Werden, G.J. and Froeb, L.M. (1994), 'The effects of mergers in differentiated products industries: logit demand and merger policy', *Journal of Law, Economics and Organization*, **10**, 407–26.

Werden, G.J. and Froeb, L.M. (1996), 'Simulating mergers among non-cooperative oligopolists', *Computational Economics and Finance: Modeling and Analysis with Mathematica*, ed. H. Varian, 2nd edn, New York: Springer-Verlag.

5. Telecommunications demand

Lester D. Taylor

INTRODUCTION

The changes that have racked the telecommunications industry during the last twenty years make this a particularly problematic chapter to write. With deregulation, competition, and rapid technological change, telecommunications demand analysts have been forced to contend with rapidly shifting industry boundaries, appearance of new products and services, firm (as opposed to market) demand functions, explosive growth of mobile and wireless telephony, disappearance of traditional sources of information and data, and, of course, emergence of the Internet. As a consequence, an overview, of the traditional sort, of the state of the art of telecommunications demand analysis (such as was presented in my 1980 and 1994 books) would be of limited relevance for the questions that are likely to be nascent in the years immediately ahead. Instead, my approach in this chapter will be to focus on fundamentals, and to describe in detail several studies that, in my view, illustrate the principles involved, with regard to both theory and technique, and transcend their particular applications. This will occupy the next three sections. An overview of the telecommunications demand literature then follows, while a list of challenges concludes the chapter.

BASIC FEATURES OF TELECOMMUNICATIONS DEMAND

The feature that most distinguishes telecommunications demand from demand for most other goods and services is that telecommunications services are not consumed in isolation. A network is involved. This gives rise not only to certain interdependencies and externalities that affect how one models consumption, but also creates a clear-cut distinction between access and usage. There must be access to the network before it can be used.

The interdependency that has received the most attention in the literature is what is usually referred to as the network (or subscriber) externality.

This externality arises when a new subscriber joins the network. Connection of a new subscriber confers a benefit on existing subscribers because the number of ports that can now be reached is increased. As this benefit is shared in common by all existing subscribers, the network externality accordingly has the dimension of a public good. The importance of this externality lies in its potential to generate endogenous growth. Because a larger network is more valuable to belong to than a smaller network, an increase in network size may increase the willingness-to-pay of existing extra-marginal subscribers to the point where they become actual subscribers, which in turn may induce a succeeding set of extra-marginal subscribers to subscribe, and so on and so forth.[1] However, the possibility of this phenomenon being of any present importance in the mature telephone systems of North America and Western Europe is obviously remote.[2]

Additional externalities are associated with usage. A completed telephone call requires the participation of a second party, and the utility of this party is thereby affected. The gratuitous effect, which falls on the second party, represents what is usually referred to as the call (or use) externality. Unlike the network externality, the call externality has never received much attention, most probably because it has never been seen as creating a problem for pricing. Assuming, as is usual, that the externality is positive, recipients of calls (since most calls are paid for by the originator) receive unpaid for benefits. The existing view in the literature pretty much has been that, since most calling is bi-directional, the externality is internalized over time as today's recipients of calls become tomorrow's originators. This balances out obligations *vis-à-vis* one another. For the system as a whole, the benefits from incoming calls can be seen as increasing willingness to pay to have a telephone, which means that a telephone system will be larger when call externalities are present than what it would be in their absence.

The foregoing conventional view of the call externality is uncontroversial.[3] But this view is not empirically helpful as its only implication is that a telephone system will be larger with call externalities than without. Nothing is said or implied about usage, which seems strange since the source of the externality is in fact usage. The problem is that an important aspect of the interdependency between callers is being overlooked. To see what is involved, consider two individuals, A and B, who regularly communicate with one another by telephone and for each of whom the call externality is positive. That is, both A and B look forward to receiving a call from the other. Two circumstances can be identified, one in which the number of calls between A and B is determined 'exogenously' as a function (say) of their respective incomes and the price of calls, and a second in which the number of calls between the two depends not only on income and price, but also on the number of calls that A makes to B and the number of calls that B makes to A. Calling is clearly endogenous in the second circumstance.

In the first circumstance, one can easily imagine the appearance of an implicit contract in which A calls B half of the time and B calls A the other half.[4] This is the situation envisioned in the conventional interpretation of the call externality, and simply involves the internalization of the externality through an implicit contract. The total number of calls is determined by income and price, and the only effect of the externality is to help determine how the cost of the calls is distributed.

The second circumstance is clearly much more complex, because not only can there be an implicit contract defining 'turn' but it is a situation in which a call can create the need for a further call. The call externality in this case not only helps to determine who pays for calls but also leads to an actual stimulation of calls. An example of this would be where A's call to B increased B's utility so much that she called A back to tell him so, or else called C to tell her that A had called. This is probably a common event in real life, but there is another situation that is probably even more common (especially among business users) in which calling triggers further calling not rooted in the call externality (as conventionally defined). This is the situation in which an exchange of information creates the need for further exchange. Several authors collaborating on a paper provide an obvious example. For lack of a better term, I have previously referred to this phenomenon as the dynamics of information exchange.[5]

The dynamics of information exchange may not only involve B returning a call to A as a result of an initial call from A, but may create a need for B to call C to corroborate something that A has told B, which in turn may lead to C calling D, and so on. Essentially the same dynamics can operate when, by whatever means, a piece of information is injected into a group that forms a community of interest. An obvious example is the death of one of the members of the group. A learns that B has died, calls C, and the grapevine kicks into operation.[6] A generic model that takes into account both this and the more conventional form of the call externality will be discussed below.

A further complication in modeling telecommunications demand arises from the fact that benefits arise not only from completed communications, but also from those that may not be made. Subscribing to the telephone network can be viewed as the purchase of options to make and receive calls. Some of the options will be exercised with certainty, while others will not be exercised at all. This is because many calls will only be made contingent upon particular states of nature whose realization is random and thus not known at the time that access is purchased. Emergency calls, such as to fire, police, or ambulance services, are obvious cases in point, but compelling urgency is not the only determinant. Many calls arise from nothing more than mood or whimsy. Option demand is thus an important characteristic of telecommunications demand.[7]

Reference to this point has for the most part been with respect to households and residential demand. Business demand is another matter, and unfortunately much more complicated. The complications include:[8]

1. The many forms of business access to the telecommunications network. Fact one would seem to be that no business in this day and age can be without telephone service. Accordingly, to approach business in an access/no access framework as with residential demand is simply not relevant.[9] What is relevant is the type of access that a business demands, together with the number of lines.
2. Internal versus external communications needs. Unlike a household, much of business telecommunications demand is driven by internal needs. These needs vary depending upon the size and type of business, and increase rapidly with the number of locations. For large national and international corporations, the demand for internal communications probably outweighs the external demand. Large private networks are a reflection of this. The presence of multiple locations, perhaps more than any other factor, is what makes modeling business demand so difficult, for one must focus on communications between and among locations as well as between the firm and the outside world.
3. The importance of telecommunications in marketing. Still another factor that complicates analysing business telecommunications demand is the heavy use of telecommunications in marketing. For many businesses, the telephone is a primary instrument of selling, whether it is in terms of direct communications with customers or maintaining contact with a sales force in the field.
4. Customer-premise equipment. Prior to the introduction of competition into the telecommunications industry, modeling the demand for customer-premise equipment was not a priority. Equipment was basic, it had to be rented, and its price was for the most part bundled into the monthly service charges. Access and equipment were essentially one and the same, and to explain the demand for one was pretty much to explain the other. Technological change and competition changed this. Access and equipment were unbundled, and customers were allowed to purchase or rent equipment from whomever they chose. As a consequence of this change, customer-premise equipment has become one of the most problem-plagued areas to model. Equipment is no longer slave to the choice of access, but is a decision variable of stature commensurate with access and usage.

5. The opportunity cost of time. An obvious consequence of economic growth is that time becomes more valuable. Business (as well as individuals acting as consumers) seeks ways to make more efficient use of time, and often increases its use of telecommunications to do so. While this may seem paradoxical, since time spent on the telephone is real time generally unavailable for any other use, the point is obvious when one considers the savings that can arise from replacing sales people in the field with advertising, facsimile, and 800 service, reduction of travel through teleconferencing, use of the now near-universal ability to communicate with anyone from anywhere, and of course telecommuting. In most situations the relevant cost of increased telecommunications usage is not the out-of-pocket cost of the increased usage, but the cost of the labor and other resources that are replaced. Identifying and measuring these costs can, of course, be a difficult task.

6. Emergence of the Internet. Whatever the underlying structure of business telecommunications demand might be it has almost certainly been balanced in recent years by the emergence of the Internet. While it is obvious that many telephone communications are lost to e-mail, it is not clear that the ultimate impact of e-mail on telephone traffic is necessarily negative. Similar opaqueness applies to e-commerce and e-business. In my view, the best strategy for dealing with the impact of the Internet in modeling business telecommunications demand will be to encompass the Internet within an expanded definition of telecommunications.

THEORETICAL CONSIDERATIONS

Let me now turn to a brief review of the theoretical considerations that have guided the modeling of telecommunications demand since the mid-1970s. As noted in the preceding section, the point of departure is the distinction between access and usage.

A Generic Model of Access Demand

For notation, let a denote access, q a measure of calling activity, and CS a measure of the net benefits from usage. For convenience, assume that usage is measured by the number of calls at a price of p per call. The conventional approach to modeling the demand for access in the literature is in a discrete choice framework, in which δ is postulated to be a zero-one random variable whose value depends upon the difference between the net benefits from usage of the telephone network (conditional on access) and the price of access, which will be denoted by π. The probability that a household will demand access can accordingly be written as

$$P(\text{access}) = P(\delta = 1)$$
$$= P(CS > \pi). \tag{5.1}$$

Net benefits, CS, in turn, are usually measured in the literature by the consumer's surplus associated with the consumption of q, that is, by

$$CS = \int_p^\infty q(z)dz. \tag{5.2}$$

Perl (1983) was one of the first researchers to apply this framework empirically, doing so by assuming a demand function of the form:[10]

$$q = Ae^{-\alpha p}y^\beta e^u, \tag{5.3}$$

where y denotes income and u is a random error term with distribution $g(u)$. CS will then be given by

$$CS = \int_p^\infty Ae^{-\alpha z}y^\beta e^u dz$$
$$= \frac{Ae^{-\alpha p}y^\beta e^u}{\alpha} \tag{5.4}$$

With net benefits from usage and access expressed in logarithms, the condition for demanding access to the telephone network accordingly becomes

$$P(\ln CS > \ln \pi) = P(a - \alpha p + \beta \ln y + u > \ln \pi)$$
$$= P(u > \ln \pi - a + \alpha p - \beta \ln y), \tag{5.5}$$

where $a = \ln A/\alpha$.

The final step is to specify a probability law for consumer surplus, which in view of the last line in (5.5) can be reduced to the specification for the distribution of u in the demand function for usage. An assumption that u is distributed normally leads to a binary probit model, while an assumption that u is logistic leads to a binary logit model. Empirical studies exemplifying both approaches will be reviewed below.[11]

Models of Toll Demand

By the late-1970s, a generic model for toll demand had been developed and used extensively in the (then) Bell System, which included price, income, market size, and habit as predictors. Market size was usually measured by the number of telephone access lines and habit was represented by the lagged value of the dependent variable.[12] In the early-1980s, a more sophisticated version of this model was developed by Bell Canada for use in hearings before the Canadian Radio, Television and Telecommunications Commission (CRTC). The ideas underlying the Bell Canada model are as follows.

Assume that there are two exchanges, A and B, that are not part of the same local-calling area, so that calls between the exchanges are toll calls. Let the number of telephones contained within the exchanges be T and R, respectively. The total possible connections between the two exchanges will therefore be $T \cdot R$. Let M denote the number of calls that are 'sent paid' from A to B during some period of time (a quarter, say), and let θ denote the proportion of potential connections $(T \cdot R)$ that are realized, so that

$$M = \theta(T \cdot R). \tag{5.6}$$

Equation (5.6) might describe the relationship that would be observed in a static world, that is, where income, price, and all other factors that affect toll calling are constant. However, these other factors do not remain constant, and we can allow for them through the value of θ. In particular, let us suppose that income (Y) and the price (P) of a toll call from A to B affect θ according to

$$\theta = a Y^\beta P^\gamma, \tag{5.7}$$

where a, β and γ are constants. The relationship in (5.6) accordingly becomes

$$M = a Y^\beta P^\gamma (T \cdot R). \tag{5.8}$$

Finally, let us suppose that M is affected by a change in potential toll connections that may be either more or less than proportional, in which case the model becomes

$$M = a Y^\beta P^\gamma (T \cdot R)^\lambda, \tag{5.9}$$

where λ is a constant, presumably of the order of one. Taking the logarithms of both sides of (5.9), we obtain

$$\ln M = \alpha + \beta \ln Y + \gamma \ln P + \lambda \ln(T \cdot R), \qquad (5.10)$$

where α is $\ln a$. With addition of a random error term, this equation represents the model that was used by Bell Canada in a variety of hearings before the CRTC in the 1980s and early-1990s.[13]

Point-to-point Toll Demand

Competition in the interLATA toll market in the USA in the 1980s quite naturally stirred interest in disaggregation, and the development of toll demand models that were route-specific.[14] Models were developed that distinguished traffic in one direction from that going in the reverse direction. These models also allowed for a call-back effect, in which calls in one direction affected the volume of calling in the reverse direction.[15] Among other things, these models can be viewed as the initial attempts to give coherence to the call externalities described in the following section.

As before, the concern is with two areas, A and B, which are not part of the same local-calling area. Let Q_{AB} denote calls from A to B, Y_A income in A, P_{AB} the price from A to B, T_A the number of telephones in A, and T_B the number of telephones in B. With this notation, the model of the preceding section becomes

$$\ln Q_{AB} = \alpha_0 + \alpha_1 \ln Y_A + \alpha_2 \ln P_{AB} + \alpha_3 \ln(T_A \cdot T_B). \qquad (5.11)$$

We now make two modifications to this model. The first is to allow for telephones in the sending area and receiving areas to have different elasticities, while the second modification is to allow for a reverse-traffic effect, whereby calling from A to B is affected by the volume of traffic from B to A. Expression (5.11) accordingly becomes

$$\ln Q_{AB} = \alpha_0 + \alpha_1 \ln Y_A + \alpha_2 \ln P_{AB} + \alpha_3 \ln T_A + \alpha_4 \ln T_B + \alpha_5 Q_{BA}. \quad (5.12)$$

Because the reverse-traffic effect is reciprocal, there will also be an equation for the traffic from B to A:

$$\ln Q_{BA} = \beta_0 + \beta_1 \ln Y_B + \beta_2 \ln P_{BA} + \beta_3 \ln T_B + \beta_4 \ln T_A + \beta_5 Q_{AB}. \quad (5.13)$$

Equations (5.12) and (5.13) form a simultaneous system and they must be estimated jointly.

EMPIRICAL STUDIES

As noted in the introduction, deregulation and competition have had mixed effects on telecommunications demand research. While the literature has continued to grow, its nature has changed, and additions to it are now less informative about the empirical structure of the industry than prior to the breakup of AT&T in 1984. Price elasticities are now viewed as important trade secrets, and few studies become openly available for rapidly growing new products and services, such as cellular telephony and Internet access. As a consequence, empirical information about the structure of telecommunications demand is becoming increasingly dated and incomplete. With this in mind, my approach in this section will be to describe three studies which, in my view, ably summarize the present state of the art in telecommunications demand modeling. In doing this, the focus will be primarily on methodology rather than on empirical results, *per se*.

Choice of Class of Local Service: Train, McFadden and Ben-Akiva (1987)

The first study to be discussed was undertaken in the mid-1980s by Train et al. (1987) for a large east coast local exchange company. Unlike a typical access demand study, where the choice is access/no access, this study focuses on the choice of class of service, given that some form of access is already demanded. What sets the study apart is that the class of service is determined jointly with a portfolio of calls. Calls are distinguished according to their number, duration, distance and time of day. A portfolio of calls is accordingly seen as consisting of a particular number of calls of a particular duration to a set of specific locations at a specific time of the day. Households in the data set analysed faced a variety of service options, ranging from budget-measured to metropolitan-wide flat-rate service, so that the cost of a portfolio varies with the particular service option selected. A conditional, that is, nested, logit framework is employed, in which the choice of service is assumed to be conditional on the portfolio of calls that is selected. With this structure, the probability of a household choosing a particular service option can be interpreted as depending upon the household's expected portfolio of calls (reflecting, for example, the tendency of households who place a lot of local calls to choose flat-rate service). Since usage is in turn conditional on access, however, the portfolio that a household will actually choose in a month will depend upon the service option that has been selected, as this determines the cost per call.

Turning to a description of the model, let the distinct times of the day that calls can be made be denoted by $t = 1, ..., T$. Further, let the geographic areas (or zones) to which calls can go be denoted by $z = 1, ..., Z$. Let N_{tz}

represent the number of calls to zone z during time-period t and let D_{tz} represent the average duration of these calls. A portfolio, accordingly, can be written as the vector with elements $(N_{11}, ..., N_{TZ}, D_{11}, ..., D_{NT})$. Denote the set of all possible portfolios by A and a particular portfolio by $i \in A$. Finally, index the available service options by $s = 1, ..., S$. In the data set analysed, three service options were available to all households, and two additional services were available to some households. Portfolios were defined for 21 time zones.

The probability of observing a particular (s, i) service–portfolio combination, given available service and portfolio options, is assumed to be nested logit. In the nesting, alternatives with the same portfolio are grouped together, while different service options define the branches.[16] Let P_{is} denote the probability of observing the combination (s, i), let $P_{s|i}$ denote the conditional probability of choosing service option s, given portfolio i, and let P_i denote the marginal probability of selecting portfolio i, so that (from the definition of conditional probability):

$$P_{is} = P_{s|i} P_i . \tag{5.14}$$

Specifically, it is assumed that

$$P_{is} = \frac{e^{y_{is}}}{\sum_j e^{y_{ij}}} \left(\frac{\sum_j e^{y_{ij}}}{\sum_k \sum_j e^{y_{kj}}} \right)^{\lambda}, \tag{5.15}$$

where:

$$P_{s|i} = \frac{e^{y_{is}}}{\sum_j e^{y_{ij}}}, \tag{5.16}$$

$$P_i = \left(\frac{\sum_j e^{y_{ij}}}{\sum_k \sum_j e^{y_{kj}}} \right)^{\lambda}, \tag{5.17}$$

and where y_{is} is a parametric function of factors specific to service option s and portfolio i. With loss of generality, y_{is} can be written as the sum,

$$y_{is} = w_{is} + v_i / \lambda, \tag{5.18}$$

where w varies with both i and s, but v varies only with i. Consequently, we will have for $e^{y_{is}}$,

$$e^{y_{is}} = e^{(w_{is} + v_i/\lambda)}$$
$$= e^{v_i/\lambda} e^{w_{is}}. \tag{5.19}$$

Similarly, for $\sum_j e^{y_{ij}}$,

$$\sum_j e^{y_{ij}} = \sum_j e^{(w_{ij} + v_i/\lambda)}$$
$$= e^{v_i/\lambda} \sum_j e^{w_{ij}} \tag{5.20}$$

so that

$$P_{s|i} = \frac{e^{w_{is}}}{\sum_j e^{w_{ij}}}. \tag{5.21}$$

Next, let

$$I_i = \ln\left(\sum_j e^{w_{ij}}\right) \tag{5.22}$$

$\sum_j e^{w_{is}}$ can then be written as

$$\sum_j e^{w_{is}} = e^{I_i} \tag{5.23}$$

Consequently, for P_i:

$$P_i = \left(\frac{\sum_j e^{y_{ij}}}{\sum_k \sum_j e^{y_{kj}}}\right)^\lambda$$
$$= \left(\frac{e^{v_i/\lambda} \sum_j e^{w_{ij}}}{\sum_k e^{v_k/\lambda} \sum_j e^{w_{kj}}}\right)^\lambda$$
$$= \frac{e^{v_i + \lambda I_i}}{\sum_k e^{v_k + \lambda I_k}} \tag{5.24}$$

The term I_i is interpreted as the 'inclusive price' of portfolio i, and its coefficient λ measures substitutability across portfolios. For $0 < \lambda < 1$, substitution is greater within nests than across nests, while for $\lambda > 1$, substitution is greater across nests than within nests. Given the nesting structure, the parameter will be less than one if households shift to different service options more readily than they shift to different portfolios. It will be greater than one if they shift to different portfolios more readily than they shift to different service options.

It remains to specify forms for v_i and w_{is}. As the only difference among service offerings is in the billing procedure, w_{is} is assumed to depend only on the cost to the customer of portfolio i under option s and option-specific constants. The specification of v_i is more complex. To begin with, it is assumed that a portfolio yields benefits through the information transmitted in calls and extracts opportunity costs through the time spent on the telephone. Suppressing for the moment variations in coefficients across time and zone categories and households, v_i for a portfolio of N_i calls of average duration D_i (with the index i deleted to simplify notation) is specified as

$$V = \theta N \ln\phi D - \alpha ND$$

$$= \theta N \ln D - \gamma N - \alpha ND, \tag{5.25}$$

where $\alpha = -2\ln\phi$.

The first term on the right-hand side in (5.25) measures the benefits of the portfolio. Each call provides a benefit, $2\theta\ln\phi D$, which is assumed to increase at a decreasing rate with duration. The parameter ϕ can be interpreted as measuring the rate of information transfer. It can be expected to be positive, but can be either greater or less than one. As θ can be interpreted as measuring the benefits from the information that can be transferred by a call, θ will obviously be positive. Consequently, γ can be of either sign. The model assumes that the benefits obtained from N calls is N times the benefit from a single call. The term αND measures the opportunity cost of the portfolio. ND represents the total amount of time spent in calling, while α measures the opportunity cost per minute. The parameters are allowed to vary in two ways. Benefits depend upon the destination of a call, while the opportunity cost depends upon the time of the day. θ, accordingly, is allowed to vary over distance zones and α over time periods. In addition, θ is assumed to vary with income and the number of telephone users in a household. Income is also assumed to affect α, the time opportunity cost of making a call.

Estimation of the model is complex because of the large number of possible portfolios. As enumeration of every possible portfolio is clearly infeasible, estimation proceeds on the basis of a sample of 10 portfolios for each household. The sample includes the portfolio actually chosen, plus a subset of nine portfolios, chosen at random from the universe of total portfolios that the household did not choose. Let B denote the sample of portfolios constructed for a particular household. Also, let $\pi(B|i)$ denote the conditional probability of constructing subset B, given that the chosen alternative is i. Since B necessarily includes the portfolio actually chosen, $\pi(B|i)=0$ for $j \notin B$. The joint probability of drawing a chosen alternative i and subset of alternatives B is

$$\pi(B, i) = \pi(i|B)P_i. \tag{5.26}$$

Consequently, from Bayes' theorem, the conditional probability of i being chosen, given B, is

$$\pi(i|B) = \frac{\pi(B|i)P_i}{\sum_{j \in B} \pi(B|j)P_j}, \tag{5.27}$$

which exists if $\pi(i|j) > 0$ for all $j \in B$. Rewritten in logit form, with P_i given by (5.24), (5.27) becomes

$$\pi(i|B) = \frac{e^{v_i + \lambda I_i + \ln \pi(B|i)}}{\sum_{j \in B} e^{v_j + \lambda I_j + \ln \pi(B|j)}}. \tag{5.28}$$

McFadden (1978) has shown that under usual regularity conditions maximizing the conditional logarithmic likelihood function,

$$L = \sum_{h=1}^{H} \ln \pi_h (i|B), \tag{5.29}$$

yields consistent estimators of the unknown parameters. Here h denotes particular households in a sample of H households.

Bypass via Extended Area Service: Kridel (1988)

The extant toll-to-local subsidy was 'justified' historically by allocating a substantial portion of the costs of local plant to toll service. At the time of the AT&T divestiture, these non-traffic-sensitive 'costs' totaled about USD16 billion, of which about USD9 billion was the interstate portion and about USD7 billion the intrastate portion. With divestiture, most of these non-traffic-sensitive costs were to be recovered through traffic-sensitive per minute charges at both ends of a toll call. Since these charges were seen as 'payment' to the toll carrier's switch (or point-of-presence), those at the sending end could be avoided if the customer were to bypass the local exchange company by connecting directly with its toll carrier's point-of-presence. As bypass facilities are not inexpensive, for direct bypass to pay requires a large volume of toll calling, and accordingly is not feasible for the vast majority of residential customers. However, in a circumstance in which there is a large volume of toll calling between adjacent regions, bypass could be affected by merging the regions into a local-calling area through Extended Area Service (EAS).[17] Bedroom communities adjacent to large central cities provide cases in point. In 1987–8 several of these communities in Texas were using the regulatory forum to demand lower toll rates via optional EAS calling plans.

In anticipation of building the necessary infrastructure for provision of the EAS calling plans, a study (Kridel, 1988) was undertaken at Southwestern Bell to assess likely demand. The point of departure for this analysis was the standard utility maximization framework, in which, given a choice between two calling plans (toll and EAS), the customer is assumed to choose the plan which yields the greater utility (as measured by consumer surplus). Specifically, if the increase in consumer surplus associated with EAS is greater than the subscription price, the customer will choose EAS. The increase in consumer surplus consists of: the toll savings due to the decrease to zero in the price of calls, and the benefits that arise from additional calling. Let the former be denoted TS and the latter V_s. EAS will accordingly be purchased when

$$\Delta CS = TS + V_s$$
$$> p_{EAS} \tag{5.30}$$

where ΔCS denotes the change in consumer surplus and p_{EAS} denotes the EAS subscription price.

Since ΔCS is not directly observable, the choice criterion in (5.30) is reformulated in probability terms as

$$\Delta CS = \Delta CS^* + u$$
$$> p_{EAS}, \tag{5.31}$$

where ΔCS^* represents the observed or deterministic portion of the increase in consumer surplus and u is a well-behaved error term. The probability of a customer choosing EAS will then be given by

$$P(EAS) = P(u > \gamma(p_{EAS} - \Delta CS^*)), \tag{5.32}$$

where γ is a parameter that can be thought of as the inclination of a subscriber to take advantage of a given net benefit (or loss) resulting from the selection of EAS.

As no historical data were then available, the model was estimated using purchase intentions data obtained in a survey of customers in the areas affected. Customers were queried whether EAS would be purchased at a given subscription price, and this price was varied across respondents. The resulting price variation allowed for predictions to be made of take-rates at various subscription prices. The customer, when providing a rational response to the EAS purchase query, is (at least subconsciously) estimating the benefits to be derived from additional calling (V_s), for the customer was informed of his/her toll savings (TS) just prior to the purchase question. Since the benefit in question is directly related to the amount of call stimulation, it is natural to treat the customer's new usage level (q_N) as a parameter to be estimated.

For the choice model represented in (5.32) to be estimated, the term ΔCS^* must be specified in terms of observable quantities. Accordingly, Kridel assumed that the demand function for q had the 'Perl' form,

$$q = Ae^{-\alpha p}y^\beta, \tag{5.33}$$

where q represents the minutes of toll usage into the EAS area, p is the toll price, y is income and A, α and β are parameters to be estimated.

With this functional form, the consumer surplus under each option will be given by:

$$CS_0^* = \int_{p_0}^{\infty} Ae^{-\alpha p}y^\beta dp$$

$$= q_0/\alpha \qquad \text{non-EAS,} \tag{5.34}$$

$$CS_N^* = \int_{p_0}^{\infty} ae^{-\alpha p}y^\beta dp$$

$$= q_N/\alpha \qquad \text{EAS} \tag{5.35}$$

where p_0 represents current toll price, q_0 the observed usage at p_0 and q_N the satiation level of usage with EAS (since $p_N = 0$).

From the expressions (5.34) and (5.35), ΔCS^* will be given by

$$\Delta CS^* = CS_N^* - CS_0^*$$

$$= (q_N - q_0)/\alpha. \tag{5.36}$$

While the satiation level of usage q_N is unobserved, it can be calculated from the demand function in (5.33) as

$$q_N = q_0 e^{-\alpha p_0}. \tag{5.37}$$

At this point, the choice model is seen to involve the parameters α and γ. The parameter α is related to the price elasticity of demand for toll, namely, $-\alpha p$, while γ (as noted earlier) can be thought of as the inclination of the subscriber to take advantage of a given net benefit resulting from selection of EAS. The higher (in absolute value) is γ, the more likely (everything else remaining constant) a customer will be to purchase EAS. Kridel extends the model to allow for individual customer effects by assuming that the parameters α and γ are functions of income and subscriber demographics. This allows all customers to have the same form of demand function, but to have different price elasticities and inclinations to purchase EAS. Demand-type variables (income, household size, and so on) are included in α, while choice-influencing variables (perceptions, education and so on) are included in γ.

The stochastic term u in expression (5.32) is assumed to have a logistic distribution, thus implying a logit form for the choice model, so that the probability of a customer choosing EAS will be given by

$$P(u > \gamma(p_{EAS} - \Delta CS^*)) = \frac{1}{1 + \exp(\gamma(p_{EAS} - \Delta CS^*))}. \tag{5.38}$$

The model was estimated for a sample of 840 households drawn at random from suburban communities using a hybrid Berndt–Hausman–Hall–Hall and quasi-Newton optimization routine.

Once estimated, the model was used to predict the take-rate and usage stimulation corresponding to EAS subscription prices of USD5, USD20 and USD25 per month. At a subscription price of USD5, the model predicted a take-rate of nearly 76 per cent and an increase of nearly 93 per cent in usage. At a price of USD25, the predicted take-rate is a little over 50 per cent and an increase in usage of about 65 per cent. The interesting (and significant) thing about these stimulation estimates is that they were about twice as large as would have been generated using Southwestern Bell's (then) existing price elasticities for toll calling.[18]

Point-to-Point Toll Demand: Larson, Lehman and Weisman (1990)

Larson et al. (1990) were the first to formulate a model incorporating reverse (or reciprocal) calling. They estimated their model using city-pair data from Southwestern Bell's service territory. The data used in estimation were quarterly time-series for the period 1977:Q1 through 1983:Q3. Nine city pairs were analysed. The city pairs were defined a priori, and for the most part were selected on a natural community of interest, based on nearness. The larger cities were designated as points A and the smaller cities as points B. The nine city pairs thus form 18 routes to be analysed, nine AB routes and nine BA routes.

The two-equation model estimated by Larson et al. was as follows:

$$\ln Q_{it} = \alpha_0 + \sum_{j=1}^{g} \alpha_i \delta_{ij} + \sum_{j=1}^{h} \beta_j \ln P_{it-j} + \sum_{j=1}^{l} \gamma_j \ln Y_{t-j} + \lambda \ln POP_t$$
$$+ \xi \ln Q_{it}^* + u_{it} \tag{5.39}$$

$$\ln Q_{it}^* = a_0 + \sum_{j=1}^{g} a_i \delta_{ij} + \sum_{j=1}^{h} b_j \ln P_{it-j} + \sum_{j=1}^{l} c_j \ln Y_{t-j} + d \ln POP_t$$
$$+ e \ln Q_{it}^* + v_{it} \tag{5.40}$$

where Q_{it} is total minutes of intraLATA long-distance traffic between the ith city pair. Traffic refers to DDD calls only, aggregated over all rate periods that originate in city A and terminate in city B. Q_{it}^* is defined similarly, and represents the reverse traffic from B to A. P_i represents the real price of an intraLATA toll call on the ith route, defined as a fixed-weight Laspeyres price index, expressed as average revenue per minute, and deflated by the appropriate state or SMSA CPI. Y_i is real per capita personal income in originating city. POP is a market size proxy, defined as the product of the populations in cities A and B. Finally, δ_{ij} is a dummy variable, which takes a value of one for $i=j$, and 0 otherwise.

The models were estimated with pooled data for the nine city pairs by two-stage least squares (2SLS). The error terms u_i and v_i were assumed to be homoscedastic within a route, but heteroscedastic across routes. Contemporaneous covariance across routes is allowed for, as is first-order autocorrelation within routes. Finally, it is assumed that $E(u_{it} v_{jt'}) = 0$ for all i, j, t, and t'. Estimation proceeded as follows.[19] To begin with, reduced-form equations were estimated using least squares dummy variables (LSDV) methods, corrected for autocorrelation and heteroscedasticity, and tested for functional form using the Box–Cox format. Fitted values from these equations were then used in direct 2SLS estimation of the structural equations. The 2SLS residuals were tested for autocorrelation and heteroscedasticity, and the structural equations were re-estimated with appropriate corrections for autocorrelation and heteroscedasticity, and then tested for correct functional form. Final estimation of the structural equations was performed by LSDV (using the initial fitted values for the endogenous variables) with an additional correction for contemporaneous covariances (across routes but not across equations).

The Larson et al. study was a pioneering effort, and is accordingly subject to the usual problems of something being cut from whole cloth with only a vague pattern for guidance. The use of per capita income, while the dependent variable is not per capita, is questionable, and almost certainly contaminates the elasticities for market size, and probably for income as well. There may also be a problem in constraining the elasticities for the magnitudes of the calling areas to be equal. Such a constraint makes sense when calling between the areas is aggregated, but probably not when the model is directional, and feedbacks from reverse traffic are allowed for. Traffic from the sending area will continue to depend upon the number of telephones in that area, but the reverse-traffic effect will probably swamp the effect of the number of telephones in the receiving area, if for no other reason than that reverse traffic is a better proxy for community of interest. Finally, use of the population as a proxy for the number of telephones may be a problem.

These criticisms, however, in no way undermine the study's central result, which is that calls in one direction tend to generate calls in return that are independent of price and income. This result is statistically strong, and has been confirmed in subsequent studies at Bell Canada and elsewhere.[20] The reverse-calling phenomenon is clearly most relevant in situations in which the same telephone company does not serve both directions of a route. In this situation, a decrease in the price on the BA route will lead to the stimulation of the traffic on the AB route in the absence of a price change on that route. Failure to take this into account can lead to serious biases in the estimates of price elasticities.[21]

Two-stage Budgeting and the Total Bill Effect in Toll Demand: Zona and Jacob (1990)

I now turn to what has been a recurring question in analysing telecommunications demand, whether households approach their expenditures for telecommunications services in terms of a total bill. The usual assumption in demand analysis, and the one that has been implicit so far in this discussion, is that telecommunications services (local, toll, features and so on) compete along with all other goods and services in a household's market basket for the household's income. In this framework, short-haul toll calls can be either a substitute or a complement for long-haul calls, and an increase in access charges need not impinge upon either. An alternative view is that households budget for a total amount to be spent on telecommunications services, so that an increase in access charges (say), will necessarily lead to less being spent on usage. At one level, the debate can be cast in terms of what is the appropriate budget constraint to be used in telecommunications demand functions: income, or a budgeted total expenditure for telecommunications? Should the household's preferences be appropriately separable (whether weakly or strongly need not concern us), the approaches are of course equivalent, as the optimization problem can be formulated in terms of a two-stage budgeting process. In the first stage, the household decides how to allocate its income between telecommunications and all other goods and services. Second-stage optimization allocates stage-one total telecommunications expenditure among the different telecommunications services (including access).

Whether or not a household's preferences are appropriately separable is ultimately an empirical question, and was the primary focus in an unpublished paper by Zona and Jacob (1990). The motivation for the Zona–Jacob study was to test a hypothesis, stronger than separability that had been current for some time, namely the view that all that matters is the total bill.[22] Households are assumed to alter their demand in response to changes in the telecommunications prices in order to keep the total bill unchanged. The practical difference between this assumption and separability (or two-stage budgeting) is in the total-bill response to price changes in individual telecommunications services. With two-stage budgeting, a household will reassess its total phone bill in response to a price change, whereas with the other assumption, the total bill will not be reassessed, but demand composition will be altered so that total spending remains the same.

The total-bill hypothesis was tested by Zona and Jacob using the Almost Ideal Demand System (AIDS) of Deaton and Muellbauer (1980). The AIDS specification is a budget share model which relates expenditure shares to prices and the level of total expenditure. The AIDS system involves flexible functional forms and does not limit substitution, but it easily allows in estimation for the restrictions across elasticities that are imposed by the two hypotheses. The Zona–Jacob procedure is to estimate three models, an unrestricted model, in which the budget constraint is income and no restrictions are imposed on the cross elasticities, and two restricted models corresponding to the two-stage budgeting and total-bill hypotheses, respectively. The unrestricted model has four share equations, for each of interLATA toll, intraLATA toll, local service (access, custom-calling features, inside wiring and so on), and all other goods. For good i, the AIDS specification is given by

$$w_i = \alpha_i + \sum_j \gamma_{ij} \ln p_j + \beta_i \ln(x/P), \qquad (5.41)$$

where w_i denotes the budget, p_j is the price of good j, x is income, and P is the quadratic price deflator defined by Deaton and Muellbauer. General theoretical restrictions are imposed on the parameters across equations in order to ensure consistency with standard demand theory.[23] Under two-stage budgeting, the second-stage conditional budget share for good *i* will be given by

$$w_i^* = \alpha_i^* + \sum_j \gamma_{ij}^* \ln p_j + \beta_i^* \ln(B/P^*), \qquad (5.42)$$

where w_i^* is the bill share for service, B is the total telecommunications bill, and P^* is a suitably defined quadratic price deflator.

In the first stage, households are assumed to allocate income between telecommunications and all other goods. Zona and Jacob again assume an AIDS specification,

$$w^G = \alpha + \gamma \ln (P_B/CPI) + \beta \ln(x/P^{**}), \qquad (5.43)$$

where w^G is the telecommunications budget share, P_B is the telecommunications group price index, CPI is the price of other goods, and P^{**} is once again a suitably defined quadratic deflator.

The data used by Zona and Jacob in estimation is obtained from four distinct areas served by the (then) United Telecommunications System.[24] Each area was in a different state and had different tariffs. The time frame for the data was June to December 1989. The telephone data were aggregated to the three services, local, intraLATA toll and interLATA toll. The variables developed included the average total bill (total revenue divided by estimates of the number of households in the areas), bill shares for the three services, price indices for the three services (average revenue per unit), a state-specific price index, and estimates of average household income. The temporal unit of observation was a month, so that the data set consisted of 28 observations (7 months, 4 areas). A fixed-effects – that is LSDV – pooled time-series, cross-section framework was used in estimation.

The results obtained by Zona and Jacob show little difference in the own- and cross-price elasticities for inter- and intraLATA tolls for the unrestricted and the two-stage budgeting models, but a big difference between these two models and the bill-targeting model. Lagrange multiplier tests fail to detect a difference between the unrestricted and the two-stage budgeting models, but reject the bill-targeting model in favor of two-stage budgeting. The results accordingly support two-stage budgeting, but not bill-targeting.

AN OVERVIEW OF THE EMPIRICAL LITERATURE

At the time of my 1980 book, we knew a lot more about the demand for usage than the demand for access, a lot more about toll usage than local usage, a lot more about residential demand than business demand, and we probably had more precise knowledge about price elasticities than income elasticities. There was some knowledge of international demand, a little about business toll demand, even less about business WATS – that is, 800 service – and private-line demand, and virtually nothing about business demand for terminal equipment. Also, virtually nothing was known regarding cross-elasticities between local and toll usage or between other categories of service. We knew that network and call externalities were important conceptual elements of telecommunications demand, but there was little empirical information as to their magnitudes. Option demand was also known to be a potentially important component of access demand, but again virtually nothing was known of its empirical importance. A continuum of price elasticities had emerged, with access at the bottom (in terms of absolute value) and long-haul toll and international calls at the top. Price elasticities clearly appeared to increase with length of haul. There was some evidence of elastic demand in the longest-haul toll and international markets, but there seemed little question that the toll market as a whole was inelastic. Income elasticities, for the most part, appeared to be in excess of one.

During the 1980s and 1990s, research on telecommunications demand added substantially to our knowledge. The small, but nevertheless non-zero, access elasticities obtained by Perl (1978) have been confirmed and sharpened in the studies of Perl (1983), Taylor and Kridel (1990), and Cain and MacDonald (1991) for the USA and Bodnar et al. (1988) and Solvason (1997) for Canada.[25] Also, progress has been made in quantifying the structure of business demand, especially with regard to standard forms of access and the demands for key, PBX and centrex systems.[26] Toll studies confirm previous estimates of the price elasticities for medium- to long-haul toll usage,[27] and provide greatly improved estimates of the price elasticity for short-haul toll calling.[28] Also an important start was made on obtaining estimates of cross-price elasticities between measured- and flat-rate local service, local service and toll, local usage and toll, and key, PBX and centrex systems.[29] Finally, starts have also been made in the analysis of the demand for custom-calling features, both singularly and in bundles. Further, the demand for access to cellular and wireless networks, and the demand for access to the Internet have begun to receive attention.[30]

On the modeling front, three innovations stand out in the literature of the 1980s and 1990s: the widespread use of quantal choice models; the analysis of toll demand on a point-to-point basis; and the application of random-coefficients methods to a system of toll demand equations. With the much greater emphasis on access demand in the 1980s, it is natural that quantal-choice models would find much greater application. In the pre-1980 literature, the only study that used a quantal-choice framework was Perl (1978). In the years since, studies using this framework include Perl (1983), Train et al. (1987), Bodnar et al. (1988), Kridel (1988), Train et al. (1989), Ben-Akiva et al. (1990), Taylor and Kridel (1990), Kridel and Taylor (1993), and Solvason (1997). The most sophisticated application is by Train et al. (1987) in which (as seen above) a nested logit model is used to analyse the choice among standard flat-rate, EAS, and measured services and the choice of particular portfolios of calls by time of day and length of haul.

As was noted in the preceding section, an important impetus to modeling toll demand on a point-to-point basis came in 1985 with the Larson et al. study at Southwestern Bell using city-pair data. While there had been earlier point-to-point studies (including Larsen and McCleary 1970; Deschamps 1974; and Pacey 1983), the Larson et al. study was the first to allow for reverse-traffic effects. This framework, as we have seen, was subsequently applied to inter-provincial toll demand in Canada by Telecom Canada (1988), and to international calling by Acton and Vogelsang (1992). While both the Larson et al. and Telecom Canada studies employed pooled time-series, cross-section data sets in estimation, the most sophisticated application of varying- and random-coefficients methodology is that of Gatto et al. (1988).[31]

Progress has also been made in clarifying understanding of the network and call externalities. In great part, this occurred as a result of protracted debate before the CRTC in the mid- to late-1980s concerning the size of the coefficients for market size in the Bell Canada toll demand models.[32] While precise estimates of the magnitudes of the externalities are unfortunately still lacking, the relevant questions are now much sharper. In 1980, the biggest concern regarding consumption externalities was whether the network externality might be large enough to justify continuing the toll-to-local subsidy. Little attention was given to the call externality because it was thought that, even if one of any magnitude should exist, it would be internalized between calling parties through call-back effects. What such a view overlooked is the likelihood that calls give rise to further calls, quite independently of price and income. Strong empirical evidence that such is the case is given in the point-to-point toll demand models of Southwestern Bell and Telecom Canada, both of which indicate that a call in one direction stimulates something like one-half to two-thirds of a call in return. The call externality thus seems to have a dynamic dimension that goes beyond a simple cost sharing of calls between two parties. As noted above, it is best, in my view, not to attribute all of the call-back effect to call externalities, but to see part of it as reflecting an independent phenomenon, or what I have called the dynamics of information exchange.

In light of all that has gone on in the telecommunications industry since 1980, an obvious question to ask is whether the changes and upheavals that have taken place have caused similar upheavals in the structure of telecommunications demand.[33] In my view, there is little evidence that this has been the case, at least for the large categories like residential and business access and medium- to long-haul toll. I find no evidence that the toll market as a whole has become either more or less elastic than in 1980.[34] Residential access price elasticities, in contrast, appear to have become smaller,[35] but this would seem to be a consequence of higher residential penetration rates.[36]

In all candidness, however, the real concern at the moment with structural change in telecommunications is not with traditional categories of wire-line access and usage, but with new forms of access and usage that are associated with technological change and the emergence of the Internet.[37] The list includes:

1. A worldwide explosion in cellular/mobile telephony. Is wireless telephony a substitute for traditional wire-line communication or a complement? What are the price and income elasticities for wireless? Has the widespread availability of wireless access increased the total market for telecommunications? These are questions for which there are virtually no answers at present, at least in the open literature.[38]

2. The demand for access to the Internet. Among other things, this has led to a sharp acceleration in the demand for second lines.[39]

3. The emergence of Internet telephony. The question here is not whether, but how and how much, traditional circuit-switched telephony will be affected. At present Internet telephony is essentially a 'new' service, for which there is no empirical information at all about its likely formal structure. Indeed, one of the biggest challenges facing the empirical analysis of Internet telephony is how to go about collecting relevant data to begin with.

4. The demand for broadband access to the Internet. There are two questions related to broadband access: what will be the size of the market; and which technology is going to win out on the supply side – cable, traditional wire-line (via DSL or similar technology), or wireless?[40]

Returning to more traditional concerns, the post-1980 literature confirms the finding that the absolute value of the price elasticity increases with length of haul. Indeed, all considered, this probably remains the best-established empirical regularity in telecommunications demand. Interestingly enough, however, no convincing explanation as to why this ought to be the case has ever been offered. Some have suggested that it reflects primarily a price phenomenon: since price elasticities (for non-constant elasticity demand functions) in general vary with the level of price, elasticities for longer-haul calls will be larger simply because they cost more.[41] On the other hand, an argument can also be made that the phenomenon represents a genuine distance effect, in that (for whatever reason) calls with a longer haul are simply more price-elastic.[42]

Finally, there is international demand. While a number of interesting and useful studies of international demand were undertaken in the 1980s and 1990s, it nevertheless remains an under-researched area. Studies of note include Telecom Canada (Appelbe et al., 1988), Acton and Vogelsang (1992), Hackl and Westlund (1995, 1996), Garin-Muñoz and Perez-Amaral (1999), Cracknell (1999), and Karikari and Gyimah-Brempong (1999). The price elasticities in the Telecom Canada study are (as noted above) based on a point-to-point model and are reported only on a unidirectional basis, that is, for a change in Canada–US rates only. The elasticities are estimated to be of the order of –0.5, which are consistent with bi-directional Canada–US elasticities of –0.9 or larger. The Acton–Vogelsang study is based upon annual data for minutes of calling between the USA and 17 Western European countries from 1979 to 1986. The volume of calling both originating and terminating in the USA was examined as a function of the price in the sending country, price in the receiving country, together with economic and conditioning variables such the volume of trade between the countries, the structure of employment, the number of telephones, and the price of telex services. Models are estimated for 1979–82 and 1983–6. Estimates of the unidirectional elasticities vary from insignificant to near –1.[43]

To conclude this overview of the empirical literature, Table 5.1 provides a guide to the post-1980 literature, organized in terms of residential and business access, short- and long-haul toll, customer-premise equipment, Internet access, and so on.

CHALLENGES FOR THE FUTURE

Unfortunately, much of what has been discussed in this chapter is dinosaur-ish. Competition, deregulation and new services made available by the convergence of computers, wireless, cable and the Internet with conventional wire-line telephony have made telecommunications demand analyses of the traditional type essentially phenomena of the past. Even if they were desirable, comprehensive industry studies of the type undertaken by Gatto et al. (1988) are impossible to pursue, for it is no longer feasible to organize the required data.[44] The distinction between access and usage is still the place

*Table 5.1   Abridged guide to the post-1980 literature on
             telecommunications demand*

Residential Access Demand	Network Externalities
Cain and McDonald (1991)	Perl (1983)
Kling and Van Der Ploeg (1990)	Taylor and Kridel (1990)
Perl (1983)	Peak/Off-Peak
Taylor and Kridel (1990)	Appelbe et al. (1988)
Train et al. (1987)	Bell Canada (1984, 1989)
Train et al. (1989)	Colias and Maddox (1990)
Business Access Demand	Gatto et al. (1988)
Ben-Akiva and Gershenfeld (1989)	Extended-Area Service
Local Calling	Kridel (1988)
Colias and Maddox (1990)	Martins-Filho and Mayo (1993)
Hobson and Spady (1988)	Optional Calling Plans
Kling and Van Der Ploeg (1990)	Watters and Roberson (1992)
Train et al. (1987)	Bypass
Train et al. (1989)	Watters and Grandstaff (1988)
Short-Haul Toll	WATS/Private Line
Duncan and Perry (1992)	Griffin and Egan (1985)
Hausman (1991)	Hausman (1991)
Kling and Van Der Ploeg (1990)	Nieswadomy and Brink (1990)
Train et al. (1987)	Watters and Grandstaff (1988)
Train et al. (1989)	Watters and Roberson (1991)
Zona and Jacob (1990)	Custom-Calling Features
Medium-Haul Toll	Kridel and Taylor (1993)
Bell Canada (1984, 1989)	Customer-Premise Equipment
Larson et al. (1990)	Ben-Akiva and Gershenfeld (1989)
Zona and Jacob (1990)	Internet/Broadband Access
Long-Haul Toll	Eisner and Waldon (1999)
Appelbe et al. (1988)	Kridel et al. (2002)
Zona and Jacob (1990)	Madden and Simpson (1997)
Gatto et al. (1988)	Madden et al. (1999)
	Rappoport et al. (1998)

to begin in studying telecommunications demand, but the purview of the subject has clearly become much larger. No longer is it adequate to look simply at wire-line access, intraLATA, interLATA, and international toll calling, 'vertical services', and customer-premise equipment. There are now many other types of access to consider, including broadband access to the Internet, and a variety of new forms of telephony, including wireless, fixed-wireless and Internet telephony. Indeed, in my view, the Internet itself is now part and parcel of telecommunications, and should be encompassed in its framework.

The models described still provide cogent points of departure. Indeed, this is the principal reason for their presentation in detail. However, their application must now be to services and markets that reflect present telecommunications structures, for suppliers as well as for users. Clearly, one of the biggest current challenges facing telecommunications demand modelers is to identify structures that can reasonably be assumed to be stable. Almost certainly, access to the telecommunications network can still be assumed to be stable, but most probably not with respect to the traditional definition of access in terms of fixed wire-line access. A broader definition of access is required. Similar considerations apply to usage. While it is probably still reasonable to attribute stability to telecommunications at different distances, times of day, and days of the week, this is probably not the case with respect to conventional definitions of intraLATA and interLATA fixed-line toll calling. Definitions must be expanded to include e-mail and other forms of Internet communication, as well as wireless-to-wireless connections.[45] Also, as has been emphasized throughout this chapter, there are major challenges concerning data. Traditional sources for telecommunications data clearly no longer suffice, or indeed are even relevant. Increasingly, data are going to have to be obtained directly from telecommunications users, rather than suppliers or from companies whose businesses are simply to monitor and count.

Finally, one of the failings of the traditional corporate decision-making structure in the telephone industry has been that demand analysis has always played a spot role, used and useful for calculating repression/stimulation effects of proposed tariff changes in regulatory proceedings, but never employed in a systematic way in budgeting, forecasting, and marketing.[46] Its home in the old Bell System was always revenue requirements, and as deregulation and competition have forced telephone companies to slim down, demand analysis has tended to be treated as regulatory overhead, no longer necessary. In truth, some slimming-down was probably in order, but to allow demand analyses to languish because of the introduction of deregulation, as has happened with US and Canadian telecommunications companies' is a mistake. One of the biggest current challenges for demand analysts in the telecommunications companies is to forge links with marketing departments, and to become integrated in company budgeting and forecasting processes. Applied demand analysis has a strategic role to play in a competitive environment, ranging from conventional types of elasticity estimation in traditional markets to the identification of new markets.[47] The possibilities are vast. It only requires imagination, hard work and some humility on the part of economists and econometricians.

NOTES

1. The treatment of consumption externalities in telecommunications literature long predates the emergence in the 1990s of a voluminous literature on the economics of networks. The seminal papers are Artle and Averous (1973), Rohlfs (1974), Von Rabenau and Stahl (1974), and Littlechild (1975).

2. In the USA and Canada, the time (if ever) for network externalities to have been a factor in fueling endogenous growth in the telephone system would have been in the first decades of the twentieth century. On the other hand, they may have been a factor in France in the 1970s, and are almost certainly of prime importance in some present-day undeveloped countries. The Internet, of course, is another matter, and there seems little question but that network externalities have been (and continue to be) of great importance in its phenomenal growth.

3. See Squire (1973) and Littlechild (1975).

4. See Larson and Lehman (1986).

5. Taylor (1994, Chapter 9).

6. The externalities associated with the dynamics of information exchange may be of even greater importance on the Internet, whereby a visit to one website may create the need for visits to a whole string of other websites.

7. The concept of option demand is well known in the literature on conservation and exhaustible natural resources (see Weisbrod, 1964; Kahn, 1966), but apart from brief mention in Squire (1973), it is absent from the pre-1980 literature on telecommunications demand. In recent years, its relevance has come to the fore in analysing the pricing of default capacity in the context of carrier-of-last-resort obligations and in the demand for optional extended-area service. See Weisman (1988) and Kridel et al. (1991).

8. See Chapter 5 of Taylor (1994).

9. This is not to argue that all businesses necessarily have telephones, but just as there are high-income households without telephone service, there are undoubtedly successful businesses for which the same is true. The numbers cannot be large, however, and the reasons are almost certainly non-economic.

10. At the time that Perl's study was commissioned by the (then) Central Services Organization of (the about to become independent) Regional Bell Operating Companies, there was a lot of concern about the possible negative effects on telephone penetration of measured local service. This function, which has subsequently become known as the 'Perl' demand function, was motivated by the fact that it can accommodate a zero, as well as a non-zero, price for usage. There was also a lot of concern regarding the effects that higher local-service charges might have on subscription rates of non-white households and households headed by females. These effects were allowed for through a bevy of socio-demographic dummy variables.

11. Empirical studies employing the probit framework include Perl (1983) and Taylor and Kridel (1990), while studies using the logit framework include Train et al. (1987) and Bodnar et al. (1988). Most empirical studies of access demand that employ a consumer surplus framework focus on local usage, and accordingly ignore the net benefits arising from toll usage. Hausman et al. (1993) and Erikson et al. (1995) are exceptions.

12. In the late-1970s, various versions of this generic model had been estimated by local Bell Operating Companies for use in rate hearings in 34 states. A tabulation of the models can be found in Appendix 2 of Taylor (1994).

13. See Bell Canada (1984, 1989), also Taylor (1984).

14. Pacey (1983) was the first attempt to model toll demand on a route-specific basis.

15. The first modeling efforts taking reciprocal calling and call-back effects into account were undertaken at Southwestern Bell in the mid-1980s. The results are reported in Larson et al. (1990). See also Appelbe et al. (1988) and Acton and Vogelsang (1992).

16. Train et al. provide an excellent discussion of the considerations that determine an appropriate nesting. The key consideration is the pattern of correlations among the omitted factors. For local calling, most of the specific factors that determine portfolio

choice, such as where the household's friends and relatives live, the time and location of activities that require use of the telephone, are not observed. These factors, however, are similar across all service offerings for any portfolio. This leads to nesting together alternatives with the same portfolio, but different service options. On the other hand, the primary factor affecting the choice of service is price, which is observable. The price of a portfolio under a particular service option is simply the cost of the portfolio under the rate schedule associated with that option. Since alternatives with the same service option, but not the same portfolio of calls, can be similar with respect to observed factors, but not (at least relatively) with respect to unobserved factors, these are not nested.

17. Other studies of the demand for EAS include Pavarini (1976, 1979) and Martins-Filho and Mayo (1993).
18. This should not be taken to mean that the existing elasticities were wrong, but only that they were not applicable to the question that was being asked. One can reasonably infer that the households whose data were being analysed exhibited above-average price responses because they were drawn from suburban communities that were already petitioning for EAS. The price elasticities obtained by Kridel were applicable to such circumstances, but almost certainly not to short-haul toll calls in general.
19. Full details of the estimation are given in Larson (1988).
20. See Appelbe et al. (1988) and Acton and Vogelsang (1992).
21. In particular, one must be careful to distinguish between unidirectional elasticities and bi-directional elasticities. Unidirectional elasticities allow for a change in price in one direction only, while bi-directional elasticities allow for price changes in both directions. In both cases, calculations must be made from the reduced forms of the models. For the formulae involved, see Appelbe et al. (1988).
22. See Doherty and Oscar (1977).
23. See Deaton and Muellbauer (1980, p. 316).
24. Since United Telecommunications was not part of the Bell System, it could continue to offer both local and long-distance services (both intra- and interLATA) after the AT&T divestiture. The Zona–Jacob study was accordingly of great interest, not only because of its focus on the bill-targeting hypothesis, but also because of the uniqueness of its data set. GTE was in similar circumstances as United in providing both local and long-distance services, but I am not aware that a Zona-Jacob type study has ever been undertaken using GTE data.
25. See also Kaserman et al. (1990) for the US and Rodriguez-Andres and Perez-Amaral (1998) for Spain.
26. See Ben-Akiva and Gershenfeld (1989) and Watters and Roberson (1991).
27. See the Bell-Intra models of Bell Canada (1984, 1989), the Telecom Canada models (Appelbe et al., 1988, Gatto et al., 1988, Larson et al., 1990, Zona and Jacob, 1990).
28. See Train et al. (1987), Kridel (1988), Train et al. (1989), Kling and Van der Ploeg (1990), Zona and Jacob (1990), Watters and Roberson (1992) and Martins-Filho and Mayo (1993).
29. See Train et al. (1987), Kridel (1988), Ben-Akiva and Gershenfeld (1989), Train et al. (1989), Kling and Van der Ploeg (1990), Watters and Roberson (1991), and Zona and Jacob (1990).
30. See Kridel and Taylor (1993), Madden and Simpson (1997), Ahn and Lee (1999), Kridel et al. (1999b, 2002) and Madden et al. (1999).
31. Interstate toll demand in the Gatto et al. study is approached in a multi-equation framework in which toll demand is disaggregated by mileage bands, time of day, and non-operator/operator handled calls. The model is estimated in a random coefficients variance-components format using a pooled data set consisting of quarterly observations for the 48 contiguous US states and the District of Columbia. Two interesting new concepts were introduced into applied demand analysis in this study, namely, stochastic symmetry and stochastic weak separability. Stochastic symmetry is taken to mean that the Slutsky symmetry conditions across commodities in a cross-section are only satisfied stochastically, while stochastic weak separability means that the price elasticities in a cross-section are assumed to be stochastically proportional to the income elasticity.

32. See Taylor (1984), Bell Canada (1984, 1989), Breslaw (1985) and Globerman (1988). See also de Fontenay and Lee (1983), and Taylor (1994, Appendix 4).
33. This is for the market as a whole. Price elasticities for individual interexchange carriers are another matter, as are also elasticities for certain submarkets within a given market. Because of a decrease in market share, the price elasticities that AT&T faces are clearly larger now than (say) in 1990. For MCI and Sprint, however, they are almost certainly smaller. For attempts to estimate firm-specific demand elasticities, see Ward (1994), Kahai et al. (1996) and Kaserman et al. (1999).
34. One of the more interesting results from research during the 1980s relates to the price elasticity for short-haul toll calls of (say) 15 to 40 miles. In 1980, the view was that this elasticity was quite small, of the order of –0.2. There was no real evidence in support of this number; it simply fitted into the flow of elasticities as one went from shorter to longer lengths of haul. The evidence now points to a value of this elasticity as being in the range of –0.25 to –0.40. See Train et al. (1987), Kridel (1988), Train et al. (1989), Kling and Van der Ploeg (1990), Zona and Jacob (1990), Hausman (1991), Martins-Filho and Mayo (1993), Duncan and Perry (1995), Rappoport and Taylor (1997) and Levy (1999). Nevertheless, it should be emphasized that this should not be viewed as an instance of structural change, but rather the emergence, for the first time, of credible evidence as to what the value of this elasticity actually is. For an account of the consequences of the failure on the part of the Regulatory Authority in California to take this value seriously, see Hausman and Tardiff (1999).
35. See, however, Cain and MacDonald (1991), who obtain residential access price elasticities that are about double Perl's, especially at very low income levels. As their results are based upon data from the mid-1980s, Cain and MacDonald interpret these higher elasticities as reflecting structural change. Whether this represents genuine structural change, or is simply a reflection of sharp shifts in the ethnic and social composition of households, though, is not clear.
36. Price elasticities calculated from the probit/logit models that are used in estimating access demand vary inversely with the level of penetration. A point that is worthy of mention is that, in comparing the results using data from the 1970 Census with those with data from the 1980 Census, Perl (1983) found an upward shift in the willingness to pay of households to have telephone services that could not be attributed to income or to changes in socio-demographic factors. It would be interesting to know whether data from the 1990 Census would show the same phenomenon vis-à-vis 1980. However, I am not aware that such a study has been undertaken.
37. For an informed discussion of some of the problems involved, see Cracknell and Mason (1999).
38. The 'closed' literature is obviously another matter, for it is precisely this type of information that has become a closely guarded proprietary asset. At the annual International Communications Forecasting Conference in San Francisco in June 1998, two analysts from SBC discussed a study of cellular demand using data from Cellular One. Their presentation was methodologically extremely informative, but devoid of numerical information about elasticities.
39. For recent studies of the demand for additional lines, see Cassel (1999), Duffy-Deno (2001) and Eisner and Waldon (1999). For a pre-Internet study of the demand for second lines in Canada, see Bodnar et al. (1988). For an initial effort to estimate the price elasticity for Internet access, see Kridel et al. (1999b).
40. The only broadband access studies that I am presently aware of are Madden and Simpson (1997), Madden et al. (1999), and a current one of my own that focuses on the demand for access to the Internet through cable modem (Kridel et al., 2002).
41. This follows from the definition of price elasticity as $(\partial q/\partial p)(p/q)$, where q and p denote quantity and price, respectively. Strong evidence against the price-elasticity/length-of-haul relationship being just a price phenomenon is provided by results from the study of Gatto et al. (1988) for interstate toll traffic in the USA. The price elasticity obtained in that model, –0.72, agrees closely with pre-1980 price elasticities for long-haul calls (see Taylor 1994, Appendix 1), but it is estimated from a data set in which the real price of

toll calling was sharply lower (because of competition) than the levels of the 1970s. Also, it should be noted that the emergence of 'postalized' pricing, in which toll rates do not vary by distance, pretty much makes the price/distance relationship a moot point.
42. For a discussion of factors that might account for the price-elasticity/length-of-haul relationship, see Taylor (1994, Chapter 11).
43. The models specified by Acton and Vogelsang allow for arbitrage and call externalities, but not for reverse-traffic effects of the type analysed in the Southwestern Bell and Telecom Canada studies. Also, the models include only the number of telephones in the European countries. For a tabulation of studies of international demand, together with estimated price and income elasticities, undertaken prior to 1980, see Taylor (1994, Appendix 2).
44. Because of fragmentation and proprietary concerns on the part of service providers, it has become increasingly necessary for analysts to work with data that are obtained directly from telecommunications users, such as from the Bill HarvestingTM and RequestTM surveys of PNR & Associates of Jenkintown, PA. For studies based on the PNR data, see Rappoport and Taylor (1997), Rappoport et al. (1998), and Kridel et al. (1999, 2002).
45. In my view, the ultimate place to seek stability in household consumption patterns is in broad consumption categories. I believe that there is basic stability in the structure defining expenditure for telecommunications in relation to other expenditure categories, as well as for functional categories of expenditure within telecommunications. The challenge, of course, is to define the functional categories in meaningful ways.
46. The reference here is to US telephone companies. Demand analysis and forecasting for years was well integrated into the corporate decision-making structure at Bell Canada. Indeed, from the mid-1970s through the mid-1990s, Bell Canada was one of the world's major centers of innovative telecommunications demand analysis activity..
47. For an innovative framework for assessing the value of new services, see Hausman and Tardiff (1997). For overviews and frameworks for forecasting the demand for new products, see Palombini and Sapio (1999), Saether (1999) and Tauschner (1999).

REFERENCES

Acton, J.P. and Vogelsang, I. (1992), 'Telephone demand over the Atlantic: evidence from country-pair data', *Journal of Industrial Economics*, **40** (3), 305–23.
Ahn, H. and Lee, M.H. (1999), 'An econometric analysis of the demand for access to mobile telephone networks', *Information Economics and Policy*, **11** (3), 297–306.
Appelbe, T.W., Snihur, N.A., Dineen, C., Farnes, D., and Giorano, R. (1988), 'Point-to-point modeling: an application to Canada–Canada and Canada–U.S. long-distance calling', *Information Economics and Policy*, **3** (4), 358–78.
Artle, R. and Averous, C. (1973), 'The telephone system as a public good: static and dynamic aspects', *Bell Journal of Economics and Management Science*, **4** (1), 89–100.
Bell Canada (1984), 'Econometric models of demand for selected intra-Bell long-distance and local services', Attachment 1 to Response to Interrogatory Bell (CNCP) 20 February 84-509 IC (Supplemental, July), CRTC, Ottawa, Ontario.
Bell Canada (1989), 'Intra-Bell customer-dialed peak and off-peak model updates,' Attachment 1, CRTC Public Notice 1988-45; Bell Canada and British Columbia Telephone-Company Review of Methodologies Used to Model Price Elasticities, 3 March.
Ben-Akiva, M. and Gershenfeld, S. (1989), 'Analysis of business establishment choice of telephone system', Cambridge Systematics, Inc., Cambridge, MA.

Bodnar, J., Dilworth, P. and Iacono, S. (1988), 'Cross-section analysis of residential telephone subscription in Canada', *Information Economics and Policy*, **3** (4), 311–31.

Breslaw, J.A. (1985), 'Network externalities and the demand for residence long distance telephone service,' Working Paper No. 1985-13, Concordia University, Montreal, Quebec.

Breslaw, J.A. (1989), 'Bell Canada and British Columbia Telephone Company: review of methodology used to model price elasticity', evidence on behalf of the Government of Quebec before the Canadian Radio-Television and Telecommunications Commission, CRTC Telecom 1988-45 (April), Ottawa, Ontario.

Cain, P. and MacDonald, J.M. (1991), 'Telephone pricing structures: the effects on universal service', *Journal of Regulatory Economics*, **3**, 293–308.

Cassel, C.A. (1999), 'Demand for and use of additional lines by residential customers', in *The Future of the Telecommunications Industry: Forecasting and Demand Analysis*, ed. Loomis, D.G. and Taylor, L.D., Boston: Kluwer Academic Publishers.

Colias, J. and Maddox, L. (1990), 'Analysis of demand for Kentucky local telephone usage', attachment to testimony of Lorraine Maddox, 'An investigation into the economic feasibility of providing local measured service telephone rates in Kentucky', Administrative Case No. 285.

Cracknell, D. (1999), 'The changing market for inland and international calls', in *The Future of the Telecommunications Industry: Forecasting and Demand Analysis*, ed. Loomis, D.G. and Taylor, L.D., Boston: Kluwer Academic Publishers.

Cracknell, D. and Mason, C. (1999), 'Forecasting telephony demand against a background of major structural change', in *The Future of the Telecommunications Industry: Forecasting and Demand Analysis*, ed. Loomis, D.G. and Taylor, L.D., Boston: Kluwer Academic Publishers.

Deaton, A.S. and Muellbauer, J. (1980), 'An almost ideal demand system', *American Economic Review*, **70** (3), 312–26.

de Fontenay, A. and Lee, J.T. (1983), 'BC/Alberta long distance calling', in *Economic Analysis of Telecommunications: Theory and Practice*, ed. Courville, L. et al., Amsterdam: North-Holland.

Deschamps, P.J. (1974), 'The demand for telephone calls in Belgium', presented at Birmingham International Conference in Telecommunications Economics, Birmingham, England, Catholic University of Louvain, Belgium.

Doherty, A.N. and Oscar, G.M. (1977), 'Will the rates produce the revenues?', *Public Utilities Fortnightly*, **99**, 15–23.

Duffy-Deno, K. (2001), 'Demand for additional telephone lines: an empirical note', *Information Economics and Policy*, **13** (3), 265–6.

Duncan, G.M. and Perry, D.M. (1995), 'Intralata toll demand modeling: a dynamic analysis of revenue and usage data', *Information Economics and Policy*, **6** (2), 163–78.

Eisner, J. and Waldon, T. (1999), 'The demand for bandwidth: second telephone lines and on-line services', Federal Communications Commission, Washington, DC.

Erikson, R.C., Kaserman, D.L., and Mayo, J.W. (1998), 'Targeted and untargeted subsidy schemes: evidence from post-divestiture efforts to promote universal service', *Journal of Law and Economics*, **41**, 477–502.

Garin-Muñoz, T. and Perez-Amaral, T. (1999), 'A model of Spain–Europe telecommunications', *Applied Economics*, **31**, 989–97.
Gatto, J.P., Kelejian, H.H. and Stephan, S.W. (1988), 'Stochastic generalizations of demand systems with an application to telecommunications', *Information Economics and Policy*, **3** (4), 283–310.
Gatto, J.P., Langin-Hooper, J., Robinson, P. and Tryan, H. (1988), 'Interstate switched access demand', *Information Economics and Policy*, 3(4), 333–58.
Globerman, S. (1988), 'Elasticity of demand for long-distance telephone service', report prepared by Steve Globerman Associates for the Federal-Provincial-Territorial Task Force on Telecommunications, Ottawa, Ontario.
Griffin, J.M. and Egan, B.L. (1985), 'Demand system estimation in the presence of multi-block tariffs', *Review of Economics and Statistics*, **67** (3), 520–24.
Hackl, P. and Westlund, A.H. (1995), 'On price elasticities of international telecommunications demand', *Information Economics and Policy*, **7** (1), 27–36.
Hackl, P. and Westlund, A.H. (1996), 'Demand for international telecommunication time-varying price elasticity', *Journal of Econometrics*, **70**, 243–60.
Hausman, J.A. (1991), 'Phase II IRD testimony of Professor Jerry A. Hausman on behalf of Pacific Bell Telephone Co.', before the Public Utility Commission of California (23 September).
Hausman, J.A., Tardiff, T.J. and Bellinfonte, A. (1993), 'The effects of the breakup of AT&T on telephone penetration in the United States', *American Economic Review Papers*, **83** (2), 178–84.
Hausman, J.A. and Tardiff, T.J. (1997), 'Valuation of new services in telecommunications', in *The Economics of the Information Society*, ed. Dumont, A. and Dryden, J., Luxembourg: Office for Official Publications of the European Communities, 76–80. A longer version of the paper was presented to the OECD Workshop on the Economics of the Information Society, Toronto, Canada (28 June 1995); National Economic Research Associates, Cambridge, MA (July 1995).
Hausman, J.A. and Tardiff, T.J. (1999), 'Development and application of long-distance price elasticities in California: a case study in telecommunications regulation', National Economic Research Associates, Inc., Cambridge, MA.
Hobson, M. and Spady, R. (1988). *The Demand for Local Telephone Service under Optional Local Measured Service*, Morristown, NJ: Bell Communications Research.
Kahai, S.K., Kaserman, D.L. and Mayo, J.W. (1996), 'Is the dominant firm dominant? An empirical analysis of AT&T's market power', *Journal of Law and Economics*, **39**, 499–517.
Kahn, A.E. (1966), 'The tyranny of small business decisions: market failure, imperfections, and the limits of economics', *Kyklos*, **19**, 23–47.
Karikari, J.A. and Gyimah-Brempong, K. (1999), 'Demand for international telephone services between the U.S. and Africa', *Information Economics and Policy*, **11** (4), 407–36.
Kaserman, D.L., Mayo, J.W., Blank, L.R. and Kahai, S.K. (1999), 'Open entry and local rates: the economics of intralata toll competition', *Review of Industrial Organization*, **14** (4), 303–19.
Kling, J.P. and Van Der Ploeg, S.S. (1990), 'Estimating local call elasticities with a model of stochastic class of service and usage choice', in *Telecommunications Demand Modeling*, ed. de Fontenay, A., Shugard, M.H. and Sibley, D.S. Amsterdam: North-Holland.
Kridel, D.J. (1988), 'A consumer surplus approach to predicting extended area

service (EAS) development and stimulation rates', *Information Economics and Policy*, **3** (4), 379–90.

Kridel, D.J., Lehman, D.E. and Weisman, D.L. (1991), 'Options value, telecommunications demand and policy', Southwestern Bell Telephone Co., St. Louis.

Kridel, D.J., Rappoport, P.N. and Taylor, L.D. (1999a), 'Competition in intralata long-distance: carrier choice models', Department of Economics, University of Missouri, St. Louis.

Kridel, D.J., Rappoport, P.N. and Taylor, L.D. (1999b), 'An econometric model of the demand for access to the Internet', in *The Future of the Telecommunications Industry: Forecasting and Demand Analysis*, ed. Loomis, D.G. and Taylor, L.D., Boston: Kluwer Academic Publishers.

Kridel, D.J., Rappoport, P.N. and Taylor, L.D. (2002), 'The demand for access to the Internet by cable modem', in *Forecasting the Internet: Understanding the Explosive Growth of Data Communications*, ed. Loomis, D.G. and Taylor, L.D., Boston: Kluwer Academic Publishers.

Kridel, D.J. and Taylor, L.D. (1993), 'The demand for commodity packages: the case of telephone custom calling features', *Review of Economics and Statistics*, **75** (2), 362–7.

Larsen, W.A. and McCleary, S.J. (1970), 'Exploratory attempts to model point-to-point cross-sectional interstate telephone demand', Unpublished Bell Laboratories Memorandum, Murray Hill, NJ (July).

Larson, A.C. (1988), 'Specification error tests for pooled demand models', presented at Bell Communications Research/Bell Canada Industry Forum on Telecommunications Demand Analysis, Key Biscayne, FL (25–27 January), Southwestern Bell Telephone Co., St. Louis.

Larson, A.C. and Lehman, D.E. (1986), 'Asymmetric pricing and arbitrage', presented at the Sixth International Conference on Forecasting and Analysis for Business Planning in the Information Age, Southwestern Bell Telephone Co., St. Louis.

Larson, A.C., Lehman, D.E. and Weisman, D.L. (1990), 'A general theory of long-distance telephone demand', in *Telecommunications Demand Modeling*, ed. de Fontenay, A., Shugard, M.H. and Sibley, D.S. Amsterdam: North-Holland.

Littlechild, S.C. (1975), 'Two-part tariffs and consumption externalities', *Bell Journal of Economics*, **6** (2), 661–70.

Madden, G. and Simpson, M. (1997), 'Residential broadband subscription demand: An econometric analysis of Australian choice experiment data', *Applied Economics*, **29**, 1073–8.

Madden, G., Savage, S.J. and Coble-Neal, G. (1999), 'Subscriber churn in the Australian ISP market', *Information Economics and Policy*, **11** (2), 195–208.

Martins-Filho, C. and Mayo, J.W. (1993), 'Demand and pricing of telecommunications services: evidence and welfare implications', *Rand Journal of Economics*, **24** (3), 439–54.

McFadden, D.L. (1978), 'Modeling the choice of residential location', in *Spatial Interaction Theory and Planning Models*, ed. Karquist et al., Amsterdam: North-Holland.

Nieswadomy, M.L. and Brink, S. (1990), *Private Line Demand: A Point-to-Point Analysis*, Denton, TX: Department of Economics, University of North Texas.

Pacey, P.L. (1983), 'Long-distance demand: a point-to-point model', *Southern Economic Journal*, **49** (4), 1094–107.

Palombini, I.M. and Sapio, B. (1999), 'Analysis of customer expectations for the introduction of new telecommunications services', in *The Future of the Telecommunications Industry: Forecasting and Demand Analysis*, ed. Loomis, D.G. and Taylor, L.D., Boston: Kluwer Academic Publishers.

Pavarini, C. (1976), 'The effect of flat-to-measured rate conversions on the demand for local telephone usage', unpublished Bell Laboratories Memorandum, Bell Laboratories, Murray Hill, NJ (27 September).

Pavarini, C. (1979), 'The effect of flat-to-measured rate conversions on local telephone usage', in *Pricing in Regulated Industries II*, ed. Wenders, J.T. Mountain States Telephone Co., Denver.

Perl, L.J. (1978), 'Economic and social determinants of residential demand for basic telephone service', National Economic Research Associates, Inc., White Plains (28 March).

Perl, L.J. (1983), 'Residential demand for telephone service 1983', prepared for Central Services Organization of the Bell Operating Companies, National Economic Research Associates, Inc., White Plains (December).

Rappoport, P.N. and Taylor, L.D. (1997), 'Toll price elasticities estimated from a sample of U.S. residential telephone bills', *Information Economics and Policy*, **9** (1), 51–70.

Rappoport, P.N., Taylor, L.D., Kridel, D.J. and Serad, W. (1998), 'The demand for Internet and on-line access', in *Telecommunications Transformation: Technology, Strategy and Policy*, ed. Bohlin, E. and Levin, S.L., IOS Press, 205–18.

Rodriguez-Andres, A. and Perez-Amaral, T. (1998), 'Demand for telephone lines and universal service in Spain', *Information Economics and Policy*, **10** (4), 501–14.

Rohlfs, J. (1974), 'A theory of interdependent demand for a consumption service', *Bell Journal of Economics and Management Science*, **5** (1), 16–37.

Saether, J.P. (1999), 'Limits to growth in telecommunications markets', in *The Future of the Telecommunications Industry: Forecasting and Demand Analysis*, ed. Loomis, D.G. and Taylor, L.D., Boston: Kluwer Academic Publishers.

Solvason, D.L. (1997), 'Cross-sectional analysis of residential telephone subscription in Canada using 1984 data', *Information Economics and Policy*, **9** (3), 241–64.

Squire, L. (1973), 'Some aspects of optimal pricing for telecommunications', *Bell Journal of Economics and Management Science*, **4** (2), 515–25.

Tauschner, A. (1999), 'Forecasting new telecommunication services at a "pre-development" product stage', in *The Future of the Telecommunications Industry: Forecasting and Demand Analysis*, ed. Loomis, D.G. and Taylor, L.D., Boston: Kluwer Academic Publishers.

Taylor, L.D. (1984), 'Evidence of Lester D. Taylor on behalf of Bell Canada concerning the price elasticity of demand for message toll service in Ontario and Quebec', Canadian Radio-television and Telecommunications Commission, November, Ottawa, Ontario.

Taylor, L.D. (1994), *Telecommunications Demand in Theory and Practice*, Boston: Kluwer Academic Publishers.

Taylor, L.D. and Kridel, D.J. (1990), 'Residential demand for access to the telephone network', in *Telecommunications Demand Modeling*, ed. de Fontenay, A. Shugard, M.H. and Sibley, D.S. Amsterdam: North-Holland.

Train, K.E., McFadden, D.L. and Ben-Akiva, M. (1987), 'The demand for local telephone service: a fully discrete model of residential calling patterns and service choices', *Rand Journal of Economics*, **18** (1), 109–3.

Train, K.E., Ben-Akiva, M. and Atherton, T. (1989), 'Consumption patterns and self-selecting tariffs', *Review of Economics and Statistics*, **71** (1), 62–73.

Von Rabenau, B. and Stahl, K. (1974), 'Dynamic aspects of public goods: a further analysis of the telephone system', *Bell Journal of Economics and Management Science*, **5** (2), 651–69.

Ward, M. (1994), 'Market power in long distance communications', Federal Trade Commission, Washington, DC (October).

Watters, J.S. and Grandstaff, P.J. (1988), 'An Econometric Model of Interstate Access', presented at BELLCORE/Bell Canada Telecommunications Demand Analysis Industry Forum, Key Biscayne, Florida, 25–27 January, South Western Bell Telephone Co., St Louis.

Watters, J.S. and Roberson, M.L. (1991), 'An econometric model for dedicated telecommunications services: a systems approach', Southwestern Bell Telephone Co., St. Louis.

Watters, J.S. and Roberson, M.L. (1992), 'An econometric model for Intralata MTS pricing plans', Southwestern Bell Telephone Co., St. Louis.

Weisbrod, B.A. (1964), 'Collective-consumption of individual consumption goods', *Quarterly Journal of Economics*, **78** (3), 471–7.

Weisman, D.L. (1988), 'Default capacity tariffs: smoothing the transitional regulatory asymmetics in the telecommunications market', *Yale Journal on Regulation* (Winter), 149–78.

Zona, J.D. and Jacob, R. (1990), 'The total bill concept: defining and testing alternative views', presented at BELLCORE/Bell Canada Industry Forum, Telecommunications Demand Analysis with Dynamic Regulation, Hilton Head, SC (22–25 April), National Economic Research Associates, Cambridge, MA.

6. Retail telecommunications pricing in the presence of external effects

Benjamin E. Hermalin and Michael L. Katz

INTRODUCTION

This chapter examines the retail pricing of telecommunications network services. By telecommunications, we mean any electronic or photonic transport of data from one party to another. These networks may be circuit switched or packet based, wireless or wire-line. Although our analysis applies most directly to one-to-one communications, some of our findings can be readily modified to apply to one-to-many, or broadcast, networks, such as over-the-air and cable television. Many of the pricing results stated here also are relevant to electronic payment networks.

Our central concerns are how to price access to a network and how to price the exchange of information based on some measure of use. We measure use in terms of discrete *messages*, where a message could be a telephone call, a paging message, a data file, an e-mail, or a video conference call, for example.

Telecommunications services possess several characteristics that raise interesting pricing issues. In this chapter, we focus on four characteristics and their implications for socially and privately optimal pricing to end-users:

1. *Consumption decisions have external effects.* Consumption of communications services involve two (or more) parties, both of whom take actions, receive benefits, and bear costs. The fact that multiple parties consume a single message gives rise to external effects, also referred to as network effects. Previous authors have distinguished two types of effect.[1] One is an *access externality*, whereby benefits accrue to existing members of a network when a new user joins the network and thus can receive messages that the original members value sending to her. The other is a *call externality*, which comprises the benefits enjoyed by a user who receives a message initiated by another user.

2. *The parties exchanging messages often have an ongoing relationship.* Two

parties may exchange messages repeatedly. The exchange of one message between two parties may affect their future message exchanges with each other, as well as with other parties. This linkage can come about both because of substitution and complementarities across messages, and because two parties have tacit or explicit agreements to share the responsibility for initiating messages (for example, by alternating who sends a message to whom).

3. *Service providers have the ability to identify customers.* Unlike many mass-market suppliers, telecommunications service providers typically can identify their customers or at least track customers' consumption patterns because a user 'attaches' herself to a provider's network in order to exchange messages. Even with prepaid telephone cards, carriers have the ability to track the rate or volume of calling made with any particular card. The fact that service providers often possess information about customers' consumption patterns and demographic characteristics makes it possible to implement sophisticated pricing schemes, including second- and third-degree price discrimination. For example, a carrier may engage in non-uniform pricing or offer special rate plans to low-income consumers.

4. *Incremental costs are often below average costs.* Telecommunications services are often supplied using technologies that exhibit economies of scale and scope. When incremental costs are less than average costs, incremental-cost pricing will lead to suppliers' suffering financial losses.

Loosely speaking, the presence of positive external effects creates an efficiency rationale for setting prices below incremental costs, while the fact that incremental costs are below average costs implies that prices may have to be set above incremental costs to generate sufficient revenues to cover the total costs of production. If the pricing decision were one dimensional, the choice of socially optimal prices subject to a profitability (producer breakeven) constraint would be uninteresting: the decision-maker would choose the lowest price that allowed the recovery of costs. However, there are two reasons why the design of socially optimal prices is a meaningful problem in the settings that we examine.[2] First, because suppliers have the ability to engage in sophisticated pricing, there can be multiple rate structures that satisfy the profitability constraint, and the problem is to find the most efficient one. Second, when both parties exchanging a message benefit from that exchange, there is a question of how to allocate the cost burden of the message exchange between them. This latter issue arises even if each consumer faces a simple uniform price.

This chapter presents a survey of these pricing issues in a monopoly setting. Our survey is unusual in three respects. First, it does not attempt to provide a comprehensive review of the literature. Instead, we present a coherent modeling framework that highlights the principal findings of key papers. Second, with respect to call externalities, we largely survey issues rather than literature. Because relatively little has been written on call externalities, we map out some of the intellectual terrain in this area in the hope of encouraging others to work on these issues and create the literature for future surveys. A third unusual feature is that most of the articles surveyed were written in the 1970s. This pattern in large part reflects the fact that the literature of the 1980s and 1990s focused on non-telecommunications 'virtual' networks (such as computer hardware and software) in which the lack of compatibility among competing networks is a central issue.[3] We chose to review the analysis of monopoly networks because, while competition is important, there is much to learn from an analysis of socially and privately optimal monopoly pricing, particularly when different carriers are interconnected and thus offer consumers a common network.

Following much of the literature on telecommunications pricing, in the next section we assume that exchange of a message generates benefits only for its sender. In this setting, access externalities can arise, but there is no possibility of call externalities. Moreover, there is no scope for charging a consumer for receiving a message; she would refuse to accept it. The pricing issue is one of choosing the structure of charges levied on the parties that send messages. Our analysis focuses on the use of two-part tariffs and the optimal trade-off between the lump-sum charge, or *access fee*, and the marginal charge, or *per-message fee*. Summarizing and extending the literature, we characterize the socially optimal two-part tariff, where the social optimum is defined in terms of total surplus maximization.

The central findings are that, in the absence of a profitability constraint, the socially optimal pricing scheme entails a per-message fee equal to the marginal message cost and an access fee set below the marginal cost of connecting a user to the network. In the presence of a profitability constraint, deviations from marginal cost pricing depend critically on consumer heterogeneity and the degree of economies of scale. In particular, suppose there are constant returns to scale. In this case, the socially optimal per-message charge is above the marginal message cost and the optimal access charge is below the marginal access cost when subscribers who are on the margin of disconnecting send fewer messages than the average sent by all subscribers. Opposite relations between these prices and costs hold if the average consumer sends fewer messages than the marginal consumer. We also examine socially optimal pricing when it is possible to charge differential prices to consumer groups that differ in terms of the costs of serving them or the external benefits they confer on others. We furthermore examine profit-maximizing pricing schemes, which, as expected, have many of the characteristics of socially optimal prices in the presence of a profitability constraint.

In the following section, we turn to consideration of call externalities. In the absence of a binding profitability constraint, the price for sending a message should be below the marginal message cost as a means of internalizing the call externality. Moreover, when access externalities are present as well, the access price should never be set above the marginal access cost absent a profitability constraint.[4] When both parties benefit from a message exchange, it is feasible to charge a party for receiving a message. Higher receiving prices reduce the set of messages accepted, which harms the message senders. Thus, there is a message acceptance externality that is similar to the access externality that arises from connection choices. In the presence of a binding profitability constraint, socially optimal pricing entails trading off the effects on the origination and acceptance of messages as the sending and receiving prices are varied along an iso-profit line.

We analyse two classes of model that extend the existing literature on call externalities. Principal findings include: it often is optimal to have the receiving party bear a positive fraction of the cost of exchanging a message; there is value to providing consumers a menu of pricing options (where different elements of the menu apportion the cost burdens differently between sender and receiver); when either of two parties can initiate a message exchange, setting the send price greater than the receive price can lead to socially wasteful strategic delay by the parties; and repeated interaction among consumers can internalize call externalities in some circumstances. We also examine the sources of inefficiency in the equilibrium outcome with call externalities.

ACCESS EXTERNALITIES

In this section, we focus on the implications of access externalities for socially and privately optimal pricing. To do so, we assume that all of the benefits of a message accrue to the sender. To simplify the analysis, we address neither general equilibrium effects nor second-best issues that can arise from distortions in other markets.

Our canonical model assumes that an individual places a monetary value of $v(n,\mathbf{z})$ on being connected to a network of size n, where the vector $\mathbf{z}$ is a placeholder for other variables, endogenous or exogenous, that might also affect its value. We ignore income effects by assuming that consumers have quasi-linear utility functions: that is, a consumer's utility is $\chi v(n,\mathbf{z}) + y$, where y is her consumption of a composite commodity comprising all other goods and services, and χ is equal to one if she is connected to the network and zero otherwise.[5] In addition to ignoring income effects triggered by changes in communications costs, we also assume that the marginal social utility of a dollar of income to a user is equal across all users. Thus, we do not consider the use of telecommunications pricing as a means of achieving socially desirable redistribution of income among households. We take total surplus, the sum of producer and consumer surplus, as our measure of welfare.

Throughout the examination of access externalities, the analysis is simplified by assuming that the size of the network is a sufficient statistic for determining the benefit function for network connection.[6] In practice, there may be communities of interest whose members more highly value exchanging messages among each other than with non-members. In this case, the size of the network would not be a sufficient statistic for determining message sending benefits and users might join a network in clusters.[7] We assume that an increase in the size of the network increases the benefit to access that is, $\partial v/\partial n > 0$.[8] This condition reflects the benefits the sender gains from a greater variety of potential message recipients.[9] We also assume that a network with no one on it provides no value to a consumer who might join; that is $v(0, \mathbf{z}) \equiv 0$.[10]

A Reduced-form Model

To give more structure to this model, assume that a user's gross benefit of connecting to the network depends on only her type and the total number of users connected to the network. Assume there is a continuum of consumer types spread uniformly on the interval $[0, N]$ with unit density. Let θ denote a consumer's (unidimensional) type. Her value of access is $v(n, \theta)$. We label types such that $\partial v/\partial \theta < 0$; that is, the greater the type index, the less the value of connecting to the network.

Because a user's value of connecting to the network depends on the number of other users who do so, consumer expectations about the number of other users connecting to the network are critical to the determination of equilibrium. In our static models, we focus on *fulfilled-expectations equilibria,* in which consumer beliefs are correct in equilibrium.

Let a be the *access* price a user pays to be hooked up to the network. Consumer optimization implies that a type-$\hat{\theta}$ consumer purchases access in a fulfilled-expectations equilibrium with a network size of n if and only if $v(n, \hat{\theta}) \geq a$. Because $v(n, \theta) > v(n, \hat{\theta})$ for all $\theta < \hat{\theta}$, it follows that all $\theta < \hat{\theta}$ must also purchase access in equilibrium. Thus, the size of the network and the greatest index of any consumer on the network must be the same in equilibrium; if the network has size $n < N$ in equilibrium, then $v(n, \theta) < a$ for all $\theta > n$ and $v(n, n) = a$.

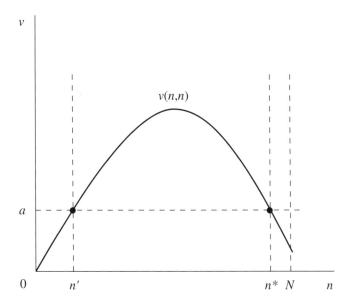

Figure 6.1   Possible equilibrium network sizes when access price is a

Network models typically possess multiple equilibria. This model is no exception. For instance, suppose that every potential user expects no other potential user to subscribe to the service. Then there is no value to purchasing access and this outcome is an equilibrium at any positive price of access. If, however, every potential user expects a large number of other users to subscribe, then there is value to subscribing and there can be equilibria with positive network subscribership. Figure 6.1 illustrates one possible situation.

As shown in the figure, there are three possible equilibrium network sizes when the access price is a: $n=0$, $n=n'$, and $n=n*$ Only the first and third equilibria are stable. If $n*$ were greater than or equal to N, then the third equilibrium would be replaced by a stable equilibrium at $n=N$.

Finding 1:   Access externalities typically give rise to multiple equilibrium network sizes for a given set of prices.[11]
There are three possible equilibrium network sizes when the access price is a: $n=0$, $n=n'$, and $n=n*$.

In what follows, we generally do not concern ourselves with the multiplicity of equilibria. Instead, we assume users are able to coordinate, and we focus on the most efficient equilibrium associated with any given pricing policy. Implicitly, we are assuming the carrier can cheaply engage in expectations management or temporary pricing strategies, for example, penetration pricing, to induce the most efficient equilibrium.

A second central feature of network models illustrated by this simple model is the efficiency of subsidizing access in the presence of externalities. The problem of finding socially optimal prices can be framed in terms of looking for the socially optimal network size and then setting the access price, a, accordingly. We begin by considering the social optimum without concern as to whether the network provider earns non-negative profits.[12] The total surplus associated with a network of size n is

$$\int_0^n v(n, \theta)d\theta - hn - F, \tag{6.1}$$

where h is the cost of hooking up the consumer to the network, which is initially assumed to be the same for all users, and F denotes the fixed cost of operating the network. The hookup cost, h, may include both network equipment and customer premises equipment, such as a telephone, a modem or a television set.

So that the problem of maximizing surplus is not trivial, assume that there exist values of $n>0$ such that (6.1) is positive.[13] In order to explore pricing issues better, we also assume that $n=N$ is not a maximum. The solution to (6.1) must therefore satisfy the corresponding first-order condition,

$$v(n, n) - h + \int_0^n \frac{\partial v(n,\theta)}{\partial n} d\theta = 0. \tag{6.2}$$

Because a larger network benefits each subscriber, the integral in (6.2) is positive, implying that $v(n^w, n^w) < h$, where n^w is the socially optimal network size: that is, at the social optimum, the marginal consumer values her connection to the network less than the cost of connecting her. Because of the externality she provides the infra-marginal consumers, however, it is nevertheless optimal to connect her to the network.[14]

One can now derive the socially optimal access price by using the fact that $a=v(n, n)$ in a size-n equilibrium. From this fact, it follows that the first-order condition (6.2) can be rewritten as

$$a - h = \int_0^n \frac{\partial v(n,\theta)}{\partial n} d\theta < 0.$$

In words, the social optimum can be achieved only if the access price is set *below* the hookup cost by an amount equal to the incremental benefits enjoyed by other subscribers to the network when the marginal subscriber connects. To summarize:

Finding 2: Except when all consumers are induced to join the network, any equilibrium associated with an access price greater than or equal to the access cost entails too little connection to the network from the perspective of social welfare. That is, $a^w < h$, where a^w is the welfare-maximizing access fee absent a network profitability constraint.

Turn next to characterization of the profit-maximizing outcome. In addition to being of interest in its own right, characterization of this outcome provides insights into the socially optimal outcome in the presence of a network profitability constraint. As before, one can express the problem in terms of network size. The profit-maximizing monopolist's optimization problem is to

$$\max_{n} nv(n, n) - nh - F.$$

If $nv(n, n) - nh < F$ for all n, then profits are maximized by shutting down. If profits are positive for some network sizes, then a profit-maximizing network size must satisfy the first-order condition

$$v(n, n) - h + n[v_1(n, n) + v_2(n, n)] = 0. \quad^{15} \tag{6.3}$$

Using the consumer's optimization condition, this equation implies that

$$a^\pi - h = -n[v_1(n, n) + v_2(n, n)],$$

where a^π denotes the profit-maximizing access fee.

How do socially and privately optimal prices and network sizes compare? Because the firm has the option of shutting down and $F \geq 0$, it must be the case that $a^\pi - h$ is non-negative whenever the profit-maximizer makes positive sales. Hence, $a^\pi > h > a^w$.

One can gain further insight into the distortion by examining consumer surplus because it is the wedge between total surplus and profits. Taking into account the monopolist's pricing rule, consumer surplus equals

$$\int_0^n v(n, \theta)d\theta - nv(n, n). \tag{6.4}$$

The derivative of consumer surplus with respect to n is

$$n\left[\frac{1}{n}\int_0^n v_1(n, \theta)d\theta - v_1(n, n) - v_2(n, n)\right].$$

The first two terms in the square brackets capture the difference between how the average subscriber and the marginal subscriber value the addition of one more subscriber to the network. Pricing is driven by the marginal subscriber's valuation. Hence, when the marginal subscriber values increased network size by less than does the average subscriber, this component of consumer surplus rises with network size, which biases the profit-maximizing supplier toward setting n too low. If the marginal consumer values network size by more than average, then this component falls with network size, which biases the profit-maximizer toward setting n too high. This distortion is similar to the one identified by Spence (1975) for a profit-maximizing monopolist's choice of product quality in a non-network market. The third term, $-v_2(n, n)$, which is positive, represents the fact that price must fall to attract the marginal subscriber. This is the usual distortion that leads to undersupply of service by a profit-maximizing monopolist. Because $n^\pi < n^w$, we know that this third term dominates when the sum of the first two is negative.

Lastly, consider the constrained (Ramsey) problem, where surplus is maximized subject to the constraint that the network at least break even:

$$\max_n \int_0^n v(n, \theta)d\theta \tag{6.5}$$

subject to

$$nv(n, n) - nh - F \geq 0. \tag{6.6}$$

Let n^c denote the optimal constrained network size and $a^c = v(n^c, n^c)$ denote the optimal constrained access price.[16]

Finding 3: $n^w > n^c \geq n^\pi$ *and* $a^w < h \leq a^c \leq a^\pi$.

Proof: Recall that $a^w < h$. Expression (6.6) could not hold if $a^c < h$, hence $a^c \geq h > a^w$.

Next we show that (6.6) is a binding constraint in program (6.5). Suppose it were not. Then, by continuity, there would exist an $\tilde{n} > n^c$ such that $\tilde{n}[v(\tilde{n}, \tilde{n}) - h] \geq F \geq 0$. Moving from n^c to $\tilde{n}$ changes welfare by

$$\left[\int_0^{\tilde{n}} v(\tilde{n}, \theta)d\theta - h\tilde{n} - F \right] - \left[\int_0^{n^c} v(n^c, \theta)d\theta - hn^c - F \right] \tag{6.7}$$

$$= \underbrace{\int_0^{n^c} [v(\tilde{n}, \theta) - v(n^c, \theta)]d\theta}_{>0} + \underbrace{\int_{n^c}^{\tilde{n}} [v(\tilde{n}, \theta) - h]d\theta}_{\geq 0} > 0 \tag{6.8}$$

which contradicts the optimality of n^c. Hence, constraint (6.6) must bind.

Now, if shutting down maximizes profits, then $n^\pi = 0 \leq n^c$. If operating maximizes profits, then

$$n^\pi[v(n^\pi, n^\pi) - h] - F \geq 0 = n^c[v(n^c, n^c) - h] - F. \qquad (6.9)$$

Hence, n^π satisfies (6.6). But if $n^\pi > n^c$, then the argument represented by expressions (6.7) and (6.8) would imply that n^π yielded greater welfare than n^c, contradicting the definition of n^c. Therefore, $n^c \geq n^\pi$. If $a^c > a^\pi$, then profits under Ramsey pricing exceed unconstrained profits from monopoly pricing, a contradiction of (6.9). Thus, $a^c \leq a^\pi$

The last step is to establish that $n^w > n^c$. Consider the derivative of welfare with respect to $\tilde{n}$, expression (6.8), evaluated at $\tilde{n} = nc$:

$$\int_0^{n^c} v_1(n^c, \theta)d\theta + v(n^c, n^c) - h,$$

which is *positive* because the integrand is positive – gross benefits increase in network size – and the Ramsey access fee, $v(n^c, n^c)$, cannot be less than the access cost. Hence, increasing the network from n^c must increase welfare and so, $n^w > n^c$. QED.

A Richer Model: Per-message Charges

With most communications networks, a user chooses both whether to obtain access and, if so, how many messages to send, for example, how many phone calls to make or e-mails to send.[17] The model in this section better allows for the richness of real-world pricing by considering per-message or per-call pricing.[18] Let p denote the price paid by the message sender or calling party. Let m denote the marginal cost of transmitting a message.[19] Let a type-θ consumer's gross benefit be $v(n, x, \theta)$ where x denotes use of the network, for example, number of messages sent. We assume positive but diminishing marginal benefits of sending messages: $\partial v/\partial x > 0$ and $\partial(\partial v/\partial x)/\partial x < 0$. We also assume the value of the marginal call increases with network size: $\partial(\partial v/\partial x)/\partial n > 0$. This benefit is conditional on network size, n, because we presume that the consumer makes her highest value calls first and that she can find a higher value person to call the larger is the network. As before, $\partial v/\partial n > 0$ and $\partial v/\partial \theta < 0$.[20]

Let $x^*(p; n, \theta)$ maximize a type-θ consumer's surplus gross of the access fee. That is, $x^*(p; n, \theta)$ maximizes

$$v(n, x, \theta) - px. \qquad (6.10)$$

Let $\sigma^*(p; n, \theta)$ denote the maximized value of (6.10):

$$\sigma^*(p; n, \theta) \equiv v[n, x^*(p; n, \theta), \theta] - px^*(p; n, \theta).$$

Because $\partial v/\partial \theta < 0$ it follows from the envelope theorem that $\partial \sigma^*/\partial \theta < 0$. Hence, if the $\hat{\theta}$-type consumer purchases access to the network, that is, if singular $\sigma^*(p; n, \hat{\theta}) - a \geq 0$, then all types θ such that $\theta < \hat{\theta}$ also purchase access. This last result means that, at prices a and p, the marginal type is defined by

$$\sigma^*(p; n, n) - a = 0 \tag{6.11}$$

in equilibrium.[21]

Absent a network profitability constraint, the social optimization problem can be posed as choosing p and n to maximize social welfare,

$$W = \int_0^n (\sigma^*(p; n, \theta) + x^*(p; n, \theta)(p - m))d\theta - hn - F. \tag{6.12}$$

Using the fact that $\dfrac{\partial \sigma^*}{\partial p} = -x^*$, the first-order conditions are

$$\frac{\partial W}{\partial p} = (p - m) \int_0^n \frac{\partial x^*(p; n, \theta)}{\partial p} d\theta = 0 \text{ and} \tag{6.13}$$

$$\frac{\partial W}{\partial n} = \sigma^*(p; n, n) + x^*(p; n, n)(p - m) - h$$

$$+ \int_0^n \left(\frac{\partial \sigma^*(p; n, \theta)}{\partial n} + \frac{\partial x^*(p; n, \theta)}{\partial n}(p - m) \right) d\theta = 0. \tag{6.14}$$

From equation (6.13), we get the usual surplus-maximizing result that price should be set equal to marginal cost, $p = m$ Consequently (6.14) reduces to

$$\frac{\partial W}{\partial n} = \sigma^*(p; n, n) - h + \int_0^n \frac{\partial \sigma^*(p; n, \theta)}{\partial n} d\theta = 0,$$

and, therefore,

$$a^w - h = - \int_0^n \frac{\partial \sigma^*(p; n, \theta)}{\partial n} d\theta < 0.$$

Finding 4: Suppose all the benefits of a message exchange accrue to the sender. Absent a profitability constraint, the welfare-maximizing message price is equal to the marginal cost of a message, and the welfare-maximizing access charge is below the marginal hook-up cost by an amount equal to the incremental surplus gained by the other network subscribers when the marginal subscriber connects.[22]

If the network must be self-supporting, then the supplier has the profit constraint:

$$(p-m)\int_0^n x^*(p; n, \theta)d\theta + (a-h)n - F \geq 0. \tag{6.15}$$

In equilibrium, the marginal consumer is defined by (6.11), so one can rewrite (6.15) as

$$(p-m)\int_0^n x^*(p; n, \theta)d\theta + (\sigma^*(p; n, n) - h)n - F \geq 0. \tag{6.16}$$

The constrained problem is to maximize surplus, expression (6.12), with respect to p and n subject to (6.16). Let λ be the Lagrange multiplier on the constraint. Then the first-order conditions for the Lagrangean, $\mathcal{L}$, are

$$\frac{\partial \mathcal{L}}{\partial p} = (1+\lambda)(p-m)\int_0^n \frac{\partial x^*(p; n, \theta)}{\partial p}d\theta$$

$$+ \lambda n\left(\frac{1}{n}\int_0^n x^*(p; n, \theta)d\theta - x^*(p; n, n)\right) = 0, \tag{6.17}$$

$$\frac{\partial \mathcal{L}}{\partial n} = (1+\lambda)(\sigma^*(p; n, n) + x^*(p; n, n)(p-m) - h) + n\lambda\frac{\partial\sigma^*(p; n, n)}{\partial n}$$

$$+ \int_0^n \left(\frac{\partial\sigma^*(p; n, \theta)}{\partial n} + (1+\lambda)\frac{\partial x^*(p; n, \theta)}{\partial n}(p-m)\right)d\theta = 0, \tag{6.18}$$

and condition (6.16). From the analysis of the unconstrained first-order conditions, (6.13) and (6.14), we know the profit constraint must bind, hence $\lambda > 0$. The bottom term in (6.17) compares average consumption with that of the marginal consumer. If the former is greater than the latter, then the bottom term is positive. The top term of (6.17) must then be negative, which, because demand curves slope downward, implies $p > m$ In other words, when the marginal subscriber sends fewer messages than the average subscriber, the optimal per-message price exceeds the cost of a message. Moreover, because the profit constraint is binding, it follows that $a < h$ when $F = 0$; in the absence of economies of scale at the network level, the access price is less than the hook-up cost. In contrast, if the marginal subscriber sends more messages than average, then the per-message price is less than cost and the access price is greater than cost.

To develop intuition, suppose $F=0$ and that both prices are initially set equal to their respective costs. Consider the welfare effects of raising the per-message price slightly and lowering the access price so as to be revenue neutral. Starting from $p=m$, the slope of the profit constraint is

$$\frac{da}{dp} = -\frac{1}{n}\int_{o}^{n} x^{*}(p; n, \theta)d\theta.$$

From (6.12), this change is also welfare neutral *except* to the extent that it changes the network size (starting from $p=m$ a small change in the prices has no effect on welfare for fixed n because the social value of the marginal message is just equal to its cost). Welfare rises if the network grows and falls if the network shrinks. The sign of the change in network size is given by the sign of

$$\frac{d}{dp}(\sigma^{*}(p; n, n) - a) = -x^{*}(p; n, n) - \frac{da}{dp}$$

$$= -x^{*}(p; n, n) + \frac{1}{n}\int_{o}^{n} x^{*}(p; n, \theta)d\theta. \qquad (6.19)$$

The sign of (6.19) is just the sign of the bottom term of (6.17). In words, if the marginal user sends fewer messages than the average, raising the per-message fee generates sufficient additional funds to more than compensate the marginal user by lowering the access charge, thus attracting new users and raising welfare through network effects.

To summarize:[23]

Finding 5: Suppose that all of the benefits of message exchange accrue to the sender and the network must be self-supporting. If the marginal subscriber sends fewer messages than the average subscriber, then the socially optimal per-message price exceeds the marginal message cost to subsidize access and/or cover network fixed costs. If the marginal subscriber's use is greater than the average, then the socially optimal per-message price is below cost, with the shortfall made up through setting the access charge above the marginal hook-up cost.[24]

Now, consider a profit-maximizing service provider. Recalling that the marginal consumer's surplus equals the access price, the monopolist's profit is:

$$n(\sigma^{*}(p; n, n) - h) + (p - m)\int_{o}^{n} x^{*}(p; n, \theta)d\theta. \qquad (6.20)$$

The first-order condition for maximizing this expression with respect to the message price is

$$(p-m)\int_o^n \frac{\partial x^*(p; n, \theta)}{\partial p}d\theta + n\left(\frac{1}{n}\int_o^n x^*(p; n, \theta) - x^*(p; n, n)\right) = 0. \quad (6.21)$$

From (6.21), whether the per-message price exceeds the per-message cost depends on who consumes more, the average consumer or the marginal consumer. If the former does, then $p > m$ if the latter does, then $p < m$ In the second case, it must be that $a > h$ otherwise the firm would do better to shut down. If, however, $p > m$ then whether the firm charges an access fee greater or less than the hook-up cost is indeterminate.[25]

Finding 6: Suppose all of the benefits of a message exchange accrue to the sender. If the marginal subscriber sends fewer messages than the average subscriber, then the profit-maximizing per-message price exceeds the marginal message cost and it can – but need not – be privately optimal to set the access charge below the hook-up cost. If the marginal subscriber's use is greater than the average, then the profit-maximizing per-message price is below the per-message cost and the access charge is above the hook-up cost.[26]

Observe from the first-order conditions (6.17) and (6.21) that, were there *no* heterogeneity in types, $p = m$ would maximize both profits and social welfare. Moreover, this logic implies the following: if access charges and message prices can be conditioned on type, that is, $a(\theta)$ and $p(\theta)$, then to maximize either total surplus or profits the message price should be set equal to marginal cost for all types, that is, $p(\theta) = m$ for all θ. The following access price schedule would support an equilibrium with n consumers on the network:

$$a(\theta) \leq \sigma^*(m; n, \theta) \text{ if } \theta \leq n$$

and

$$a(\theta) = \infty \text{ if } \theta > n.$$

Intuitively, setting $p = m$ maximizes the total surplus generated by any consumer's use of the network and personalized access charges can be used to appropriate the surplus for the supplier to the extent needed to cover costs or maximize profits. Setting $a(\theta) = \sigma^*(m; n, \theta)$ extracts all of the surplus for the network supplier in a non-distorting way. Under these conditions, a profit-maximizing monopolist can engage in perfect price discrimination and – because the firm appropriates all of the surplus – will act to maximize total surplus.

Summarizing this discussion,

Finding 7:   Suppose all of the benefits of a message exchange accrue to the sender. If access prices can be conditioned on consumer type, then it is socially and privately optimal to set the per-message price equal to marginal cost and use type-dependent access charges to raise needed revenue or desired profits. In particular, if all consumers are identical, then it is privately and socially optimal to set the per-message price equal to per-message cost, that is, $p^w = p^c = p^\pi = m$.[27]

Identifiably Different User Groups

In practice, personalized prices are likely to require more information than a service provider will have available. In this subsection, we examine pricing when users can be placed in sets with different average characteristics. For example, rural and urban consumers likely differ along one or more dimensions, such as the costs of providing them access. And they may differ in terms of the average level of external benefits they confer on others or their willingness to pay for access. In this section, we examine simple models of pricing in response to differences in access costs and differences in the generation of externalities.

(i)   Differential access costs

Wire-line communications technologies typically exhibit economies of density and, consequently, access costs to serve users in rural areas are high relative to the costs of connecting urban users. An important question for telecommunications policy is whether urban users should subsidize rural users. Suppose that there are two groups with N consumers each. It costs h_L to provide access to a member of the low-cost group and h_H to provide access to a member of the high-cost group, where $h_L < h_H$. As before, a user of type θ enjoys dollar benefits $v(n, \theta)$ from being connected to a network with a total of n subscribers, where n equals the sum of n_L and n_H, the numbers of subscribers in the two groups. This functional form embodies the assumption that a user derives equal benefits from being able to exchange messages with a member of either group. We also assume that the distribution of θ is identical in the two groups.

Suppose the network provider can identify to which group any given user belongs and arbitrage across groups is infeasible. Then the social problem is to choose group-specific prices (a_L, p_L) and (a_H, p_H) to maximize welfare. As before, this problem can be restated in terms of the number of connected users, n_L and n_H. The social problem is to

$$\max_{\{n_L, n_H, p_L, p_H\}} \int_0^{n_L} (\sigma^*(p_L; n, \theta) + x^*(p_L; n, \theta)(p_L - m))d\theta - h_L n_L$$

$$+ \int_0^{n_H} (\sigma^*(p_H; n, \theta) + x^*(p_H; n, \theta)(p_H - m))d\theta - h_H n_H.$$

Finding 8:   Suppose all of the benefits of a message exchange accrue to the sender and there are two identifiable groups of consumers that are identical except for the fact that the access costs for members group H are higher than those for members of group L. Under socially optimal pricing with or without a network profitability constraint: (a) $n_L < n_H$; and (b) $p_H > p_L$ or $a_H > a_L$ or both.

Proof: Let (p_L^*, a_L^*) and (p_H^*, a_H^*) denote the welfare-optimal prices. Suppose that under these prices $n_L < n_H$. Reverse the prices, so that type-L consumers now face the type-H prices and *vice versa*. Under these new prices, $n_L' = n_H$ and $n_H' = n_L$; that is, the size of each population on the network also reverses. Total gross consumption benefits are unchanged. However, the change in costs is $(n_L - n_H)(h_H - h_L) < 0$. This contradicts the optimality of the original prices. Hence, one can conclude that $n_L \geq n_H$ under welfare-optimal pricing.

Suppose $n_L = n_H$. Then there is no loss in using the same maximizing (p, a) pair for both segments: $a_L^* = a_H^* \equiv a^*$ and $p_H^* = p_L^* \equiv p^*$. Consider the effects of lowering a_L by da and raising a_H by da. n_L increases by $\frac{dn_L}{da_L} da$ and n_H falls by $\frac{dn_H}{da_H} da$. Starting from equal prices for the two groups, $\frac{dn_L}{da_L} = \frac{dn_H}{da_H} \equiv \frac{dn}{da}$ and the total number of subscribers is unchanged. Similarly, gross consumption benefits are unaffected. The resulting change in welfare is

$$\left[\sigma^*\left(p; n, \frac{n}{2}\right) + x^*\left(p; n, \frac{n}{2}\right)(p - m) - h_L \right.$$

$$\left. - \sigma^*\left(p; n, \frac{n}{2}\right) - x^*\left(p; n, \frac{n}{2}\right)(p - m) + h_H \right] dn$$

$$= (h_H - h_L)dn > 0.$$

This contradicts the hypothesized optimality of (p_L^*, a_L^*) and (p_H^*, a_H^*). QED.

The intuition underlying this result is easiest to see when the carrier is subject to a profit constraint. By assumption, any two consumers generate the same access externality. Hence, given the tradeoff between connecting a member of one group or the other, it is efficient to connect a member of the low-cost group. Moreover, in the presence of a profitability constraint, it may be efficient to set prices charged some users above marginal cost to generate revenues to subsidize other users. Indeed, by the logic above, in some cases it is constrained efficient to raise prices to the high-cost users above costs and use the resulting profit to subsidize the connection of low-cost users. This pattern of socially optimal cross-subsidy is the opposite of the one commonly advocated in policy debates! For instance, the subsidization of high-cost rural areas has been a central tenet of public policy in the United States.

(ii)   Differential externalities

Suppose that consumers in one group generate lower externalities on a per-user basis than do members of another group. For example, rural callers may be more isolated from other subscribers to the telephone network and may generate fewer benefits for others. Or middle- and high-income households may be more valuable to telemarketers than are low-income households. It is easy to show that if two groups are otherwise identical, then it is socially optimal to subsidize the high-externality group relative to the low-externality group.

Dynamic Pricing Models

To this point, we have considered only static models. The existence of multiple equilibrium levels of subscribership raises the question of what determines the realized equilibrium level, including any dynamics. In this section, we briefly consider dynamic models, primarily to identify this area as one in which further research is needed.

The process by which consumers form expectations about network size is a critical part of equilibrium determination. Thus far, we have examined fulfilled-expectations equilibria and – in cases of multiple equilibria – have selected the outcome Pareto preferred by users. In such markets, there are no meaningful dynamics because consumers form expectations based on their knowledge of current conditions, for example, the current price and the distribution of consumer demands, without reference to the history of the market. However, consumers often are not well informed, and they may not be fully rational calculators and coordinators. In these situations, history may matter and interesting dynamics can develop.

Consider the following simple model of adaptive expectations that, to some extent, captures the opposite pole from fulfilled expectations. There are multiple discrete periods indexed by t. A user's actual benefits in period t are $v(n_t, \theta)$ The user bases his or her subscription decision on the expected network size. With adaptive expectations, all users making connection decisions at time t expect the network size in that period to be equal to the actual network size in the previous period. Under these expectations, a user connects to the network at time t if and only if $v(n_{t-1}, \theta) \geq a_t$.

By themselves, even adaptive expectations are not enough to generate interesting dynamics in the unconstrained welfare maximization problem. Absent a profit constraint, welfare is maximized by setting the access price low enough in each period to induce the network size that maximizes single-period welfare. There may be a large first-period subsidy required to compensate for user expectations. Once the optimal network size has been established for one period, the network supplier can support the welfare optimum by charging the same access price as it would under a one-period, fulfilled-expectations equilibrium.

If the network provider faces a sufficiently strict profitability constraint, such pricing is infeasible. Suppose, for example, that the firm's pricing is constrained to setting $a_t \geq h$ for all t.[28] It is easy to demonstrate that it is socially optimal to set $a_t = h$ for all t such that $n_{t-1} \leq n^w$ where as before n^w is the network size that maximizes welfare in a one-period, fulfilled-expectations equilibrium without a profitability constraint.

Before characterizing the dynamics of this simple model, one more assumption is needed. Let n_0 denote consumers' (exogenously given) initial expectations for the size of the network in period 1. The network size in period t, n_t, solves

$$v(n_{t-1}, n_t) = a_t. \qquad (6.22)$$

Because $v(n, \cdot)$ is monotonic, equation (6.22) has a unique solution for each n_{t-1}.

The behaviour of this system depends on the nature of the $v(n, n)$ as n varies because $v(\hat{n}, \hat{n}) = \hat{a}$ implies that $\hat{n}$ is a steady state when the access price is $\hat{a}$. Figure 6.2 illustrates one interesting case for a fixed access price of $a = h$.[29]

Panel A of Figure 6.2 reproduces Figure 6.1 reflected about the 45° line. The three equilibria in Panel A correspond to three steady states of the dynamics. There are stable steady states at $n_t \equiv 0$ and $n_t \equiv n^*$. There is also an unstable steady state at $n_t \equiv n'$.

Panel B shows three possible dynamic paths depending on the initial network size, n_0. If $n_0 > n^*$ then, because $v(n_0, n_0) < a$, there is a neighborhood of types, $\theta \in (n_0 - \varepsilon, n_0]$, who would prefer *not* to remain on the network. However, $v(n_0, n^*) > v(n^*, n^*) = a$, and thus $\varepsilon < n_0 - n^*$; that is, the

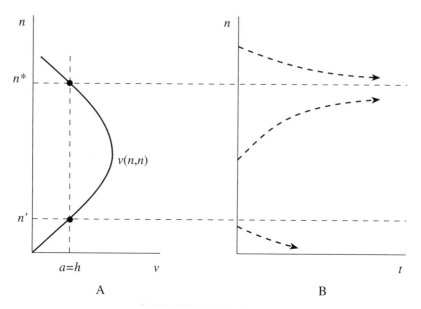

Figure 6.2   Possible dynamics under adaptive expectations

network does not immediately shrink to n^* but rather to some intermediate size between n^* and n_0. Repeating the argument with n_{t-1} rather than n_0 demonstrates that the network shrinks toward but does not actually reach n^*. When $n' < n_0 < n^*$, $v(n_0, n_0) > a$ and there is a neighborhood of types, $\theta \in [n_0, n_0 + \varepsilon]$, who want to join the network. But, because $a = v(n^*, n^*) > v(n_0, n^*)$, $\varepsilon < n^* - n_0$. Thus, the network does not immediately jump to n^* but approaches it from below. Repeating the argument with n_{t-1} rather than n_0 indicates that the network grows toward but does not reach n^*. Finally, if $n_0 < n'$, then $v(n_0, n_0) < a$ and there is a neighborhood of types, $\theta \in (n_0 - \varepsilon, n_0)$, who would then prefer not to remain on the network. Because $v(n_0, 0)$ can be greater or less than a, the network can immediately disappear or, in other cases, disappear gradually.

The assumption that the network provider can never price below cost is a strong one. In both theory and practice, providers engage in *penetration pricing* to launch networks. Under such a strategy, the provider prices below cost to promote network growth.[30] To see the power of penetration pricing, suppose that $n_0 = n'$ in Figure 6.2. Consider what would happen if the network provider were to incur a small loss in the first period by setting a_1 just less than h and then set price equal to the marginal access cost in all subsequent periods. The network size would grow over time until it approached n^*. Total surplus would be considerably higher than if the network size had remained at n' Moreover, given the increased network size, the firm could recoup its losses during the periods following the initial price cut. Hence, in this extreme situation, a small amount of penetration pricing would give rise to large welfare gains. While the effects of penetration pricing are less dramatic in other cases, this pricing strategy can promote welfare with or without a profit constraint.

Penetration pricing may also be used by a profit-maximizing network provider. Faced with $n_0 = n'$,[31] a profit-maximizing firm may choose to subsidize first-period consumption to put the network on a trajectory that will increase both the number of subscribers and the price that can be charged per subscriber.[32]

This simple model is just a start. It would be useful to explore more general expectations processes and profit constraints. For example, it would be useful to examine a model in which expected network size in period t is an increasing function of the previous period's network size and a decreasing function of the present period's price. Such an expectations process would be a (very) reduced-form representation of a situation in which consumers estimated demand to form a rational forecast of current consumption decisions.

Turning to budget constraints, a broad family to examine is

$$\sum_{t=0}^{T} \pi(t)\delta^t \geq -L(T) \text{ for all } T, \tag{6.23}$$

where $L(T)$ is the absolute present value of the maximal allowable accumulated losses to time T and δ is the one-period discount factor. For example, if the provider receives a government subsidy of ℓ per period, then $L(T) = \ell \sum_{t=0}^{T} \delta^t$. If the provider is given a one-time lump-sum subsidy of L_0, then $L(T) = L_0$.

Although we have only scratched the surface here, it is useful to summarize our broad findings:

Finding 9: In the presence of access externalities, expectations play a critical role in determining the equilibrium outcome. The combination of adaptive expectations and a binding budget constraint can give rise to meaningful dynamics in a multi-period, welfare-optimizing equilibrium. When consumers have adaptive expectations, it can be socially and privately optimal to engage in penetration pricing, with or without a profitability constraint.

Carrier Competition

Throughout, this section has examined socially and privately optimal monopoly pricing. In practice, many telecommunications services are provided by competing suppliers. Moreover, these competitors may or may not have networks that are compatible with one another in the sense of being able to exchange messages. To simplify the problem, the analysis below takes the degree of compatibility as exogenous. In practice, compatibility may be an endogenous choice made by the carriers.[33]

(i) Compatible networks
Consider a situation in which there are two suppliers offering identical services over interconnected networks. A type-θ consumer's gross benefits from sending x messages are $v(n, x, \theta)$, where n is the total number of distinct subscribers to the networks. Supplier i has costs $mX^i + hn^i$ where X^i is the total number of messages carried by firm i and n^i is the total number of its subscribers. We illustrate several important issues by considering the case of static Bertrand competition. To avoid the well-known non-existence of pure-strategy Bertrand equilibrium in the presence of economies of scale, assume $F=0$.

Typically, the fact that competition drives price toward incremental costs is a virtue. However, Finding 5, indicates that marginal-cost pricing is typically *not* the efficient solution even when pricing is constrained to ensure that service providers do not suffer losses.[34] Indeed, as we will now show by example, there is no guarantee that a competitive (that is, marginal-cost-pricing) Bertrand equilibrium is more efficient than the profit-maximizing monopoly equilibrium, even in the absence of economies of scale.

Consider the following example.[35] There are two types of consumer, A and B, of sizes n_A and n_B, respectively, with the following characteristics. Let $N=n_A+n_B$. If N customers are on the network, then type j has demand:

$$x_j(p, N) = \begin{cases} \bar{x}_j & \text{if } p \leq \bar{v}_j \\ 0 & \text{if } p > \bar{v}_j \end{cases},$$

where

$$(\bar{v}_A - m)\bar{x}_A > h > (\bar{v}_B - m)\bar{x}_B. \tag{6.24}$$

The second inequality in (6.24) captures the possibility that a B type either has too low a value per call or originates too few calls to gain much surplus from connecting to the network.

Consistent with our earlier analysis, the quantity demanded at p is less if fewer than N consumers connect. If only the n_A type-A consumers are on the network, a type-A subscriber has demand:

$$x_A(p, n_A) = \begin{cases} \bar{x}_A & \text{if } p \le \underline{v}_A, \\ 0 & \text{if } p > \bar{v}_B, \end{cases}$$

where $m < \underline{v}_A < \bar{v}_A$ and $(\underline{v}_A - m).\bar{x}_A \ge h$.

It is readily shown that both firms' setting access charges and per-message fees equal to their respective marginal costs is an equilibrium. Note that, in this equilibrium, only type-A consumers are served and on the network.

There cannot be any other equilibrium. To see why, observe that we can quickly rule out there being any other equilibrium in which B types are on the network. From expression (6.24), type-B users are not willing to pay enough for a provider to cover its costs of serving them. Hence, neither firm could serve B types exclusively. Suppose firm 1 serves both A and B types in equilibrium. What of firm 2? Type-A consumers must be indifferent between purchasing from firm 1 or firm 2:

$$(\bar{v}_A - p^1).\bar{x}_A - a^1 = (\bar{v}_A - p^2).\bar{x}_A - a^2,$$

where p^i and a^i are the per-message price and access charge, respectively, set by carrier i. Hence

$$a^1 + p^1.\bar{x}_A = a^2 + p^2.\bar{x}_A.$$

Therefore,

$$a^1 + (p^1 - m).\bar{x}_A = a^2 + (p^2 - m).\bar{x}_A;$$

firm 2 must make the same profit per type-A customer as does firm 1. Moreover, because firm 1 must make a positive profit on type-A consumers to offset its losses on type-B consumers, firm 2 makes a positive profit on its type-A consumers as well. Now, from (6.24), it must be that (i) $\bar{x}_B < \bar{x}_A$ or (ii) $\bar{v}_B < \bar{v}_A$ or both. If (i), then firm 2 could lower p^2 by a bit and raise a^2 by a bit such that: (a) all A s strictly prefer firm 2 to firm 1; (b) all B s strictly prefer firm 1 to firm 2; and (c) firm 2's profit per A is only slightly less than under its 'equilibrium' prices. But since firm 2 now corners the A market, while avoiding the B market, and makes almost the same per type-A consumer, this must be a profitable deviation for firm 2. If only (ii) holds, then firm 2 could raise p^2 by a bit and lower a^2 by a bit such that (a), (b) and (c) are satisfied. Again, because firm 2 would corner the A market, while avoiding the B market, and make almost the same per type-A consumer, this would be a profitable deviation for firm 2. By contradiction, it is established that only type-*A* consumers can be served in equilibrium. But if only *A* s are served, then Bertrand competition implies prices must equal marginal costs in equilibrium.

It has been shown that only type-*A* consumers are served under the competitive equilibrium (a *B* type's surplus even with $p=m$ is less than $a=h$ so she won't connect). Total welfare is

$$n_A[(\underline{v}_A-m).\bar{x}_A-h].   \qquad (6.25)$$

A profit-maximizing monopolist, however, may wish to sell to both types. Let $p=\min\{\bar{v}_A, \bar{v}_B\}$ and $a=(\bar{v}_B,-p).\bar{x}_B$. If $\bar{v}_A$ is sufficiently greater than $\underline{v}_A$, then total welfare will be greater than (6.25) because the positive externality of connecting *B*-types more than offsets the *B*-types' negative *direct* contribution to welfare. Furthermore, if $\bar{v}_B$ is sufficiently greater than $\underline{v}_A$, then the monopolist's profits are greater than if it just sold to type-*A* consumers, that is, greater than (6.25). In other words, unlike the Bertrand duopolists, a monopolist is able to capture – and thus internalize – some of the network externality that the type-*B* consumers contribute. Hence, the monopolist can be willing to subsidize type-*B* consumers to induce them to connect to the network, while the Bertrand duopolists cannot.

To summarize:

Finding 10: Bertrand competition among rival suppliers with interconnected networks can result in an equilibrium outcome that yields strictly lower social welfare than either the Ramsey or unregulated monopoly outcomes.

The sub-optimality of marginal-cost pricing raises the issue of whether the market outcome would be improved if the carriers charged each other an above-cost interconnection fee for exchanging traffic. Doing so might be seen as a way of elevating the equilibrium per-message charge above social marginal cost and subsidizing access. The reason is that a carrier would take the interconnection charge for outgoing traffic into account in setting *p*, and the carriers would compete by setting their access charges below the marginal cost of access because each carrier would recognize that a new subscriber generates interconnection profits from incoming traffic. Such interconnection pricing, however, creates an artificial advantage of network scale because a carrier benefits from keeping traffic 'on net'. Further analysis is needed.[36]

(ii) Incompatible networks

Now consider competition between two incompatible networks, 1 and 2. Total surplus is equal to

$$\sum_{i=1}^{2}\left(\int_{\Theta^i}[v^i(n^i, x^i(p^i; n^i, \theta), \theta) - mx^i(p^i; n^i, \theta)]d\theta - hn^i - F\right),$$

where Θ^i is the set of consumers connected to the ith network, v^i is the benefit function for service on the ith network, and x^i is the corresponding demand function.

If the networks provide otherwise identical services, that is, $v^1(n, x, \theta) = v^2(n, x, \theta)$ $\forall n, x, \theta$, then it can never be first best to have two incompatible networks because of the loss of network effects from fragmentation and the duplication of any network fixed costs.

Suppose the two networks are differentiated and each user subscribes to at most one network. Conditional on the existence of two networks, the social problem can be thought of as having two components: whether a consumer connects to a network and, if so, which one. If a user joins network 1 instead of network 2, then users on the first network gain, but there is a social opportunity cost of the benefits forgone by the users on the second network. The individual user, however, maximizes $\sigma_i(p_i; n_i, \theta) - a_i$, and ignores the effects on other users. Thus, prices have to play a coordination role to internalize the balance of the opposing externalities. It might even be optimal to tax subscription to one network in order to induce users to join the other network.[37] Not surprisingly, competition among profit-maximizing firms may also fail to allocate users optimally across networks.[38] On the other hand, it is possible that having competition between incompatible networks is a second-best regulatory solution *vis-à-vis* unfettered monopoly.

CALL EXTERNALITIES

The maintained assumption that the receiving party enjoys no benefits from a message exchange clearly is unrealistic in many cases. This section relaxes this assumption to examine the pricing implications of call externalities. To simplify the analysis, we begin by assuming that the set of parties connected to the network is invariant with respect to pricing. A form of access externalities still arises, however, because the receiver must choose whether to accept a message. Throughout most of this section we also abstract from any repeated-play considerations (see below for discussion of the consequences of repeated interaction). In particular, each party is motivated only by his or her private benefit and the price he or she faces for a single potential message exchange.

Consider two individuals, A and B, who may wish to communicate. The value or benefit of this communication to party j is v_j, which may be stochastic. The value or benefit of no communication is normalized to zero for each party. The cost of exchanging a message between A and B is $m > 0$. Welfare is maximized if all messages for which $v_A + v_B \geq m$ are exchanged and no messages for which $v_A + v_B < m$ are exchanged.

When message exchange generates benefits and costs for both parties, there are important differences between situations in which either party can initiate a message exchange ('two-way calling') and those in which only one party can do so ('one-way calling'). One-way calling has several interpretations. The first is that message origination is literally one-sided. Many telecommunications technologies, such as paging and pay telephones, are inherently one-way technologies. Other technologies are two-way, but in many instances only one of the two parties knows there is value in communicating. For instance, A could wish to announce some news to B. Alternatively, A could be a diner calling a restaurant, B, to make a reservation. In such situations, it is reasonable to view only one of the two parties as the potential message initiator. Other situations, in which the parties are more symmetrically informed about the value of communicating and it is technically feasible for either party to send a message to the other, are two-way calling situations.

An alternative interpretation of the distinction between the one-way and two-way calling models is the following. In a two-way calling model, a party may strategically delay sending a costly message in anticipation that the other party will initiate the exchange instead. For low-cost messages, this type of strategic behavior may be implausible, and the situation may better be approximated by a pair of one-way calling models in which a party sends a message if and only if the expected value of the message exceeds the cost.

By and large, economists have ignored the fact that a given message or call generates benefits at both ends. Previous theoretical work on communications pricing has tended to note the possibility of call externalities and then assert that the effects between two users largely offset one another.[39] Four exceptions are Squire (1973), Srinagesh and Gong (1996), Acton and Vogelsang (1992), and Hahn (2000).[40] Each of these papers implicitly examines the one-way calling case. While their models allow any user both to send and receive messages, the models assume that the sets of messages sent and received are independent of each other. In effect, each paper examines a pair of one-way calling models.

Squire (1973) considers socially optimal pricing under the assumption that all users have the same expected benefits from receiving a message.

Srinagesh and Gong (1996) consider more general distributions and identify implicit welfare tradeoffs for two cases. In the first, receivers cannot refuse to accept messages, while in the second receivers can selectively block low-value messages. As the authors note, these are polar cases and in many situations are unrealistic. For instance, a consumer receiving a voice telephone message can refuse to answer but cannot make this decision based on the value of the call. Acton and Vogelsang (1992) assume that the receiver gets some number of calls and his decision is simply how many to answer. The incremental value of a call to the receiver depends solely and deterministically on the order in which it is received.[41] Lastly, Hahn (2000) characterizes the profit-maximizing non-uniform price schedule for sending messages. He does not, however, allow for non-zero receive prices.

The analysis of this section extends the literature by characterizing optimal send and receive prices for a variety of situations, including two-way calling and realistic settings in which receivers can refuse to accept messages but must do so on the basis of limited information about incoming messages.

The analysis first considers situations in which a given message can originate from only one of the two parties. The analysis then turns to situations in which either party can send a message.

One-way Calling Decisions

Suppose that only party A can send a message. Consider first situations in which the network provider is restricted to uniform pricing: if a message is exchanged, the sender pays p and the receiver pays r. Otherwise no payments are made, for example, there is no charge if the receiver refuses to accept the message.

As noted earlier, the parties' values of communicating, v_A and v_B, could be stochastic at the time the parties make their send and receive decisions. Specifically, each individual has some prior knowledge (type, signal, and so on), $\omega_j \in \Omega_j$. The vector (ω_A, ω_B) has some joint distribution $\Psi(\omega_A, \omega_B)$, which is common knowledge. Each (ω_A, ω_B) pair defines a joint distribution over (v_A, v_B): $G(v_A, v_B | \omega_A, \omega_B)$.

Among the many possible consequences of this structure is 'learning'. A's calling B could provide B information that B then uses to estimate his value of accepting the message. Likewise, B's accepting the message could provide information to A. To illustrate this latter point, suppose that $v_A = v_B$, B knows these values, that is, $\omega_B = v_A = v_B$, but A does not. If $p = r = m/2$, then A always calls, and B answers if and only if $v_B \geq m/2$ (equivalently, if and only if $v_A \geq m/2$). A relies on B to receive the message if and only if $v_A \geq p$.

Because of such complex inferences, solving a general model is difficult, and the present analysis is restricted to limited cases. The initial case is circumscribed by two simplifying assumptions. First, there is only one type of receiver, that is, there is a single value of ω_B. Under this assumption, the sender cannot learn from the receiver's actions. Thus, it is a weakly dominant strategy for A to send a message whenever $E\{v_A|\omega_A\} \geq p$, and we assume she plays this strategy.

The second initial simplifying assumption limits the nature of learning by the receiver from the sender's actions. It is assumed that, if one knows the expected value of v_A (conditional on ω_A), then learning ω_A provides *no* additional information about v_B.[42]

For notational convenience, assume that v_A is known by A with certainty when she decides whether to send a message. This assumption is without further loss of generality given the first two assumptions because a random v_A can be replaced by its expected value conditional on ω_A. Under this assumption, A's strategy is to send a message if and only if $v_A \geq p$.

B does not know v_B's realization until he accepts the message. We assume, however, that B knows the joint distribution of v_A and v_B, $G(\cdot,\cdot)$. Let $g(\cdot,\cdot)$ denote the associated density and let $g(v_s|v_t)$ denote the density of v_s conditional on v_t. To avoid certain technical matters that are cumbersome to handle, we assume the support of v_A is a compact set. Given A's strategy, it is privately optimal for B to accept a message if and only if

$$E\{v_B|v_A \geq p\} \geq r.$$

Because the receiver, B, does not know v_B until after answering, it is not generally possible to achieve a first-best efficient solution (that is, to have completed calls if and only if $v_A + v_B \geq m$). Instead, we ask whether second-best – information-constrained – efficiency obtains: a message is exchanged if and only if

$$E\{v_B|v_A\} + v_A \geq m. \qquad (6.26)$$

Before considering this model in full, two simple examples serve to sharpen intuition. First, suppose that any message of value v_A to the sender has value μv_A to the receiver. For any message exchange, A receives $1/(1+\mu)$ of the total benefits and B receives $\mu/(1+\mu)$. It is readily shown, and intuitive, that setting $p = m/(1+\mu)$ and $r = m\mu/(1+\mu)$ satisfies the profitability constraint and induces the information-constrained efficient level of message exchange: costs are shared in proportion to gross benefits and both parties choose to exchange a message if and only if the expected total gross benefits exceed the expected total costs. Moreover, B takes A's actions as a signal that B's private benefits of accepting the message exceed his costs.

Second, suppose that for any message of value v_A to the sender, the receiver's value is a constant, β. Again, it is intuitively clear that setting $p = m - \beta$ and $r = \beta$ satisfies the profitability constraint and induces efficient message exchange.[43]

Both of these examples satisfy the following condition, which is central to a more general analysis:

Condition 1. There exists a v^* such that $v_A \geq v^*$ implies $E\{v_B | v_A\} + v_A \geq m$ and $v_A < v^*$ implies $E\{v_B | v_A\} + v_A < m$ for all possible values of v_A.[44]

Because we have assumed the support of v_A is compact, we can assume v^* is an element of the support of v_A without further loss of generality.[45]

The importance of this condition can be seen intuitively as follows. Under any pricing scheme, the receiver, B, either always accepts or always rejects messages. Unless the problem is degenerate, optimality requires that B accept all messages. In the absence of a profitability constraint, B's accepting can be assured by setting a low value of r. Achieving efficiency thus reduces to inducing A to initiate a message if and only if it is information-constrained efficient. Being able to induce such a calling pattern is equivalent to being able to divide the support of the sender's message values into an increasing set that corresponds to efficiency. This is what Condition 1 ensures.

If there is a profit constraint, then Condition 1 is insufficient to guarantee that an information-constrained efficient outcome is supportable through simple pricing even if $F=0$. By Condition 1, efficiency can be achieved only if $p = v^*$, but B may not accept messages when $r = m - v^*$. The following condition assures that she will:

Condition 2. $E\{v_B | v_A \geq v^*\} \geq E\{v_B | v^*\}$.

Finding 11: Suppose that: (a) if one knows the expected value of v_A (conditional on ω_A), then learning ω_A provides no additional information about v_B, and (b) there is a single type of receiver. Absent a network profitability constraint, information-constrained efficiency can be achieved if and only if Condition 1 holds. If the network provider is subject to a profit constraint and $F=0$, then Conditions 1 and 2 are sufficient for simple pricing to support the information-constrained optimum.

Proof: First, suppose there is no profit constraint. To demonstrate sufficiency, let $p = v^*$ and set $r = \inf\{v_B\}$. A sends a message if and only if $v_A \geq p = v^*$. That is, communication is initiated if and only if it is information-constrained efficient. Because $v_B \geq r$ for all v_B, all sent messages are accepted. To demonstrate necessity, suppose p achieves information-constrained efficiency. Then $v_A \geq p$ implies $E\{v_B|v_A\} = v_A \geq m$ and $v_A < p$ implies $E\{v_B|v_A\} + v_A < m$; p satisfies the definition of v^*.

Now, suppose there is a profit constraint and $F = 0$. The following prices support the information-constrained optimum and satisfy the profit constraint:

$$p = v^* \text{ and } r = m - v^*. \tag{6.27}$$

Observe that A sends a message if and only if $v_A \geq v^*$. Hence all sent messages are second-best efficient and, hence, all should be accepted. Because B knows $v_A \geq v^*$ if called, he will answer if $E\{v_B|v_A \geq v^*\} \geq r$. We have

$$
\begin{aligned}
E\{v_B|v_A \geq v^*\} &\geq E\{v_B|v^*\} \\
&\geq m - v^* \text{ (because } E\{v_B|v^*\} + v^* \geq m \tag{6.28} \\
&= r,
\end{aligned}
$$

which establishes the result.[46] QED.

If Condition 2 did not hold, then inequality (6.28) would hold strictly in reverse. For those cases in which $E\{v_B|v^*\} + v^* = m$ (as would be true, for instance, if v^* were in the interior of a continuous support over v_A), then reversing (6.28) would mean that, at any price such that A would send a message if and only if it were second-best efficient, B would not accept the messages sent. That is, in those cases, Condition 2 would also be necessary.

Condition 2 establishes that A's willingness to call is considered 'good news' by B about the value of the message to him. In other words, it is the assumption that the sender's valuing the call by more makes it more likely that the receiver will too. Indeed, if v_A is good news about v_B in the more formal sense of Milgrom (1981), then Condition 2 follows. Milgrom's definition of good news is

$$g(v_A|v_B)g(\tilde{v}_A|\tilde{v}_B) > g(v_A|\tilde{v}_B)g(\tilde{v}_A|v_B) \tag{6.29}$$

whenever $v_A > \tilde{v}_A$ and $v_B > \tilde{v}_B$, that is, $g(\cdot|\cdot)$ satisfies the strict monotone likelihood ratio property. Milgrom's Proposition 4 can then be invoked to establish Condition 2 *and* to prove that $E\{v_B|v_A\} + v_A$ is increasing in v_A. The latter implication implies Condition 1, and thus the information-constrained optimum can be achieved even when the firm is subject to a profit constraint.

Observe too that the inequality between the expectations in Condition 2 is a weak inequality. This means that second-best efficiency can also be achieved even if A's willingness to call is not informative about B's value.[47]

Summarizing this analysis:

Finding 12: Suppose that: (a) if one knows the expected value of v_A (conditional on ω_A), then learning ω_A provides no additional information about v_B, and (b) there is a single type of receiver. If either: (i) $g(\cdot|\cdot)$ satisfies the strict monotone likelihood ratio property, or (ii) the receiver's value is distributed independently of the sender's value, then the information-constrained optimum can be achieved while satisfying the profit constraint when $F=0$.

We close the discussion of single receiver types by briefly examining the profit-maximizer's problem. The firm sets $r = E\{v_B|v_A \geq p\}$ – because there is only one type of receiver, the network provider can fully extract surplus from that user. Hence, any distortion must arise in pricing to the sender. To see why, consider the expressions for expected profits and expected surplus. The network's expected profits are

$$G_A(p)\,[p + E\{v_B|v_A \geq p\} - m],$$

where $G_A(\cdot)$ is the distribution function for v_A. Total expected surplus is

$$G_A(p)\,[E\{v_A + v_B|v_A \geq p\} - m].$$

Using these two expressions, a revealed-preference argument readily shows that the profit-maximizing price weakly exceeds the surplus-maximizing price.[48]

So far, we have assumed that there is only one type of receiver. Not surprisingly, achieving the information-constrained efficient outcome is less likely in the more common situation in which a single pricing scheme applies to heterogeneous receivers. Condition 1 remains a sufficient condition (but is no longer necessary) for attainment of the first best in the absence of a network profitability constraint.[49] However, it appears unlikely that a single value of p^* would correctly divide the space into efficient and inefficient calls for all receiver types.

In a different context, Diamond (1973) examined the problem of congestion externalities when users are heterogeneous. The agents creating externalities in his model are similar to the senders in our model. If one assumes that r is set low enough that all messages are accepted, his results can be reinterpreted for positive consumption externalities with heterogeneous users.[50] Restated, he derived the following results. First, when all consumers give rise to the same externality, a uniform price below marginal cost by the amount of the externality restores efficiency. Second, when individuals differ in terms of the externalities they generate and demands depend only on price, the socially optimal uniform price is less than marginal cost by an amount equal to the weighted average of the externalities generated, with the weights equal to price derivatives of demand. When individuals are heterogeneous and demands for sending messages depend on the quantities of messages received, the rule has to be modified to account for the sensitivities of consumers who generate large externalities to both price and incoming message volumes. The direct applicability of these results is limited by the fact that it typically is not optimal to set r less than or equal to the lowest possible value of $E\{v_B\}$, which could be negative. Diamond's model, however, identifies considerations that continue to be relevant, if not determinant, in a general setting.

Additional results in the presence of a budget constraint can be obtained by assuming that v_A and v_B are stochastically independent. Although independence is a strong assumption, it is a useful case to consider. The importance of this assumption is that neither party draws inferences about his or her value of message exchange from the behavior of the other party. This assumption strikes us as reasonable in many situations where B potentially receives messages from a large number of heterogeneous senders and he does not know the identity of the sender before accepting a message, for example, when answering the telephone.

Observe that independence makes the most sense with truly one-way technologies. Specifically, the earlier story about two-way technologies in which only one party knows that the call is worth completing becomes problematical when $r > p$. In this case, every user would rather initiate calls than accept them, which can lead to strategic considerations similar to those discussed in the next subsection.

Assuming independence, interpret v_A as the sender's (expected) value of conveying a message and v_B as the receiver's (expected) value of getting a message. A weakly dominant strategy for the sender is to send if and only if $v_A \geq p$. Likewise a weakly dominant strategy for the receiver is to accept the message if and only if $v_B \geq r$. Figure 6.3 illustrates. As the left panel shows, it is efficient to exchange all messages above the line $v_B = m - v_A$.[51] The right panel of Figure 6.3 shows the set of messages exchanged when users face prices p and r. In the case illustrated, p and r sum to m and thus satisfy the network profitability constraint when $F = 0$. As is clear from a comparison of the shapes of the message exchange areas in the two panels of the figure, it is impossible to achieve efficient message exchange – with or without a network profit constraint – using only simple pricing. The next result helps characterize the optimal second-best pricing in the presence of a network profitability constraint.

As a further simplifying assumption, assume a *common* distribution

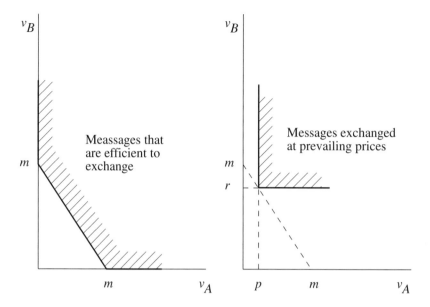

Figure 6.3   Messages that are efficient to exchange and those that are exchanged in equilibrium

function $G(\cdot)$ for v_A and v_B.[52] Let $S(v) = 1 - G(v)$ be the survival function. If $g(\cdot)$ is the density associated with $G(\cdot)$, then $S'(v) = -g(v)$

Consider prices p and $r = m - p$, which satisfy the breakeven constraint. Expected welfare, EW, is

$$EW = \int_{m-p}^{\infty} \left(\int_{p}^{\infty} (v_A + v_B - m) \, g(v_A) dv_A \right) g(v_B) dv_B$$

$$= [1 - (m-p)] \int_{p}^{\infty} v_A \, g(v_A) dv_A$$

$$+ [1 + G(p)] \int_{m-p}^{\infty} v_B \, g(v_B) dv_B - m[1 - G(p)][1 - G(m-p)].$$

Recalling the definition of the survival function, $S(\cdot)$, we see that

$$EW = -S(m-p) \int_{p}^{\infty} vS'(v) dv - S(p) \int_{m-p}^{\infty} vS'(v) dv - mS(p)S(m-p).$$

Integrating by parts,

$$EW = S(m-p)\left[pS(p) + \int_{m-p}^{\infty} S(v)dv\right]$$

$$+ S(p)\left[(m-p)S(m-p) + \int_{m-p}^{\infty} S(v)dv\right] - mS(p)S(m-p)$$

$$= S(m-p)\int_{p}^{\infty} S(v)dv + S(p)\int_{m-p}^{\infty} S(v)d(v). \tag{6.30}$$

We can now establish:

Finding 13: Suppose $F=0$ the network is subject to a profitability constraint, and v_A and v_B are independently and identically distributed according to the differentiable distribution function $G(\cdot)$ If $G(\cdot)$ is weakly convex,
then the socially optimal prices divide the cost of message exchange evenly
between sender and receiver.

Proof: Differentiating (6.30) with respect to p yields the first-order condition:

$$S'(p)\int_{m-p}^{\infty} S(v)dv - S'(m-p)\int_{p}^{\infty} S(v)dv = 0$$

Clearly, $p = m/2$ solves the first-order condition. By construction,
$r = m-p = m/2$.

  If (6.30) is concave, then the second-order conditions are met as well. The
second derivative of (6.30) is

$$\int_{m-p}^{\infty} S(v)dv + S'(p)S(m-p) + S'(m-p)S(p) + S''(m-p)\int_{p}^{\infty} S(v)dv. \tag{6.31}$$

$G''(\cdot) \geq 0$ implies expression (6.31) is negative, because the integrals are positive, $S''(\cdot) = -G''(\cdot) \leq 0$, $S'(\cdot) < 0$, and $S(\cdot) > 0$. Expression (6.31) therefore
establishes that (6.30) is concave if $G(\cdot)$ is a convex distribution. QED.

Examples of weakly convex distributions include uniform distributions
and certain other power-function distributions.[53]

Figure 6.4 illustrates the logic of Finding 13. The figure shows the effects
of shifting from unequal send and receive prices, (p_1, r), to equal prices,
$(m/2, m/2)$. Area *I* represents messages that are exchanged at the original
prices but not the equal ones. The sum of Areas *II* and *III* represents messages that are exchanged at the equal prices, but not at the original ones.
The welfare effects depend on the comparison of the messages lost and

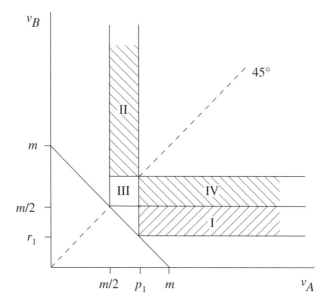

Figure 6.4   Effects of a move towards equal send and receive prices

gained from the shift to equal send and receive prices. When v_A and v_B are symmetrically distributed, the density over Area *II* is identical to the density over Area *IV*. Consider the set of points with Areas *I* and *IV* for any fixed value of $v_A \geq p_1$. Under the assumption that $G(\cdot)$ is weakly convex, the probability weight put on the points in Area *IV* is at least as great as the weight put on those points in Area *I*. Moreover, every such point in Area *IV* has a higher value of v_B than does any such point in Area *I*. Hence, Area *IV* has both greater probability weight and a higher value of gross consumption benefits than does Area *I*. It follows that the welfare gains, Areas *III* and *IV*, exceed the welfare losses, Area *I*. Therefore, the move to equal prices raises welfare. A similar argument applies to cases in which the initial send price is less than the receive price.

Interestingly, for some distributions of message valuations, how the costs of communicating are split is irrelevant to achieving optimality. Suppose $G(\cdot)$ were the exponential distribution with mean μ. Then $S(z) = e^{-z/\mu}$ and, hence, $EW = 2\mu e^{-m/\mu}$, which is independent of p; any division of the cost will achieve optimality.

What about a profit-maximizing provider of network services? It is instructive to divide the monopolist's problem into two steps. First, suppose the monopolist wishes to achieve a profit margin of γ per completed call.

This is achieved by any p and r pair such that $p+r-m=\gamma$. The monopolist wishes to choose the pair from this line that maximizes the probability of a completed call:

$$S(p)S(r)=S(p)S(\gamma+m-p). \tag{6.32}$$

The corresponding first-order condition for an interior maximum is

$$S'(p)S(\gamma+m-p)-S(p)S'(\gamma+m-p)=0, \tag{6.33}$$

of which $p=1/2(\gamma+m)$ – and, thus, $r=1/2(\gamma+m)$ – is a solution. Equation (6.33) is sufficient, as well as necessary, if expression (6.32) is a concave function of p. It is readily shown that $G(\cdot)$ convex is a sufficient condition for (6.32) to be concave.

This result can also be established by applying the graphical analysis of Figure 6.4. The only differences from the earlier welfare analysis is that the monopolist attaches equal dollar value to any two points that involve message exchange and m is replaced by $m+\gamma$. As shown earlier, when $G(\cdot)$ is weakly convex, Area *IV* gets greater probability weight than does Area *I*. Hence, the monopolist gains sales from the move to equal send and receive prices.

Once the optimal prices p and r for a given γ are found, the second step is to maximize expected profits with respect to γ. When $p=r=1/2(\gamma+m)$ is optimal, the monopolist's problem is

$$\max_{\gamma} S\left(\frac{\gamma+m}{2}\right) S\left(\frac{\gamma+m}{2}\right) \gamma$$

Because the derivative of this expression with respect to γ is positive at $\gamma=0$ (it equals S^2), the usual monopoly distortion of too few sales holds. Note, however, that conditional on this distortion, monopoly imposes no further distortion in terms of the optimal division of the cost and mark-up between sender and receiver. To summarize:

Finding 14:   Suppose that v_A and v_B are independently and identically distributed according to the differentiable distribution function $G(\cdot)$. If $G(\cdot)$ is weakly convex, then a profit-maximizing monopolist sets equal send and receive prices. The profit-maximizing price level exceeds the welfare-maximizing price level.

In other cases, there can be two types of distortion. In addition to setting the markup too high, the profit-maximizing network may distort the relative levels of the send and receive prices. This second distortion arises because – for a given margin – the profit maximizer chooses the relative prices to maximize the probability that a message will be exchanged, while the welfare maximizer takes into account the value of the messages exchanged.

(i) Menus

So far, it has been assumed that the network sets a single pair of send and receive prices. In this setting, a sender with a very valuable message has the same chances of having it accepted by the receiver as does a sender with a much less valuable message. There is no scope for the sender with a very valuable message to the volunteer to assume more of the cost in order to increase the odds that the receiver will accept the call. Giving the sender a menu of prices – and allowing the receiver to know the charge he correspondingly faces – provides a means around this problem.

To be concrete, suppose that the two parties' values of message exchange are independently distributed. Suppose too that when restricted to a single price pair such that $p+r=m$ the network optimally chooses $p^*<m$ and r^*. Now, consider a menu in which the potential sender, if she chooses to send a message, chooses between prices p^* and $\tilde{p}$, where $m \geq \tilde{p} > p^*$. If the sender elects $\tilde{p}$, then the receiver pays $\tilde{r} \equiv m - \tilde{p}$ if he accepts the message. Welfare is weakly greater under this menu: all messages that would have been exchanged under the single-price regime will be exchanged under the menu and some surplus-generating messages that would not have been exchanged may now be exchanged.[54]

Intuitively, the sender can be thought of as a monopoly producer of messages that she 'sells' to the receiver at a price r. Giving her the option of charging a lower price cannot reduce welfare, and it can raise welfare to the extent that the lower price is closer to the monopoly price for some types of sender. This monopoly metaphor also points out that adding a receive price to a menu that is greater than the lowest existing price can reduce welfare because it can give the sender added ability to exercise her market power.

To summarize:

Finding 15: Suppose that v_A and v_B are independently distributed and each party knows his or her own value of message exchange. Adding a send price to a menu that is greater than any existing send price weakly increases welfare. Adding a send price to a menu that is less than some existing send price can reduce welfare.

(ii) Simultaneous connection and acceptance externalities

Once one allows for multiple types, it is also natural to examine connection decisions as well as message decisions. Squire (1973) examined such a model and derived the finding that the socially optimal access charge might be greater than the marginal access cost even in the absence of a profitability constraint (Squire, p. 521). For the setting examined by Squire (in which $v_B \equiv \beta$) this result is false. The socially optimal prices absent a profit constraint are $p = m - \beta$, $r \leq \beta$, and $a < h$ where a is found by the methods of our earlier analysis of access externalities. This result follows from the fact that the send and receive prices perfectly internalize the call externalities and thus the access price can be determined from a model that ignores the call externalities. Optimal prices for the other settings remain to be explored.

TWO-WAY CALLING DECISIONS

Now suppose that either A or B can send a message to the other, so that the calling direction is endogenous. The main consequence of endogenizing the calling direction is that it causes each player to face the choice of either sending a message or waiting to be sent one from the other party. If the price to the initiating party exceeds the cost to the receiving party, then waiting is preferable conditional on a message's being exchanged. The danger, however, is that both wait and, thus, a valuable message may fail to be exchanged.

To model this situation, we consider a highly parameterized waiting game (see Hermalin and Katz 2001, for a more general version). Specifically, assume the values of communicating, v_A and v_B, are drawn from the same finite support. For convenience, we normalize that support to the unit interval, [0, 1]. Scale the cost of calling, m, by the same proportion. So that the analysis is non-trivial, assume $m \in (0, 1)$. Further assume that the values are drawn independently and according to the uniform distribution.

Each player knows his or her own value of communicating, but not the other player's. The value of communicating decays over time. Specifically, if a message is exchanged at time t, then it is worth $ve^{-\delta t}$, where δ is the common decay rate.

We focus here on two pricing regimes: *sender pays*, that is, $p = m$ and $r = 0$, and *equal cost sharing*, that is, $p = m/2 = r$. Under the equal-sharing regime, there is no advantage to delay and there is immediate calling if at least one party's value of communicating exceeds $m/2$. That message is accepted if the receiver's value exceeds $m/2$;[55] that is, we have immediate message exchange if and only if $\min\{v_A, v_B\} \geq m/2$. Figure 6.5 illustrates.

With sender pays, there is reason to engage in delay. Let $\tau(v)$ denote the time that a player with initial value v makes a call if he or she has not yet

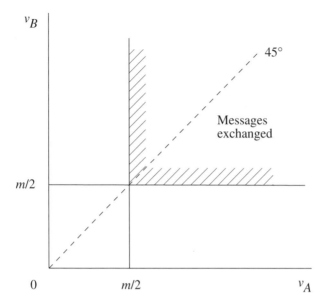

Figure 6.5 Messages exchanged when the message cost is evenly divided, that is, $p = r = m/2$

called. From the perspective of one player, there is a distribution, $Q(\cdot)$, over when he will be called. Assume the corresponding density, $q(\cdot)$, exists.[56] Then $\tau(\cdot)$ is the solution to the program

$$\max_{t}[1 - Q(t)](ve^{-\delta t} - p) + \int_{0}^{t} ve^{-\delta z}q(z)dz. \tag{6.34}$$

The first-order condition is

$$q(t)\,p - \delta ve^{-\delta t}[1 - Q(t)] = 0. \tag{6.35}$$

The cross-partial derivative of expected utility, (6.34) with respect to t and v is

$$-\delta e^{-\delta t}[1 - Q(t)] < 0$$

which, because it is negative, implies $\tau(\cdot)$ is a decreasing function.

In what follows, it is helpful to work with the survival function, $\Phi(t) \equiv 1 - Q(t)$. Using the survival function, we can rewrite (6.35) as

$$-\frac{\dot{\Phi}}{v\Phi}=\frac{\delta}{p}e^{-\delta t}. \tag{6.36}$$

If the equilibrium is symmetric, as we now assume, then

$$\Phi(\tau(v))=\Pr\{t\ge\tau(v)\}.$$

Recall $\tau(\cdot)$ is decreasing, so

$$\begin{aligned}\Phi(\tau(v_0))&=\Pr\{t\ge\tau(v_0)\}\\&=\Pr\{v\le v_0\}\\&=v_0\text{ (because }v\text{ is distributed uniformly).}\end{aligned}$$

Using this result in (6.36), the differential equation becomes

$$-\frac{\dot{\Phi}}{\Phi^2}=\frac{\delta}{p}e^{-\delta t},$$

the solution to which is

$$\frac{1}{\Phi}=K-\frac{1}{p}e^{-\delta\tau(v)}. \tag{6.37}$$

K a constant. Consider the user with $v=1$. Because $\tau(\cdot)$ is decreasing, this user calls first. Knowing that fact and given that the value of calling is decaying, there is no reason for this user to delay sending a message in equilibrium; $\tau(1)$ must be equal to 0. Plugging this value into equation (6.37), K must be $\frac{1+p}{p}$. Solving for $\tau(\cdot)$:

$$\tau(v)=\frac{1}{\delta}\ln\left(\frac{v}{v(1+p)-p}\right). \tag{6.38}$$

A consumer with the value v_0 will never send a message after time t_0 if $v_0 e^{-\delta t_0}\le p$. Consequently, the last type to call is v^ℓ, where v^ℓ solves

$$ve^{-\delta\tau(v)}=p. \tag{6.39}$$

Some algebra on (6.38) and (6.39) yields

$$v^\ell=\frac{2p}{p+1} \tag{6.40}$$

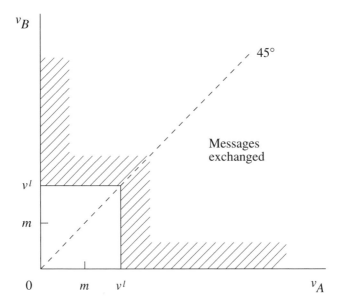

Figure 6.6   Messages exchanged under sender-pays regime

Recall that $v \leq 1$ for all realizations and, thus, calling occurs only if $p<1$. Hence, equation (6.40) implies $v^\ell > p$. In words, there exists message values such that a party values exchanging a message by more than the cost of sending one, but never sends a message because she waits to receive one at a lower price. This fact is a consequence of strategic delay. Figure 6.6 illustrates the set of values for which messages are eventually exchanged.

For future reference, observe that the time of the last possible message is

$$t^\ell \equiv \tau(v^\ell) = \frac{1}{\delta}\ln\left(\frac{2}{p+1}\right).$$

We now compare the two pricing regimes in terms of expected welfare. Under equal division, expected welfare is

$$W_E = \int_{m/2}^1 \int_{m/2}^1 (v_A + v_B - m)\, dv_A dv_B$$
$$= \frac{(2-m)^3}{8}. \tag{6.41}$$

Under the sender-pays regime, expected welfare is

$$W_S = 2 \times \int_{v^I}^1 \left(\int_0^v e^{-\delta\tau(v)}(v+u) - m) \, du \right) dv$$

$$= 2 \times \int_{\frac{2m}{m+1}}^1 \left(\int_0^v \left[\frac{v(1+m) - m}{v}(v+u) - m \right] du \right) dv \qquad (6.42)$$

$$= 1 - \frac{3m}{2} + \frac{2m^3}{(1+m)^2}.^{57}$$

A little algebra shows that

$$W_E - W_S = \frac{m^2(6 + 4m^2 - 5m - m^3)}{8(1+m)^2} > 0.$$

That is, expected welfare is greater when the message cost is evenly divided than when the sender alone pays.

Finding 16: Given a uniform and independent distribution of message values, expected welfare is greater when the sender and receiver bear equal shares of the cost of exchanging a message than when the sender bears the full cost alone.

Finding 16 is the consequence of two factors. First, for relatively large costs – that is, m near 1 – the likelihood that either player ever initiates a message becomes too small under a sender-pays regime. Second, even when costs are not that large, there is still the problem of strategic delay. To see the importance of delay, fix $\tau(v) \equiv 0$. The notional level of expected total surplus under the sender-pays regime is then

$$2 \times \int_{v^I}^1 \left(\int_0^v \left[e^{-\delta \times 0}(v+u) - m \right] du \right) dv = 2 \times \int_{\frac{2m}{m+1}}^1 \left(\int_0^v \left[v+u-m \right] du \right) dv$$

$$= \frac{1 + 2m - 6m^3 + 3m^4}{(1+m)^3}. \qquad (6.43)$$

It can be shown that (6.43) exceeds (6.41) if and only if $m > 0.513$. In words, except for delay, sender-pays would yield greater welfare than would equal division when the cost of a message is low (roughly less than the expected value of a message to either party). But there is a strategic delay, which is why sender-pays is dominated by equal division.

It is important to recognize that Finding 16 depends critically on the distributional assumptions made. Suppose, in contrast, that v_A is distributed on $\{0, m+\upsilon\}$, where $0<\upsilon<m$, and $v_B\equiv m+\upsilon-v_A$. In this case, it always is efficient to exchange a message because $v_A+v_B=m+\upsilon>m$. No message would be exchanged under equal division, that is, $p=r=m/2$, because for any message one of the parties gets no benefit from the exchange. A message would, however, always be exchanged under sender-pays.[58] Intuitively, this finding is similar to our earlier result that offering consumers a menu of options can be efficient. In this case, the options are whether to be an active sender or a passive receiver.

We close this subsection by noting that several of the results here and in the previous subsection indicate that it is not socially optimal to have one party bear the full marginal costs of exchanging a message. Yet, except in certain circumstances when someone on a wireless network receives a call, calls for which both parties pay on a traffic-sensitive basis are rare in practice.

Cost Differences across Users

The costs of exchanging a message may not be the same between all possible user pairs. Suppose that the cost of exchanging a message between users A and B is m_A+m_B. In analysing such situations, it is important to distinguish between cases in which the component costs are exogenous and those in which they are endogenous. When costs are due to consumers' locations, and communications costs are only a small part of the overall costs and benefits of locating in a particular place, these costs may be taken to be exogenous.[59] In other cases, the component costs are endogenous. For example, when a consumer subscribes to a wireless telephone service, the per-message cost is generally greater than that of a wire-line service.

Consider first the case of exogenous cost differences. In this case, there is no economic rationale for trying to assign particular costs to one user or another in a given pair. All that matters is the cost of exchanging a message between that pair. Hence, when costs are exogenous, all of our earlier results continue to apply.

Finding 17: With exogenous costs, the privately and socially optimal send and receive prices are unaffected by changes in the component prices that leave the sum of those prices unaffected.[60]

Now, consider endogenous costs. The analysis is more complex than with exogenous costs because prices can influence the choice of message exchange technology, as well as the volume of message exchange. Here, we offer two one-way calling examples to illustrate a range of possibilities. In both examples, suppose $F=0$ and A has a choice between two technologies, one of which generates greater calling benefits but is also more costly in terms of the marginal cost of message exchange. In the first example, suppose $v_B = \mu v_A$ for any call under either choice of technology (the more expensive technology might be a mobile telephone that allows the two parties to exchange messages on a more timely basis and thus benefits both). In this case, the socially optimal prices are $p = m_k/(1 + \mu)$ and $r = m_k \mu/(1 + \mu)$, where m_k is the marginal message cost of the kth technology. For any given technology, these prices satisfy the profitability constraint and induce the information-constrained efficient level of message exchange. Moreover, because A gets a constant share of the net benefits under either technology, she chooses the technology that maximizes total surplus.

When A's share of the benefits varies across technologies, optimal pricing must take that into account. Suppose, for instance, that the more costly technology generates additional benefits for A but not for B, for example, the technology also allows A to consume other services, such as cable television, that do not involve B. In this case, a pricing scheme that sets r higher for the more expensive technology in order to cover the higher marginal message cost will tend to bias A's choice toward that technology. In effect, B would be subsidizing A's consumption of the additional benefits associated with the more costly technology.

Sources of Distortion

The first best often is unattainable in the single-message models examined above, and it is useful to understand the reasons why in greater depth. One set of reasons center around transaction costs and the lack of a bargaining or coordination mechanism. There is no pre-communication communication; the parties cannot exchange information about their values of message exchange prior to exchanging a costly message and thus cannot fully coordinate their actions. Moreover, absent repeat play, the parties have no mechanism, such as side payments, through which to internalize external effects.

In theory, alternative institutional arrangements or technologies might reduce transaction costs and allow coordination and bargaining. For example, the valuable messages to be exchanged might be 'large' and the parties might be able first to exchange limited messages communicating their valuations at a near-zero relative cost. Moreover, the parties might be able to commit to making monetary transfers to one another.

Even under these alternative institutions, the fact that a party's valuation of a message is private information could still give rise to information rents and inefficient outcomes. Suppose, for example, that each party's value were verifiable. In this situation, the service provider could set the price of a message equal to m and the parties could then write an enforceable contract that required A to call B if and only if $v_A + v_B \geq m$ and set monetary transfers to divide the surplus however determined by the outcome of bargaining. Given that the message exchange values would be common information, it is reasonable to assume the bargaining would be efficient. Rubinstein bargaining, for instance, would lead to agreement in the first round (Rubinstein 1982). With low enough transactions costs, the bargaining could even take place on a per-message basis without the need for the verifiability of users' valuations. As long as the valuations were common knowledge, the parties could be expected to bargain to an efficient outcome.

When the values of message exchange are private information, however, the ability to achieve efficiency will be limited even if there is costless preplay communications and negotiation, as well as the possibility of side payments. To see why, consider the problem from the mechanism design perspective.

Assume that v_A and v_B are the private information of A and B, respectively. Assume, too, that these values are independently distributed, but that their distributions are common knowledge. Lastly, assume that A and B can commit to a mechanism *before* they learn their private information. In this context, a mechanism is an indicator function, $\chi(\hat{v}_A, \hat{v}_B) \in \{0, 1\}$ and a pair of payment functions also contingent on $(\hat{v}_A, \hat{v}_B)$, where $\hat{v}_i$ is player i's announced value of communicating.[61] Consider a mechanism under which payments to the center (telecommunications provider) are

$$\frac{m}{2}\chi(\hat{v}_Z, \hat{v}_B) + P_i(\hat{v}_A, \hat{v}_B).$$

Observe that payments have a portion that is contingent on a message's being sent ($\chi = 1$) or not ($\chi = 0$) and a non-contingent portion needed to align incentives.

We want the mechanism to be efficient, that is, $\chi(\hat{v}_A, \hat{v}_B) = 1$ if and only if $\hat{v}_A + \hat{v}_B \geq m$. In addition, the mechanism must induce truth telling, that is, $\hat{v}_i = v_i$ for all v_i in equilibrium. Finally, we want it to be balanced, that is, $P_A + P_B = 0$ for all v_A and v_B.

We can insure the mechanism is balanced if we set the non-message-contingent payments to be

$$P_A(\hat{v}_A, \hat{v}_B) = \rho_A(\hat{v}_A) - \rho_B(\hat{v}_B) \qquad (6.44)$$

and

$$P_B(\hat{v}_A, \hat{v}_B) = \rho_B(\hat{v}_B) - \rho_A(\hat{v}_A) \qquad (6.45)$$

where

$$\rho_i(\hat{v}_i) = -E_{v_j}\left\{\left(v_j - \frac{m}{2}\right)\chi(\hat{v}_A, \hat{v}_B)\right\}, \qquad (6.46)$$

for $i \neq j$ and E_{v_j} denotes expectation with respect to v_j.

Finding 18: The mechanism defined by (6.44)–(6.46) induces the efficient outcome for all values of the message.

Proof: We need to show that: (i) truth-telling is an equilibrium, and (ii) the equilibrium is efficient. To see that truth-telling is an equilibrium, suppose that A anticipates that B will tell the truth. A's best response is to choose $\hat{v}_A$ to maximize

$$E_{v_B}\left\{\left(v_A - \frac{m}{2}\right)\chi(\hat{v}_A, v_B) - P_A(\hat{v}_A, v_B)\right\},$$

which is equivalent to choosing $\hat{v}_A$ to maximize

$$E_{v_B}\left\{\left(v_A - \frac{m}{2}\right)\chi(\hat{v}_A, v_B) - \rho_A(\hat{v}_A)\right\} =$$

$$E_{v_B}\left\{(v_A + v_B - m)\chi(\hat{v}_A, v_B)\right\}.$$

Truth-telling by A clearly maximizes this last expression. Given that A and B are interchangeable, this result means that there is a truth-telling equilibrium. By construction of χ, P_A and P_B, the equilibrium is efficient and balanced. QED.

Although Finding 18 is provocative, it hinges on two assumptions that very likely limit its normative implications. First, there is no exit from the mechanism after learning one's value of communicating, v_i. Whenever $v_i < m/2$ party i would prefer not to honor her commitment to play the mechanism: she pays P_i, which yields nothing at best and *costs* her $m/2 - v_i$ at worst.[62] In most situations, it is difficult to imagine enforcing commitment to a mechanism such as this. In general, constructing a balanced mechanism that also exhibits *ex post* individual rationality – that is, no one will exit after learning his type – is not possible (see Myerson and Satterthwaite, 1983, for an analysis).

Second, Finding 18 relies on the assumption of independent values. As others have shown, correlation in values can make it impossible to define an efficient balanced mechanism, see, for example, the discussion in Hermalin and Katz (1993).

Repeat Play and Ongoing Caller Relationships

As many earlier authors have noted, repeated interaction among users can internalize call externalities in some circumstances because users can engage in tit-for-tat and similar strategies.[63] It is useful to introduce a final model to explore this idea further. Suppose that each party's private value of exchanging a message can take on a range of values on the interval $[v^L, v^H]$. Assume that

$$v^H < m < 2v^L. \tag{6.47}$$

Given assumption (6.47), a sender-pays pricing regime would lead to no calling in a one-shot setting, even though it is always efficient for a message to be exchanged. Suppose, however, that the two parties have the potential to call each other repeatedly. Specifically, each period, each caller learns his or her value of message exchange in that period. The distribution of values is common knowledge, but the realizations in any period are private information. Let $\delta \in (0, 1)$ be the common per-period discount factor.

Suppose the two callers agree to alternate calling each other. Under sender-pays, the maximal loss from calling in an assigned period is $v^L - m$. Let v^E be the expected private value of a call. Caller i's expected continuation value is

$$V_i = \frac{\delta}{1 - \delta} v^E - \frac{\delta^2}{1 - \delta^2} m.$$

The harshest possible punishment regime is one in which the parties revert to the Nash equilibrium with no calling in any period, which drives both surplus levels to zero in the continuation game. For the party whose turn it is to be the sender not to deviate from the agreement, it must be the case that $v^L - m + V_i \geq 0$, or

$$v^L + \frac{\delta}{1 - \delta} v^E \geq \frac{1}{1 + \delta^2} m. \tag{6.48}$$

Multiplying both sides of (6.48) by $1 - \delta$ reveals that the condition for cooperation is equivalent to

$$(1 - \delta)v^L + \delta v^E \geq \frac{1}{1 + \delta} m. \tag{6.49}$$

As δ goes to zero, the future does not count, and the assigned sender has incentives to deviate: the limit of the left-hand side of (6.49), v^L, is less than the limit of the right-hand side, m. As δ goes to one, the future is given a lot of weight, and the limit of the left-hand side, v^E, exceeds the limit of the right-hand side, $m/2$ (recall assumption (6.47)).

Thus, if the two callers are sufficiently patient, they can support the first-best outcome with message exchange in every period even though the highest individual value of calling is less than the price of sending a message.

A critical assumption in this illustrative model is that exchanging a message is always efficient. Generally, if there are valuation draws such that some messages are inefficient, and the realization of such draws are not common knowledge, then it will not be possible to achieve an efficient outcome: the player whose turn it is to be the receiver cannot know whether he failed to receive a message because the sender determined that exchange was socially inefficient that period or because she chose to deviate from the cooperative path. Nonetheless, repetition will still allow some internalization, so users can still be expected to do better in a repeated situation than in a one-shot situation. Summarizing these results,

Finding 19:   Repeat play can (partially) internalize call externalities.

CONCLUSION

With the rise of the Internet and ongoing development of wireless technologies, telecommunications networks continue to grow in economic and social importance. There is a clear need for additional research on the pricing of these networks. This is particularly true in the areas of call externalities and markets in which there are competing service providers. In addition to issues of competition and coordination, the presence of multiple carriers raises important issues concerning inter-carrier charges, such as interconnection charges, access fees, and reciprocal compensation. Although there is a large literature examining inter-carrier charges, it generally ignores call externalities. A notable exception is DeGraba (2000), who characterizes efficient interconnection fees under the assumption that the sender's and the receiver's message values are identically equal. While a promising start, much work remains to be done.

ACKNOWLEDGEMENTS

The authors wish to thank Professor Oystein Fjeldstad for helpful comments. They also wish to acknowledge the support of the Walter A. Haas School of Business through the Willis H. Booth and Edward J. & Mollie Arnold Chairs and the University of California through a Committee on Research Grant. The views expressed here are solely those of the authors.

As such, they do not necessarily reflect the views of the chair donors, the University of California, or consulting clients in government and the telecommunications industry.

NOTES

1. See Taylor (1994, Chapter 9) for a useful survey of telecommunications externalities, particularly empirical issues. Laffont and Tirole (2000) offer a more theoretical analysis of many of these issues.
2. To simplify the analysis and focus on the issues created by external effects, we examine only single-service markets. In multi-service markets, there can be interesting price structure issues even if each service is sold at a uniform price, as long as prices can vary across services.
3. Katz and Shapiro (1994) survey much of this literature.
4. The latter result stands in contrast to a result presented in Squire (1973), which we believe is incorrect for reasons discussed below.
5. Willig (1979) provides a rigorous justification of the use of the consumer surplus approach in the presence of income effects.
6. This property is satisfied under at least two circumstances. One is when all users are identical in terms of the benefits their subscriptions confer to any given sender (although these values may vary across senders, and sending a message to one user may be a substitute for sending it to another). In this case, consumer benefits depend on the number of users connected to the network, but not their composition. The other case is where users are heterogenous in terms of the values that their subscriptions confer to parties sending messages to them, but for any family of pricing strategies, users join the network in the same order as prices are varied. In a model with variable numbers of messages (as examined in the next subsection), being able to rank demand curves in a constant order for any quantity (that is, non-crossing) and subscriber set is a sufficient condition for this latter property to hold.
7. Rohlfs (1974) examines the effects of communities of interest on membership dynamics.
8. Metcalf's Law asserts that $v(n,\mathbf{z})$ is a linear function of n. Other analysts assume that $v(n,\mathbf{z})$ is a concave function of n, as could arise if sending a message to one subscriber is a substitute for sending a message to another. Lastly, Crémer (2000) notes that, for certain types of communications, access externalities may be convex over a range of penetration levels. In particular, at high levels of penetration a network might offer a single way to reach an entire community of people (what Crémer labels a *broadcasting service*). At low levels of penetration, it is necessary to rely on alternative media as well in order to ensure that everyone is reached, and the use of multiple media can lead to costs that reduce net benefits.
9. If programming is supplied under conditions of monopolistic competition, similar effects arise on broadcast networks because a user's decision to subscribe increases the variety of programming available to other viewers: an increased potential audience attracts more program suppliers. Under market structures in which programming suppliers earn non-zero profits, one has to take those profits into account in calculating social welfare.
10. This is a reasonable assumption for a communications network; it need not be for a hardware–software, or virtual, network. Consider the use of Microsoft Word. In that case, there are presumably some benefits to being the sole member of the 'network' of Word users.
11. For an early analysis, see Rohlfs (1974).
12. Doing so implicitly assumes that any funds needed to cover the network provider's costs can be raised in a non-distortionary manner. To the extent that subsidy funds are costly,

optimal pricing rules resemble those found below for situations in which a profit constraint must be met, except that the Lagrange multiplier for the profit constraint is replaced by the shadow price of subsidy funds.

13. Observe that $n=0$ is a local maximum of (6.1) because the derivative of (6.1) evaluated at $n=0$ is $-h<0$. Consequently, a network with a 'small' positive number of subscribers can never be efficient.

14. Artle and Averous (1973) provide an early analysis of the effects of access externalities on the optimal allocation of resources.

15. We use the convention that the partial derivative of a function, ζ, with respect to its kth argument is denoted ζ_k.

16. It is a simple matter to show that the first-order condition for Ramsey pricing is the weighted sum of the first-order conditions for unconstrained welfare optimization and profit maximization. Specifically, it entails setting the left-hand side of equation (6.2) plus the Lagrange multiplier for the profit constraint times the left-hand side of (6.3) equal to zero. Finally, we note that the problem here is among the simplest Ramsey-pricing problems, see Wilson (1993, Chapter 5) for a complete treatment of Ramsey pricing. Also see Ramsey (1927).

17. In reality, messages can be of different sizes or lengths. Variable message length is easily incorporated into the analysis of this section by interpreting a message as a time unit of calling or a number of bytes or packets of data.

18. The previous subsection can be seen as assuming a fixed per-message price and thus defining $v(\cdot)$ as an indirect utility function. Moreover, because we did not consider the producer surplus from per-message charges, the per-message price was implicitly set equal to the per-message cost.

19. One might hold the view that, for modern telephony, m is so small as to be meaningless. While $m=0$ might be a reasonable approximation for local wire-line telephone service, it is a poor approximation for wireless and long-distance calling, especially some international calling. Looking to the future, the costs associated with sending other forms of messages, such as streaming video, may be significant. Finally, even when the network's cost of transmitting a message is small, the users could still incur significant opportunity costs connected with the time expended and disruption of dealing with a message.

20. This model is similar to the one examined by Littlechild (1975). An important difference is that Littlechild assumed that consumers could be ordered in terms of quantities demanded at any given price, rather than consumer surplus as we assume.

21. As before, we restrict attention to interior equilibria in which *not* all potential consumers choose to gain access to the network.

22. This result was first derived by Littlechild (1975). Willig (1979, pp. 130–33) derived a similar result in a model allowing for income effects and multiple types of messages.

23. Ng and Weisser (1974) obtained a similar result for welfare-optimal two-part tariffs in the absence of externalities. They also extended the result to a class of models in which consumers differ along more than one dimension. Oi (1971) examined the profit-maximizing two-part tariff in a situation without externalities and reached similar conclusions about the per-message price and the per-message cost.

24. Willig (1979, pp. 134–7) also derived conditions for prices that maximize welfare subject to a profitability constraint. However, in interpreting his results, he restricted attention to the case in which the message price does not affect the access decision. Hence, he found that the optimal message price is equal to or greater than marginal cost, with the latter occurring whenever the profitability constraint is binding.

25. For p near m, the access fee must exceed the hook-up cost, for example, if the types are nearly identical. Alternatively, suppose $v(n, x, \theta)=\ln(x)e^{-\theta}\ln(n+1)$, $m=0.01$, $h=0.2$, and $N=10$. Then it can be shown that a monopolist would set $p=1.15$ and $a=-0.02$ that is, $p>m$ but $a<h$.

26. This result was first derived by Littlechild (1975).

27. This result was first derived by Littlechild (1975). Curiously, he also states a claim on page 667, without proof, that appears to contradict this result.

28. Artle and Averous (1973) considered a similar model, in which they assumed the price is

exogenously fixed at a constant level. They explored conditions under which the number of users increases over time.

29. In the interest of brevity, we do not offer a complete analysis of the model here.

30. Artle and Averous (1973) did not consider the possibility of penetration pricing; Rohlfs (1974) did.

31. For other parameter values a profit-maximizing provider may choose a *milking* strategy, whereby it sets price above *h* and runs the network size down to 0.

32. Formally, the profit-maximizer's first-order condition is similar to the one-period fulfilled-expectations condition (6.3), with $\delta n_{t+1} v_1(n_t, n_{t+1})$ replacing $n_t v_1(n_t, n_t)$, where δ is the one-period discount factor and $0 < \delta < 1$. In words, the provider takes into account the fact that a larger network size this period means it can charge a higher price next period, *ceteris paribus*, rather than contemporaneously as in the fulfilled-expectations equilibrium.

33. For a summary discussion of private compatibility choices, see Katz and Shapiro (1994).

34. Even absent network effects, the 'competitive equilibrium' may not induce Ramsey prices when there are economies of scale and carriers face profitability constraints. For discussions of multi-carrier pricing, see Braeutigam (1984) and the citations therein.

35. In the absence of network externalities, one can readily show that marginal cost pricing is the unique pure-strategy equilibrium under Bertrand competition between undifferentiated sellers with equal costs. With network externalities, one must check whether a deviation from a candidate asymmetric equilibrium that involves a price increase by one firm would shrink the network by so much as to render that deviation unprofitable. Consequently, a general analysis is difficult and we confine ourselves to an example.

36. See Chapter 5 of Laffont and Tirole (2000) for more on the economics of interconnection.

37. This case could arise when a user's private benefits were greater from joining one network, but his subscription would generate considerably greater external benefits on the other network.

38. For a survey of the relevant literature, see Katz and Shapiro (1994).

39. Willig (1979, pp. 124–5) establishes conditions under which call externalities will be internationalized in the demand for sending messages and thus can be incorporated into the standard analysis of access externalities. Essentially the condition is that sending a message triggers a set number of incoming messages.

40. In a companion paper, Hermalin and Katz (2001), we further develop the retail pricing analysis presented in this section. DeGraba (2000) investigates optimal inter-carrier prices in a multi-network model in which one-half of the benefits of each message exchange accrue to the receiver. As we show in the next subsection, first-best message exchange can be attained in this case by setting both the send and receive prices equal to one-half of the marginal cost of message exchange.

41. Rather than being concerned with the welfare effects of pricing, Acton and Vogelsang (1992) are concerned exclusively with deriving the demand for sending messages.

42. This assumption is satisfied, for example, when ω_A is unidimensional and $E\{v_A | \omega_A\}$ is strictly monotonic in ω_A.

43. Squire (1973) reached a similar finding in a model that assumes $E\{v_B | v_A\} \equiv \beta$, where β is independent of the number of messages exchanged.

44. If it is efficient to exchange all messages, that is, $G(v_A, m - v_A) = 0 \ \forall v_A$, then $v^* = \min \{v_A\}$.

45. Let V_A be the support of v_A. If Condition 1 holds, consider a v^* satisfying the condition and define $\hat{v}$ as $\min \{v_A \geq v^* | v_A \in V_A\}$. Because V_A is compact, $\hat{v}$ is defined and is an element of V_A. But observe $\hat{v}$ could replace v^* in Condition 1 and the condition would still hold.

46. In the degenerate case in which all messages are inefficient, efficiency is achieved by any $p + r \geq m$, because $v_A \geq p$ implies $v_B \leq r$.

47. When *A*'s willingness to send a message is not informative about *B*'s message value, the second-best outcome is achieved by setting the receiver's price to the lesser of his expected value of a call and the cost of a call, that is, $r = \min \{E\{v_B\}, m\}$, and the sender's price to cost minus the receiver's price, that is, $p = m - r$.

48. By revealed-preference,

$$G_A(p^w)(E\{v_A|v_A\geq p^w\}-p^w)\geq G_A(p^\pi)(E\{v_A|v_A\geq p^\pi\}-p^\pi),$$

where p^w is the surplus-maximizing price and p^π is the profit-maximizing price. This can be satisfied only if $p^w\leq p^\pi$.
49. It is no longer necessary because there may be values of r that induce some types of receiver to accept calls while other types do not. The receive price is a less blunt instrument than in the case of a single receiver type.
50. Under the assumption that r is set low enough to induce all messages to be accepted, neither a sender nor a receiver bases decisions on inferences about message values drawn from conditioning on the other party's behavior.
51. Nothing in the analysis precludes messages with negative values, although they make little sense in this context for A. For convenience, however, illustrative graphs are drawn for positive-value messages only.
52. Asymmetric cases are explored in Hermalin and Katz (2001).
53. Power-function distributions describe random variables, Z, defined on an interval $[z_1, z_2]$, such that $\Pr\{Z\leq z\}=\left(\dfrac{z-z_1}{z_2-z_1}\right)^c$. If $c\geq 1$, the distribution is everywhere weakly convex.
54. In the case of independently and identically distributed values of message exchange, the latter happens when there exists a v_A such that $G(v_A)<1$ and $(v_A-p^*)G(r^*)<(v_A-\tilde{p})G(\tilde{r})$.
55. We will not worry about who is the sender and who is the receiver when both parties' values exceed $m/2$ and, thus, both are willing to call immediately. We will simply assume the call goes through, that is, there is no 'busy signal', and the cost is split.
56. Hermalin and Katz (2001) show that this differentiability assumption is not essential for the analysis.
57. The double integral corresponds to the area beneath the 45° line in Figure 6.6. That is, we calculated the expected value when A calls first and then double it to include when B calls first.
58. We are assuming here that the distributional assumptions are common knowledge between A and B.
59. This assumption is very likely reasonable for the vast majority of households but may not be appropriate for telecommunications-intensive commercial users.
60. This result can be seen formally by substituting m_A and m_B into our earlier results for both one-way and two-way calling. The two terms always appear as a sum. This result was first identified by Srinagesh and Gong (1996).
61. By the revelation principle, there is no loss of generality in restricting attention to direct revelation mechanisms. For more on the revelation principle see, for example, Fudenberg and Tirole (1991, Chapter 7).
62. Note that the losses would be incremental to any sunk costs initially incurred to join the network.
63. Acton and Vogelsang (1992), Littlechild (1975), and Taylor (1994, Chapter 9) all mention the possible gains from repeated interaction, but do not model it.

REFERENCES

Acton, J.P. and Vogelsang, I. (1992), 'Telephone demand over the Atlantic: Evidence from country-pair data', *Journal of Industrial Economics*, **40** (3), 305–23.
Artle, R. and Averous, C, (1973), 'The telephone system as a public good: static and dynamic aspects', *Bell Journal of Economics and Management Science*, **4** (1), 89–100.

Braeutigam, R.R. (1984), 'Socially optimal pricing with rivalry and economies of scale', *RAND Journal of Economics*, **15** (1), 127–34.

Crémer, J. (2000), 'Network externalities and universal service obligation in the Internet', *European Economic Review*, **44**, 1021–31.

DeGraba, P. (2000), 'Bill and keep interconnection for competing networks', unpublished manuscript, US Federal Communications Commission.

Diamond, P. (1973), 'Consumption externalities and imperfect corrective pricing', *Bell Journal of Economics and Management Science*, **4** (2), 526–38.

Fudenberg, D. and Tirole, J. (1991), *Game Theory*, Cambridge, MA: MIT Press.

Hahn, J-H. (2000), 'Nonlinear pricing of a telecommunications service with call and network externalities', Department of Economics Working Paper, Keele University, Keele, Staffordshire, UK.

Hermalin, B.E. and Katz, M.L. (1993), 'Judicial modification of contracts between sophisticated parties: a more complete view of incomplete contracts and their breach', *Journal of Law, Economics, and Organization*, **9**, 230–55.

Hermalin, B.E. and Katz, M.L. (2001), 'Sender or receiver: who should pay to exchange an electronic message?', Working Paper, Walter A. Hass School of Business.

Katz, M.L. and Shapiro, C. (1994), 'Systems competition and network effects', *Journal of Economic Perspectives*, **8** (2), 93–115.

Laffont, J.J. and Tirole, J. (2000), *Competition in Telecommunications*, Cambridge, MA: MIT Press.

Littlechild, S.C. (1975), 'Two-part tariffs and consumption externalities', *Bell Journal of Economics and Management Science*, **6** (2), 661–71.

Milgrom, P.R. (1981), 'Good news and bad news: representation theorems and applications', *Bell Journal of Economics and Management Science*, **12** (2), 380–91.

Myerson, R. and Satterthwaite, M. (1983), 'Efficient mechanisms for bilateral trading', *Journal of Economic Theory*, **28**, 265–81.

Ng, Y.K and Weisser, M. (1974), 'Optimal pricing with a budget constraint – The case of the two-part tariff', *Review of Economic Studies*, **41** (3), 337–45.

Oi, W.T. (1971), 'A Disneyland dilemma: two-part tariffs for a Mickey Mouse monopoly', *Quarterly Journal of Economics*, **85**, 77–96.

Ramsey, F.P. (1927), 'A contribution to the theory of taxation', *Economic Journal*, **37**, 47–61.

Rohlfs, J. (1974), 'A theory of interdependent demand for a communications service', *Bell Journal of Economics and Management Science*, **5** (1), 16–37.

Rubinstein, A. (1982), 'Perfect equilibrium in a bargaining model', *Econometrica*, **50**, 97–110.

Spence, A.M. (1975), 'Monopoly, quality and regulation', *Bell Journal of Economics and Management Science*, **6** (2), 417–29.

Squire, L. (1973), 'Some aspects of optimal pricing for telecommunications', *Bell Journal of Economics and Management Science*, **4** (2), 515–25.

Srinagesh, P. and Gong, J. (1996), 'A model of telecommunications demand with call externality', unpublished manuscript, Bellcore.

Taylor, L.D. (1994), *Telecommunications Demand in Theory and Practice*, Dordrecht: Kluwer Academic Publishers.

Willig, R.D. (1979), 'The theory of network access pricing', in H.M. Trebing (ed.), *Issues in Public Utility Regulation*, East Lansing: Michigan State University.

Wilson, R.B. (1993), *Nonlinear Pricing*, Oxford: Oxford University Press.

7. Vertical integration in telecommunications

Dennis L. Weisman

INTRODUCTION

The two dominant trends in the telecommunications industry with respect to market structure are consolidation and vertical integration. These developments harbor complex public policy and regulatory issues that simultaneously renew old debates and provoke new ones. This chapter focuses on the strategic motivation for vertical integration in telecommunications markets and the corresponding regulatory and public policy issues. The unique history of the telecommunications industry in the United States (USA) provides a useful setting in which to frame these questions. The central issues under analysis, however, have broader implications that shed light on recent industry trends in Europe and throughout the world (Fraquelli and Vannoni, 2000).

In the USA, SBC led the consolidation trend among the Regional Bell Operating Companies (RBOCs), having moved first to acquire Pacific Telesis, then Southern New England Telephone, and finally Ameritech.[1] The merger between Bell Atlantic and NYNEX placed control of local exchange telecommunications along the northeastern seaboard in the hands of a single entity. The trend continues with Bell Atlantic and GTE[2] planning to merge pending approval by state and federal regulators.[3]

The two largest long-distance carriers, AT&T and MCI, led the vertical integration trend in the industry. AT&T has acquired Teleport Communications and is merging with cable giant TCI and Media-One.[4] With this latest acquisition, AT&T becomes the largest cable television provider in the USA. In 1998, regulators approved a merger between MCI and WorldCom.[5] A merger between MCI/WorldCom and Sprint is pending. These mergers provide AT&T and MCI with the potential to connect with their customers without relying on the local exchange facilities of the RBOCs. To date, Bell South and, until quite recently, US West have stayed clear of the 'merger frenzy'. US West agreed to a merger with Qwest Communications in the summer of 1999 in a negotiated settlement that also resulted in a partnership between Global Crossings and Frontier Corporation.[6] The 'musical chairs' of industry consolidation and vertical integration shows no sign of abating any time soon, nor is it confined to the USA.[7]

It is noteworthy that policymakers in the USA appear to have been caught off guard by these emerging industry trends, despite early predictions of such industry restructuring.

> The AT&T divestiture may eventually be seen as a type of experiment in industry genetics, the long run effect was to clone multiple copies of the former Bell System. Divestiture brought to an end AT&T's control over end-to-end connectivity in order to eliminate its ability to discriminate against its rivals. Some ten years after divestiture, the industry trend is one of vertically-integrated supply, as consortia are forming once again to provide for end-to-end connectivity. The Bell System is no longer, but its progeny lives on, albeit more combative, market oriented, and financially focused than its ancestry.
>
> In this sense, we have come full circle in the telecommunications industry, from end-to-end connectivity to partitioned markets, back to end-to-end connectivity. In light of this progression, the age-old question of whether a market is a natural monopoly or whether it is competitive no longer seems relevant. We have seemingly arrived at a parallel universe in which there is not one Bell System, but several. Questions of terms and timing remain, but these are merely the rules of engagement. A galactic battle of the titans seems inevitable. (Sappington and Weisman, 1996, pp. 68–9).

In contrast to the USA, most European countries chose to retain a vertically-integrated structure for the monopoly incumbent at the outset of liberalization.[8] None the less, Europe is in the midst of a major transition on two fronts. The first is the transition from public ownership to private ownership; and the second is the transition from monopoly to competition. Britain has been the leader in terms of liberalization,[9] followed by Spain, France, Germany and Italy (Waverman and Sirel, 1997). While aggressive competition policy (that is, network unbundling and interconnection) generally lags behind that in the USA, the European Commission has been the agent of change in mandating a strict timetable for introducing competition into traditionally closed markets.

Supply-side Considerations

The AT&T divestiture was nothing if not a conscious decision to superimpose an artificial market structure on the US telecommunications industry in the long-run interest of fostering competition. Writing on the fifth anniversary of the break-up of the Bell System, William Baxter, former Assistant Attorney-General for Antitrust and the architect of the divestiture decree, reflected on the rationale for the AT&T divestiture.

The decree implicitly made a wager that the regulatory distortions of those portions of the economy which could have been workably competitive, yielded social losses in excess of the magnitude of economies of scope that would be sacrificed by this approach. It was a wager, a guess. It would be absurd to pretend it was made on the basis of detailed econometric data. It was not; we did not have the data. Of course, all other courses from that point were also guesses. Clear proof was not about to become available any time soon. It was a judgment call, and I guess, in some senses, I do not yet know. Maybe we will never know whether it was right or wrong. (Baxter, 1991, p. 30).

In light of Baxter's reflections, it is tempting to conclude that the trend toward vertical integration reflects supply-side considerations – opportunities to recapture economies of scope that were sacrificed at divestiture. This is a plausible hypothesis, but the empirical support for economies of scope in telecommunications is decidedly mixed (Charnes, Cooper and Sueyoshi, 1988; Evans and Heckman, 1988; and Nadiri and Nandi, 1997). This does not necessarily imply that such economies do not exist, merely that their consistent measurement has proven elusive.

Spiller and Cardilli (1997, p. 137) make the following observations with regard to competitive entry in Chile. 'Chile's absence of unbundling requirements, combined with a strong tendency by players towards vertical integration (which suggests strong economics of scope at work), has helped unleash a remarkable level of competition in the provision of local services'.

Demand-side Considerations

It is not possible to conclude on the basis of the existing economic literature that supply-side considerations alone are responsible for the trend toward vertical integration in telecommunications markets. We are thus forced to look elsewhere for a more complete explanation.

Demand-side considerations can also provide a compelling rationale for the vertical integration trend in telecommunications markets (Ware, 1998). The RBOCs are allowed to petition for entry into the interLATA long-distance market under Section 271 of the 1996 Telecommunications Act.[10] At the time of writing, only one RBOC in one state, Bell Atlantic (New York), has been granted interLATA entry. The prospect that more of these petitions will be granted in the near future is of no small concern to the long-distance carriers, however. The pent-up demand for one-stop shopping for local telephone service, long-distance telephone service, Internet access, high-speed data transmission, wireless services and video entertainment position the RBOCs to capture a large share of the long-distance marketplace once the interLATA restriction falls.

The ability of the RBOCs to compete with AT&T and MCI requires that they be able to provide similar one-stop shopping offerings. A physical presence with customers that, for the long-distance carriers, can be obtained only through vertical integration, may be seen as essential to competing with the RBOCs once the interLATA restriction falls. In a similar vein, we should expect the RBOCs to seek out opportunities to align themselves with wholesale network capacity providers in order to lessen any ongoing dependence on their rivals (the long-distance carriers) for out-of-region network capacity.[11] In both cases, the motivation for vertical integration is the same: provide one-stop-shopping for telecommunications services while minimizing dependence on rivals for necessary or essential inputs to production.

In New Zealand and Australia vertically-integrated providers of both local and long-distance telephone services are beginning to make inroads and put downward pressure on prices (Spiller and Cardilli, 1997, p. 134). In Chile, almost all long-distance companies have adopted a strategy of becoming integrated providers, but focusing on a particular set of customers rather than services. Notably, there are no unbundling or resale requirements in Chile and competition is pronounced with at least six competing facilities-based local exchange competitors in and around Santiago (Spiller and Cardilli, 1997, p. 135).

Public Policy Issues

It is noteworthy that policymakers in the USA have reacted in disparate ways toward the two dominant emerging market trends in the telecommunications industry. The trend toward vertical integration on the part of the long-distance carriers (for example, AT&T and MCI) has been viewed positively, primarily because such vertical integration harbors the potential for increasing competition in local telephone markets – a paramount objective of the 1996 Telecommunications Act.[12] The second vertical integration trend – RBOC entry into interLATA long distance markets – is far more controversial. The reasons for this controversy are primarily twofold. First, the long-distance carriers claim that the long-distance marketplace is already competitive and that RBOC entry would not likely result in lower prices. Second, RBOC entry into interLATA markets could result in higher prices if the RBOCs leverage their local exchange monopolies and engage in discrimination (sabotage) against their long-distance rivals.

INCENTIVES FOR EXCLUSIONARY BEHAVIOR

It is common in the telecommunications industry for an upstream monopolist to supply an input essential to the production of the downstream service. This essential input is commonly referred to as access. For example, long-distance carriers rely upon the access services supplied by local exchange telephone companies for the origination and termination of long-distance messages. In the USA, these long-distance carriers, along with other prospective entrants in the market for local telephone service, also depend on the incumbent local telephone companies for unbundled network elements – the inputs required to assemble competing local telephone service offerings (Harris and Kraft, 1997; Sidak and Spulber 1997a).

The so-called access-pricing problem is a principal focus of the recent regulatory economics literature (Baumol and Sidak, 1994a, 1994b; Kahn and Taylor, 1994; Laffont and Tirole, 1994, 1996; Armstrong et al., 1996; Baumol et al., 1997; Laffont et al., 1998a, 1998b; Lewis and Sappington, 1999). An issue closely related to the access-pricing problem concerns the prospect that a vertically-integrated provider (VIP) with a monopoly in the upstream access market will engage in exclusionary behavior to impede competition in the downstream market (Bernheim and Willig, 1994). The two types of exclusionary behavior that seemingly pose the greatest concern are a price squeeze, and discrimination or sabotage.

In the case of a price squeeze, the VIP essentially charges downstream competitors a higher price for the input than it charges (imputes) to itself. This practice squeezes the profit out of the downstream market for competitors and may force them to exit the market (Perry, 1989). Discrimination occurs when the VIP intentionally degrades the quality of the access service provided to downstream competitors with the effect of raising rivals' costs (Krattenmaker and Salop, 1986; Salop and Scheffman, 1983).

Overview of Theoretical Literature – Cournot Competition

Vickers (1995) examines whether the upstream monopoly supplier of access should be allowed to enter the downstream market.[13] Allowing such entry allows for the realization of scope economies, but also creates the incentive to raise rivals' costs. He finds that when the number of downstream firms is sensitive to the price of access, vertical integration allows a given total output to be produced at lower average cost because fixed costs of production are incurred fewer times. This advantage may offset the incentive to raise rivals' costs. Conversely, if the number of firms is not sensitive to the level of the access price, then the optimal access price is optimally set below marginal cost in order to offset the imperfect competition downstream.[14] Vickers indicates that in this latter case it may be doubly undesirable to allow the monopolist into the deregulated sector.

Economides (1998) develops a formal model of non-price discrimination (sabotage) by an input monopolist. One of the primary conclusions of this research is that the incentive to discriminate exists regardless of the relative (in)efficiency of the VIP and its downstream rivals.[15] Weisman and Kang (2000) and Mandy (2000) show that, at least as a theoretical matter, this claim is incorrect. The VIP's incentive to discriminate against independent downstream rivals is not unequivocal, even in the absence of explicit costs of discrimination (for example, financial penalties, market expulsion, and so on). The economic rationale for this finding is straightforward. The experience of the independent rivals in the downstream market may confer an efficiency advantage (learning by doing) that the VIP (and access supplier) may not wish to distort. In other words, the more efficient the independent rival the larger its downstream output and hence the greater its demand for access services, *ceteris paribus*.

A necessary condition for the incentive to discriminate not to arise in equilibrium is that the independent rivals be more efficient than the VIP.[16] In general, the incentive to engage in discrimination is decreasing in the level of the access charge. Moreover, the incentive to discriminate always exists when access charges are set equal to marginal cost (Weisman and Kang, 2000; Weisman, 2000).

In a related context, Weisman and Williams (2000) explore a somewhat different question. They assume that the VIP has both the incentive and the ability to engage in discrimination against downstream rivals and measure the level of discrimination that must take place in order to just offset the consumers' surplus gains from additional entry. They conclude that the VIP would have to raise its rivals' marginal costs by 50 per cent or more in order to neutralize the expected consumers' surplus gains from RBOC entry into the interLATA long-distance market. They question whether discrimination levels of this magnitude could plausibly escape detection by the regulatory authority.

Overview of Theoretical Literature – Bertrand Competition

In a Bertrand framework, the incentive to discriminate turns on the degree of substitutability between competing downstream service offerings (Beard et al., 1999). In general, the incentive to discriminate does not arise in equilibrium when the downstream products are sufficiently differentiated. As in the Cournot case, the incentive to discriminate always exists when access charges are set equal to marginal cost.

The VIP optimally balances profit opportunities in both upstream (access) and downstream markets. Discrimination leads to decreased demand for access, and an increase in the independent rivals' price, which causes the demand for the VIP's downstream service to increase through the cross-price elasticity. For a cross-elasticity sufficiently large (respectively, sufficiently small) the gain in downstream profit resulting from discrimination dominates (respectively, is dominated by) the loss in access profits (for access prices set above marginal cost). As a result, the incentive to discriminate does (respectively, does not) arise in equilibrium (Weisman 2000). Moreover, the incentive to discriminate does not arise in equilibrium when the VIP's incentive is to reduce downstream prices (Weisman 1995, 1998).[17]

Summary and Assessment

The incentive to discriminate turns on the relative efficiency of the downstream firms, the heterogeneity of the product offerings, the level of the access charge and the explicit costs of discrimination. If we abstract from this last consideration, which by its very nature harbors profound uncertainty, and recognize that regulators are moving rapidly toward cost-based pricing of access, the incentive for VIP discrimination appears to be more likely than not.

COURNOT ANALYSIS

The principal findings of this analysis are twofold. First, abstracting from the explicit costs of discrimination, when the VIP is no less efficient than its independent rivals, the incentive to discriminate always exists in equilibrium. Second, the efficiency superiority of the independent rivals must be pronounced in order for the VIP's incentive to discriminate not to arise in equilibrium.

Basic Model

There is assumed to be a single VIP of both the essential upstream access service and the downstream service. There are also assumed to be $n-1$ identical downstream independent rivals of the VIP, where the integer $n \geq 2$. Production is of the fixed-coefficient type, requiring one unit each of access and a complementary input that may be self-supplied by one or more independent rivals. The regulated price and marginal cost of access are w and c, respectively. The cost of the complementary input is s^i, where $i = V$ or I indicates the VIP or independent rival, respectively.

The inverse market demand function is given by $P(Q)$, where $Q = q^V + (n-1)q^I$ is market demand, q^V and q^I represent the output of the VIP and the independent rival, respectively, and $P'(Q)$ is assumed to be bounded. The profit function for the VIP is given by:

$$\Pi^V = (w-c)(n-1)q^I + [P(Q) - c - s^V]q^V. \tag{7.1}$$

The profit function for the representative rival is given by:

$$\Pi^I = [P(Q) - w - s^I]q^I \tag{7.2}$$

The Nash equilibrium is implicitly defined by the solution to the following first-order conditions:

$$q^V: q^V P'(Q) + P(Q) - c - s^V = 0; \text{ and} \tag{7.3}$$

$$q^I: q^I P'(Q) + P(Q) - w - s^I = 0. \tag{7.4}$$

Primary Findings

(i) General analysis

To determine whether the VIP has incentives to discriminate, we examine whether raising rivals' costs ($\Delta s^I > 0$) leads to increased profits for the VIP. The first finding identifies a sufficient condition for the incentive to discriminate to arise in equilibrium.

Finding 1. The VIP has an incentive to discriminate in equilibrium when it is no less efficient than its independent rivals ($s^V \le s^I$).

Proof: Partially differentiating (7.1) with respect to s^I, rearranging terms, and appealing to the envelope theorem yields:

$$\frac{\partial \Pi^V}{\partial s^I} - (n-1)\frac{\partial q^I}{\partial s^I}[w - c + q^V P'(Q)]. \tag{7.5}$$

Rearranging terms in (7.3) yields

$$q^V P'(Q) = c + s^V - P(Q) \tag{7.6}$$

Substituting (7.6) into (7.5) and simplifying yields

$$\frac{\partial \Pi^V}{\partial s^I} = (n-1)\frac{\partial q^I}{\partial s^I}[w + s^V - P(Q)], \tag{7.7}$$

which is positive since $P > w + s^V \ \forall s^V \le s^I$ and $\dfrac{\partial q^I}{\partial s^I} < 0$ (Dixit, 1986, p. 120).

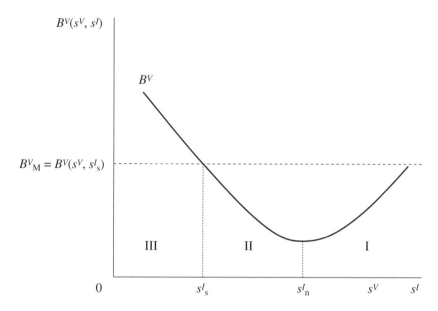

Figure 7.1   VIP's profit function

Two results follow immediately from Finding 1. First, when the VIP is no less efficient than its downstream rivals ($s^V \leq s^I$), $P > w + s^V$ which implies that the P-F constraint does not bind in equilibrium. Second, the contrapositive of Finding 1 implies that a necessary condition for the VIP not to discriminate in equilibrium is that the independent rivals be more efficient than the VIP ($s^V > s^I$).

The economic intuition for these results can be seen clearly in Figure 7.1. Holding s^V fixed for given values of $w > c$ and $n \geq 2$, the VIP's profit function is parabolic, taking on a minimum at s_n^I. It is immediately clear from (7.7) that for $s^I < s_n^I$, $P < w + s^V$, the imputation constraint is not satisfied, and the profit function is decreasing in s^I (discrimination). Conversely, for $s^I > s_n^I$, $P > w + s^V$, the imputation constraint is satisfied, and the profit function is increasing in s^I (discrimination). Precisely (and uniquely) at s_n^I, $P = w + s^V$, which implies that $P - c - s^V = w - c$, or the VIP's margin in the downstream market is equal to its margin in the upstream access market. At this point, the VIP's profit function is non-increasing in this first (incremental) unit of discrimination.

The convexity of the VIP's profit function, however, implies that while $\dfrac{\partial \Pi^V}{\partial s^I} > 0$ is sufficient to establish incentives for discrimination, the converse is not true. Figure 7.1 reveals that unless the regulator has the ability to detect discrimination perfectly, the VIP can increase profits by moving beyond this first (incremental) unit of discrimination.

The previous discussion reveals that discrimination may not arise in equilibrium when the efficiency superiority of the independent rivals is sufficiently pronounced. This result is intuitive and suggests that the experience of the independent rivals in the downstream market may confer an efficiency advantage (learning by doing) that the VIP (and access supplier) will not wish to distort. In other words, the more efficient the independent rival the larger its output and hence the greater the demand for access services, *ceteris paribus*. For the 'sufficiently inept' VIP, the efficient independent rival represents the proverbial goose that lays the golden egg, where the 'golden egg' refers to access volumes.

Table 7.1   Necessary and sufficient conditions for (non-)discrimination outcomes

Region	Condition on s_0^I	Determination
I	$s_0^I > s_n^I$	Sufficient for discrimination
II	$s_0^I \leq s_n^I$	Necessary for non-discrimination
III	$s_0^I \geq s_s^I$	Sufficient for non-discrimination

Table 7.1 reflects the three regions identified in Figure 7.1, where the default (undistorted) values of s^I are denoted by s_0^I. The VIP has an incentive to discriminate in Region I unless the regulator's ability to monitor discrimination is perfect. The VIP has no incentive to discriminate in Region III. In Region II, the VIP's incentive to discriminate turns on the degree of imperfection of the regulator's ability to monitor discrimination and the associated financial penalties, should discrimination be detected.

The analysis reveals that the rivals must be more efficient than the VIP in order for the incentive to discriminate not to arise in equilibrium. In addition, we know from (7.3) and (7.4) that if the rival's efficiency superiority is sufficiently pronounced, the VIP will not participate downstream in equilibrium. The outstanding question thus concerns whether there exists an efficiency differential sufficiently large such that the VIP participates in the downstream market; and the VIP has no incentive to discriminate against the downstream rivals in equilibrium (when there are no explicit costs of discrimination). We explore this question for the special case of linear market demand.

(ii)   Linear analysis

Finding 2.   When market demand is linear of the form $P(Q) = A - BQ$, a necessary condition for non-discrimination is that

$$ns^V - (n-1)s^I \geq [A + c - 2w]. \tag{7.8}$$

Proof: Straightforward computations based on (7.3) and (7.4) reveal that when demand is linear the downstream outputs of the VIP and the independent rivals are given respectively by

$$q^V = \frac{1}{(n+1)B}[A - n(c+s^V) + (n-1)(w+s^I)], \text{ and} \tag{7.9}$$

$$\varepsilon^I = \frac{1}{(n+1)B}[A - 2(w+s^I) + (c+s^V)]. \tag{7.10}$$

Substitute (7.9) and (7.10) into the expression for inverse market demand, $P(Q) = A - BQ$, to obtain

$$P(Q) = \frac{1}{(n+1)}[A + (n-1)(w+s^I) + (c+s^V)], \tag{7.11}$$

Recall that $P(Q) \leq w + s^V$ is a necessary condition for non-discrimination from Finding 1. Imposing this necessary condition on (7.11) and simplifying yields the result in (7.8). QED.

Example 1.   It is instructive to compute the efficiency differential, $s^V - s^I$, necessary for a non-discrimination outcome in the case of interLATA long-distance service in the USA. Based on current volumes and prices, the linear market demand function for the interLATA long-distance segment is given by $P(Q) = 26.71 - 0.0457Q$ (Weisman and Williams 2000). The consensus value for s^I is approximately one cent per minute. The current Herfindahl index for the interLATA market is 0.2640 (FCC 1999, Table 1.4), which corresponds to $n = 3.8$. Assume that allowing the RBOCs to serve this market would increase the equivalent number of equally-sized firms by 1 to $n = 4.8$. The current values for the price and marginal cost of access per minute are $w = 2.85$ cents and $c = 1$ cent, respectively. Substituting these values into (7.8) and solving for s^V yields $s^V \geq 5.38$. Hence, a necessary condition for the VIP not to discriminate in equilibrium is that $s^V \geq 5.38s^I$, or the VIP's complementary input costs are some 400 per cent higher than those of its downstream rivals. Finally, note that if the access charge is raised to $w = 5$, the critical value for s^V falls to $s^V \geq 4.48s^I$.

This gives rise to an interesting paradox: reducing access charges, to the extent that it exacerbates the VIP's incentive for discrimination, can actually serve to raise input costs for independent rivals.

Thus far in the analysis, we have treated the efficiency differential, $s^V - s^I$, as a parameter that measures innate differences in productive efficiency between the independent rivals and the VIP. An alternative interpretation is that this parameter reflects differing degrees of market power on the part of complementary input suppliers. Such differences in market power may arise because an independent rival self-supplies the complementary input or because it enjoys 'most favored nation' status with the complementary input supplier.[18]

Define the critical efficiency differential as the value of $s^V - s^I$ that solves (7.8) as an equality. The following result shows how this critical efficiency differential varies with n and w.

Finding 3.   When inverse market demand is linear of the form $P(Q) = A - BQ$, the critical efficiency differential, $s^V - s^I$, is decreasing in n and w.

Proof: A sufficient condition for (7.8) to be satisfied is that

$$(s^V - s^I) \geq \frac{[A + c - 2w]}{n}, \text{ since} \qquad (7.12)$$

$$ns^V - (n-1)s^I \geq n(s^V - s^I). \qquad (7.13)$$

The result follows from the fact that the expression on the right-hand side of the inequality in (7.12) is decreasing in n and w. QED.

It follows that the larger the number of competitors in the downstream market, the smaller the efficiency differential required to satisfy the necessary condition for non-discrimination. As the number of downstream competitors increases, the VIP's market share decreases, *ceteris paribus*. This reduced participation in the downstream market causes the VIP to place greater value on the independent rivals' relative efficiency in generating access volumes.

In the next finding, we derive sufficient conditions for non-discrimination in the case of linear demand that depend on the VIP's profits with independent rivals participating in the downstream market being no lower than the VIP's (monopoly) profit absent such participation.

Finding 4.   Let $x = c + s^V$ and $y = w + s^I$. Then for n sufficiently large and linear market demand, $P(Q) = A - BQ$, the VIP has no incentive to discriminate in equilibrium when

$$0 < y - x \leq \left[\frac{A - x}{2}\right]\left[\frac{4(w - c) - (A - x)}{4(w - c)}\right] \text{ and } w - c > \frac{(A - x)}{4}. \quad (7.14)$$

Proof: From (7.9), (7.10) and (7.11), it is straightforward to show that the profit function for the VIP is given by

$$\Pi^V(s^V, s^I) = \frac{n-1}{B(n+1)}[[A - 2y + x][w - c]] + \frac{1}{B}\left[\frac{A + (n-1)y - nx}{(n+1)}\right]^2. \quad (7.15)$$

When the VIP is a monopolist downstream, $n = 1$, and (7.15) reduces to

$$\Pi^V_M = \frac{1}{B}\left[\frac{A - x}{2}\right]^2. \quad (7.16)$$

The second term to the right of the equals sign in (7.15) is nonnegative. Hence, a sufficient condition for $\Pi^V(s^V, s^I) \geq \Pi^V_M$ is that

$$\frac{n-1}{B(n+1)}[[A - 2y + x][w - c]] \geq \frac{1}{B}\left[\frac{A - x}{2}\right]^2. \quad (7.17)$$

As n grows large, the term $\frac{n-1}{n+1} \to 1$ in (7.17). Simplifying and solving for y yields

$$y \leq \left[\frac{A + x}{2}\right] - \left[\frac{1}{2(w - c)}\right]\left[\frac{A - x}{2}\right]^2. \quad (7.18)$$

To rule out degenerate cases, it is necessary to ensure that the VIP participates in the downstream market, or $q^V > 0$. From (7.9), this requires that $A + (n-1)y - nx > 0$. Solving for y yields

$$y > \frac{nx - A}{n - 1}. \quad (7.19)$$

Letting n grow large in (7.19) implies that $y > x$, which, when appended to (7.18), yields

$$x < y \leq \left[\frac{A + x}{2}\right] - \left[\frac{1}{2(w - c)}\right]\left[\frac{A - x}{2}\right]^2. \quad (7.20)$$

Subtracting x from each part of the chain of inequalities in (7.20) and simplifying yields (7.14). The set $\{x, y\}$ that satisfies the leftmost term in (7.14) is non-empty when the rightmost term in (7.14) is strictly positive. This requires that $4(w - c) > (A - x)$, or $(w - c) > (A - x)/4$,[19] which is the condition in (7.14). QED.

The bounds on the $y-x$ term in Finding 4 have an intuitively appealing economic interpretation. The VIP must be sufficiently inept to foster a dependence on the independent rivals (for generating access volumes), but not so inept that it voluntarily retreats from the downstream market altogether. This result confirms that there exist conditions under which discrimination is inconsistent with profit-maximizing behavior on the part of the VIP, even when there are no explicit costs of discrimination. The following is an example.

Example 2. Let $A = 18$, $B = 2$, $n = 10$, $w = 5$, $c = 1$, $s^V = 7$ and $s^I = 3.5$. For this set of parameters, $x = 8$ and $y = 8.5$. It is straightforward to show using (7.9), (7.10), (7.11), (7.15) and (7.16) that $q^V = 0.66$, $q^I = 0.41$, $P = 9.32$, and $\Pi^V(s^V, s^I) = 15.6 > 12.5 = \Pi^V_M$. Hence, the VIP participates in the downstream market and yet has no incentive to discriminate in equilibrium.

Summary

This analysis explores the relationship between upstream access pricing and downstream exclusionary behavior in a Cournot framework. When incumbency confers no efficiency advantage ($s^I \geq s^V$), the VIP has unambiguous incentives to discriminate against its rivals in the downstream market. Yet, when the independent rivals enjoy a pronounced efficiency advantage, the incentive to discriminate need not arise in equilibrium. The likely source of such a cost advantage is a combination of market experience, weak vertical-integration economies, and market power for the complementary input.

These findings present policymakers with something of a dilemma: the incentives to discriminate are most pronounced precisely when the efficiency gains from vertically-integrated supply are likely to be quite large. Prohibiting the VIP from serving the downstream market solves the discrimination problem but runs the risk of foreclosing an efficient provider. Permitting the VIP to serve the downstream market may lead to gains in productive efficiency but risks allocative efficiency losses from discrimination and exclusion.

BERTRAND ANALYSIS

The principal findings of this analysis are twofold. First, the incentive for exclusionary behavior almost always arises in equilibrium under a parametric (constant) access price. An access price that is 'too high' can lead to a price squeeze and one that is 'too low' creates incentives for discrimination. Second, the access pricing rule that simultaneously eliminates the incentive for both forms of exclusionary behavior in equilibrium is unique and equal to the familiar efficient component pricing rule (ECPR).

Basic Model

There is a single VIP that serves as a monopolist in the upstream access market and a single independent downstream provider. The downstream demand functions for the VIP and the independent provider are given by $Q^V(P^V, P^I)$ and $Q^I(P^I, P^V)$, where P^i, $i = V$ and I denotes the respective downstream prices for the VIP and the independent rival. The downstream outputs of the VIP and the independent downstream provider are imperfect substitutes so that $Q^i_{pj} > 0$ for $i, j = V$ and I, $i \neq j$, where the subscripts denote partial derivatives. There are assumed to be no income effects.

As in the Cournot analysis, the regulated price of access is denoted by w and the corresponding constant marginal cost of access is denoted by c. Production is of the fixed-coefficient type: each unit of the downstream output requires one unit of access and one unit of a complementary input. The cost of each unit of the complementary input is denoted by s^i, $i = V$ and I. Let $d \geq 0$ denote the increment by which the VIP raises the per unit costs of its downstream rival through discrimination. Finally, we assume that there are no explicit costs of discrimination for the VIP.

The profit functions for the VIP and the independent rival are assumed to be concave in their own prices and are given respectively by:

$$\Pi^V = Q^I(P^I, P^V)[w - c] + Q^V(P^V, P^I)[P^V - c - s^V], \text{ and} \qquad (7.21)$$

$$\Pi^I = Q^I(P^I, P^V)[P^I - w - s^I - d]. \qquad (7.22)$$

The VIP and the independent rival are assumed to set their downstream prices simultaneously and independently. For a parametric access price, w, the Nash equilibrium is characterized by the solution to the following first-order conditions:

$$P^V: \frac{\partial Q^I}{\partial P^V}[w - c] + \frac{\partial Q^V}{\partial P^V}[P^V - c - s^V] + Q^V = 0. \qquad (7.23)$$

$$P^I: \frac{\partial Q^I}{\partial P^I}[P^I - w - s^I - d] + Q^I = 0. \qquad (7.24)$$

Definition 1.   (Displacement Ratio)

$$\sigma = -\left(\frac{\partial Q^I}{\partial P^I} \bigg/ \frac{\partial Q^V}{\partial P^I}\right).[20]$$

The displacement ratio (σ) is simply the absolute value of the change in the output of the independent rival associated with a one-unit increase in the output of the VIP.

The VIP is generally required by the regulator to satisfy a price floor (P-F) or imputation constraint (Hausman and Tardiff, 1995). The P-F constraint requires that the downstream price for the VIP be no lower than the direct cost of providing downstream output plus the net contribution forgone (opportunity cost) in not providing upstream access.

Definition 2. *(P-F Constraint)*

$$P^V \geq c + s^V + \sigma[w-c] \Leftrightarrow w \leq c + \sigma^{-1}[P^V - c - s^V]. \qquad (7.25)$$

The P-F constraint recognizes that each unit of the VIP's downstream output displaces σ units of the rival's downstream output that, in turn, causes the VIP to forgo the sale of σ units of upstream access. The latter follows directly from the assumed nature of the production technology.

The price squeeze is but one form of exclusionary behavior that is of potential concern to the regulator, however. The other is non-price discrimination or sabotage. We assume that the environment is one in which the access-pricing rule is the only instrument available to the regulator to control exclusionary behavior. Hence, the primary research question to be addressed here concerns the existence of an access-pricing rule that ensures the VIP practices neither form of exclusionary behavior in equilibrium.

Primary Findings

We begin with a formal definition of what it means for the VIP not to have an incentive to discriminate in equilibrium.

Definition 3. *(Non-Discrimination Condition)*
A necessary condition for the VIP not to have an incentive to discriminate in the equilibrium defined by (7.23) and (7.24) is that

$$\frac{\partial \Pi^V}{\partial d} \leq 0.[21]$$

Totally differentiating the VIP's profit function (7.21) with respect to d, and appealing to the envelope theorem yields

$$\frac{\partial \Pi^V}{\partial d} = \left[\frac{\partial Q^I}{\partial P^I}(w-c) + \frac{\partial Q^V}{\partial P^I}(P^V - c - s^V) \right] \frac{\partial P^I}{\partial d}. \qquad (7.26)$$

The sign of the expression in (7.26) is equal to the sign of the expression in the large brackets, since $\dfrac{\partial P^I}{\partial d} > 0$ (Dixit, 1986, p. 120).

To establish a benchmark for the analysis, we begin by investigating the incentives for discrimination under marginal cost access pricing.

Finding 5.   The *non-discrimination condition* is not satisfied when the access price (w) is set equal to the VIP's marginal cost of providing access (c).

Proof: Setting $w = c$ in (7.26) yields

$$\frac{\partial \Pi^V}{\partial d} = \left[\frac{\partial Q^V}{\partial P^I} (P^V - c - s^V) \right] \frac{\partial P^I}{\partial d} > 0, \tag{7.27}$$

since Q^V and Q^I are substitutes and $P^V - c - s^V > 0$ in equilibrium from (7.23).

The economic intuition for Finding 5 is straightforward. Under marginal cost access pricing, there is no opportunity cost of discrimination in terms of forgone upstream profits, but the VIP realizes greater downstream demand due to the higher price charged by its rival. Hence, the VIP's incentive to discriminate is unequivocal.

Finding 6.   The *non-discrimination condition* is satisfied when $w \geq c + \sigma^{-1}[P^V - c - s^V]$ in the equilibrium defined by (7.23) and (7.24).

Proof: Set the expression in large brackets in (7.26) to be non-positive. Simplifying and solving for w yields

$$w \geq c - \frac{\partial Q^V}{\partial P^I} \left[\frac{\partial Q^I}{\partial P^I} \right]^{-1} [P^V - c - s^V] = c + \sigma^{-1}[P^V - c - s^V]. \tag{7.28}$$

Finding 6 further implies that for any $w > c$, the non-discrimination condition is satisfied when the downstream outputs are highly imperfect substitutes (that is, σ^{-1} is 'sufficiently small'). Discrimination induces the independent rival to set a higher downstream price. This higher downstream price has two effects on the VIP's profits: decreased demand for the VIP's upstream access, and increased demand for the VIP's downstream output. With σ^{-1} 'sufficiently small' and $w > c$, the first effect dominates the second, which explains why the non-discrimination condition is satisfied in equilibrium.

Finding 7.   The parametric access price that simultaneously satisfies the non-discrimination condition and the P-F constraint is unique and is given by $w^* = c + \sigma^{-1}[P^V - c - s^V]$.

Proof: The non-discrimination condition requires that $w \geq c + \sigma^{-1}[P^V - c - s^V]$. The P-F constraint requires that $w \leq c + \sigma^{-1}[P^V - c - s^V]$. The unique value of w that satisfies the above requirements is given by $w^* = c + \sigma^{-1}[P^V - c - s^V]$.

Finding 7 suggests that unless the parametric access price, w, is such that it equates the margin in the upstream market with the (effective) opportunity cost of supplying access in the downstream market, one of the two forms of exclusionary behavior will arise in equilibrium. This finding suggests that in general the incentives for exclusionary behavior can be eliminated only by linking the VIP's upstream price (w) and downstream price (P^V) so as to maintain parity between the upstream margin and the (effective) downstream opportunity cost. As the following discussion makes clear, the parity-pricing principle is the defining property of the ECPR (Baumol et al., 1997).

The ECPR requires that the price of access be set equal to the direct incremental cost of providing the upstream access service plus the net contribution forgone (opportunity cost) in not providing the downstream retail service. The traditional conception of the ECPR measures opportunity costs with respect to the pre-entry price of the incumbent provider that is assumed set by the regulatory authority (Baumol and Sidak, 1994a, 1994b; Kahn and Taylor, 1994). Alternatively, opportunity costs can be measured with respect to the prevailing downstream prices. In this form, the ECPR is sometimes referred to as the M-ECPR, where the 'M' indicates a market-determined measure of opportunity cost (Sibley et al., 1998; Sidak and Spulber, 1997a, 1997b).

Definition 4.   (ECPR)
The efficient component pricing rule is given by $w(P^V) = c + \sigma^{-1}[P^V - c - s^V]$.

The ECPR has the property of preserving parity between the upstream margin and the (effective) downstream opportunity cost.

Finding 8.   The unique access-pricing rule that simultaneously satisfies the non-discrimination condition and the P-F constraint is the ECPR.[22]

Proof: The first-order conditions for profit-maximization under this non-parametric access pricing rule, $w(P^V)$, are given by:

$$P^V: \frac{\partial Q^I}{\partial P^V}[w - c] + \frac{\partial w}{\partial P^V}Q^I + \frac{\partial Q^V}{\partial P^V}[P^V - c - s^V] + Q^V = 0. \qquad (7.29)$$

$$P^I: \frac{\partial Q^I}{\partial P^I}[P^I - w - s^I - d] - \frac{\partial w}{\partial P^I}Q^I + Q^I = 0. \qquad (7.30)$$

Totally differentiating the VIP's profit function with respect to d, we obtain

$$\frac{\partial \Pi^V}{\partial d} = \left[\frac{\partial Q^I}{\partial P^I}[w - c] + \frac{\partial Q^V}{\partial P^I}[P^V - c - s^V]\right]\frac{\partial P^I}{\partial d}$$

$$+ \left[\frac{\partial Q^I}{\partial P^V}[w - c] + \frac{\partial w}{\partial P^V}Q^I + \frac{\partial Q^V}{\partial P^V}[P^V - c - s^V] + Q^V\right]\frac{\partial P^V}{\partial d}. \qquad (7.31)$$

The second term in brackets on the right-hand side of (7.31) is identically zero by the envelope theorem upon appeal to (7.29). Hence, (7.31) reduces to (7.26) and the remainder of the proof follows directly from the proof of Finding 7.

Summary

This analysis explores the relationship between upstream access pricing and downstream exclusionary behavior in a Bertrand framework. We find that the incentive for a price squeeze or discrimination almost always arises in equilibrium for any parametric access price. An access price that is 'too high' can lead to a price squeeze and one that is 'too low' creates incentives for discrimination.

A potentially important finding is that the access-pricing rule that simultaneously eliminates the incentives for both forms of exclusionary behavior is unique and equal to the familiar ECPR. This pricing rule states that the regulator should set the access price equal to the direct cost of providing access (c) plus the opportunity cost incurred in supplying the unit of access ($\sigma^{-1}[PV - c - s^V]$).

CONCLUSIONS

The trend toward vertical integration in telecommunications is both irreversible and multi-faceted, reflecting a host of demand-side and supply-side considerations. Pent-up demand for one-stop shopping and bundling of telecommunications services is driving vertical integration on the demand side; and the opportunity to realize economies of scope is driving vertical integration on the supply side. Both of these considerations must be viewed against the background of an increasingly competitive telecommunications marketplace.

In addition to the pure business motivation for vertical integration in telecommunications, there are also complex questions of strategic behavior and public policy. In the case of vertical integration by an upstream monopolist, regulators are naturally concerned about the possibility of exclusionary behavior. In the USA and other countries, these concerns have served to slow the pace of vertical integration. The analysis herein suggests that, while the incentives for exclusionary behavior are not unequivocal, such concerns on the part of policymakers are clearly warranted.

Increasing competition in both upstream and downstream markets will tend to reduce if not eliminate the opportunities for exclusionary behavior in the longer run. In the interim, a careful understanding of the costs and benefits of vertical integration and the role of efficient access-pricing rules in tempering incentives for exclusionary behavior should serve to inform the design of competition policies that strike the proper balance.

ACKNOWLEDGEMENTS

I am grateful to Jaesung Kang for expert research assistance. The usual caveat applies.

NOTES

1. *Telecommunications Reports*, 8 April 1996, 12 January 1998, and 18 May 1998, respectively.
2. *Telecommunications Reports*, 29 April 1996 and 3 August 1998, respectively.
3. Analysts are at a loss to explain why the RBOCs have to date steadfastly abstained from entering each other's territories. The FCC, however, has conditioned approval of the SBC–Ameritech merger on a strict timetable for out-of-region entry on the part of the consolidated entity. The RBOCs themselves have argued that consolidation will increase the likelihood of so-called out-of-region entry. See Weisman (1999) for an analysis of this claim.
4. *Telecommunications Reports*, 12 January 1998 and 29 June 1998.
5. *Telecommunications Reports*, 6 October 1997.
6. *Telecommunications Reports*, 21 June 1999 and 24 May 1999. See also Blumenstein (1999).
7. For two examples, BC TELECOM and TELUS have merged in Canada (*Telecommunications Reports*, 26 October 1998) and AT&T and British Telecom have embarked on a significant joint venture (*Telecommunications Reports*, 3 August 1998).
8. Waverman and Sirel (1997, p. 123) point out that until quite recently, the Italian telecommunications market had a structure similar to the United States after the break-up of AT&T with separate local and long-distance providers.
9. Following privatization, the government put in place a protected duopoly market structure comprised of British Telecom and Mercury, a subsidiary of Cable and Wireless. Mercury concentrated on providing long-distance service rather than local telephone service with the exception of its larger customers. Moreover, within the City of London,

Mercury has a very extensive local network. Following the review of the restricted duopoly policy, more than sixty new operators have applied for licenses to compete with British Telecom and Mercury. The entrants include British Rail, British Waterways, AT&T and a host of cable television companies (Armstrong, Cowan and Vickers 1994, p. 203).

10. As part of the AT&T divestiture in 1984, the United States was divided into 161 local access transport areas (LATAs). The RBOCs are permitted to provide long-distance telephone service within LATAs but not between LATAs. The 1996 Telecommunications Act allows the RBOCs to petition for entry into the in-region) interLATA market once they have satisfied a 14-point checklist that specifies the manner in which they must open their local networks to competition. See Kahn, Tardiff and Weisman (1999).

11. Examples include SBC's agreement with Williams Pipeline Company to purchase long-distance capacity (*Telecommunications Reports Daily*, 8 February 1999) and the QWEST agreements with BellSouth and US West concerning joint marketing (*Telecommunications Reports*, 12 October 1998).

12. A notable exception to this statement occurred when news reports surfaced that SBC and AT&T were contemplating a merger. Then FCC Chairman Reed Hundt stated that such an arrangement was 'unthinkable.' (*Telecommunications Reports*, 19 June 1997).

13. Riordan (1998) considers the possible anticompetitive effects of backward integration by a dominant firm. By integrating backwards, the dominant firm can increase the price of the scarce input, reduce the size of the competitive fringe and thereby gain more power in the downstream market.

14. See Panzar and Sibley (1989) for a related discussion.

15. In a similar context, Sibley and Weisman (1998b) show that in the presence of capacity constraints, the incentive to discriminate need not arise in equilibrium even when the VIP and the independent rivals are of equal efficiency downstream. Furthermore, the more high-powered the regulatory regime (that is, the smaller the degree of earnings sharing), the less likely the incentive to discriminate will arise in equilibrium. This occurs because high-powered regimes tend to increase the relative financial importance of the upstream (access) market relative to the downstream market, *ceteris paribus* (Sibley and Weisman 1998a).

16. It is noteworthy that the RBOCs claim, in part, that they should be allowed to enter the long-distance marketplace because of their ability to leverage economies of scope. Superior efficiency gives rise to countervailing effects, however. On the one hand, RBOC entry harbors the potential to reduce long-distance prices through superior technical efficiency. On the other hand, if the RBOCs are more efficient than their downstream rivals, they will have an incentive to discriminate. This discrimination may ultimately lead to higher long-distance prices.

17. Reiffen (1998) contends that the RBOCs likely have an incentive to discriminate against their rivals in the long-distance market. The rationale for this conclusion is twofold: first, the downstream products are close substitutes; second, discrimination may actually be a cost-reducing activity for the RBOCs if they supply a lower-quality access service to rivals.

18. For example, in the USA, the 4-plus backbone long-distance networks are all owned by incumbent long-distance providers. The MCI–WorldCom–Sprint merger will reduce the number of independently owned and operated long-distance networks. This consolidation may ultimately lead to increased market power for network capacity. An increase in market power gives rise to countervailing effects. Higher prices for the complementary input tend to raise downstream prices, but also reduce the VIP's incentive to discriminate by raising the effective efficiency differential.

19. This condition states that the margin in the upstream access market ($w–c$) must exceed one-fourth the difference between the maximum willingness to pay for the downstream service (A) and the VIP's marginal cost for the downstream service (x).

20. The total differential of $Q^V(P^V, P^I)$ is given by $dQ^V = \frac{\partial Q^V}{\partial P^V} dP^V + \frac{\partial Q^V}{\partial P^I} dP^I$. Set $dQ^V = 1$ and

$dP^V = 0$, then $dQ^V = \frac{\partial Q^V}{\partial P^I} dP^I = 1$. Solve for dP^I and obtain $dP^I = \left[\frac{\partial Q^V}{\partial P^I}\right]^{-1}$. The absolute

value of the change in Q^I induced by dP^I is given by $dQ^I = -\left(\frac{\partial Q^I}{\partial P^I} \bigg/ \frac{\partial Q^V}{\partial P^I}\right) = \sigma$.

21. Whereas $\frac{\partial \Pi^V}{\partial d} > 0$ is sufficient for discrimination, $\frac{\partial \Pi^V}{\partial d} \leq 0$ is only necessary for non-

discrimination when the access price is parametric. This asymmetry arises because there may be conditions under which 'marginal discrimination' is unprofitable, but 'extra-marginal discrimination' is profitable.

22. Under the ECPR, the access price, $w(P^V)$, serves to maintain parity between the upstream margin and the (effective) downstream opportunity cost in equilibrium so that

$\frac{\partial \Pi^V}{\partial d} = 0, \forall d.$

REFERENCES

Armstrong, M., Cowan, S. and Vickers, J. (1994), *Regulatory Reform: Economic Analysis and British Experience*, Cambridge, MA: MIT Press.

Armstrong, M., Doyle, C. and Vickers, J. (1996), 'The Access Pricing Problem: A Synthesis', *Journal of Industrial Economics*, **44**, 131–50.

Baumol, W. and Sidak, J. (1994a), 'The Pricing of Inputs Sold To Competitors', *Yale Journal on Regulation*, **11**, 171–202.

Baumol, W. and Sidak, J. (1994b), *Toward Competition in Local Telephony*, Washington DC: The AEI Press and Cambridge, MA.: MIT Press.

Baumol, W., Ordover, J. and Willig, R. (1997), 'Parity Pricing and Its Critics: A Necessary Condition for Efficiency in the Provision of Bottleneck Services to Competitors', *Yale Journal on Regulation*, **14**, 145–63.

Baxter, W.F. (1991), 'Questions and Answers with the Three Major Figures of Divestiture', in B.G. Cole (ed.) *After the Break-Up: Assessing the New Post-AT&T Divestiture Era*, New York: Columbia University Press, pp. 21–49.

Beard, T., Kaserman, D. and Mayo, J. (1999), 'Regulation, Vertical Integration, and Sabotage', research monograph.

Bernheim, D.B., and Willig, R.D. (1994), 'Appropriate Preconditions for Removal of the InterLATA Restrictions on the BOCs', Affidavit filed with the United States Department of Justice in support of AT&T's Opposition to Ameritech's Motions for 'Permanent' and 'Temporary' Waivers from the Interexchange Restriction of the Decree, Case No. 82-0192 (D.D.C. Feb. 15).

Blumenstein, R. (1999), 'Quest Wins US West for $35 Billion', *The Wall Street Journal*, A3.

Charnes, A., Cooper, W. and Sueyoshi, T. (1988), 'A Goal Programming/Constrained Regression Review of the Bell System Break-Up', *Management Science*, **34**, 1–26.

Dixit, A. (1986), 'Comparative Statics for Oligopoly', *International Economic Review*, **27** (1), 107–22.

Economides, N. (1998), 'The Incentive for Non-Price Discrimination by an Input Monopolist', *International Journal of Industrial Organization*, **16**, 271–84.

Evans, D. and Heckman, J. (1988), 'Natural Monopoly and the Bell System: A Response To Charnes, Copper, and Sueyoshi', *Management Science*, **34**, 27–38.

Fraquelli, G., and Vannoni, D. (2000), 'Multidimensional Performance in Telecommunications, Regulation and Competition: Analysing the European Major Players', *Information Economics and Policy*, **12** (1), 27–46.

Harris, R., and Kraft, J. (1997), 'Meddling Through: Regulating Local Telephone Competition in the United States', *The Journal of Economics Perspectives*, **11** (4), 93–112.

Hausman, J. and Tardiff, T. (1995), 'Efficient Local Exchange Competition', *The Antitrust Bulletin*, 529–56.

Kahn, A.E, and Taylor, W. (1994), 'The Pricing of Inputs Sold to Competitors: A Comment', *Yale Journal on Regulation*, **11**, 225–40.

Kahn, A.E., Tardiff, T.J. and Weisman, D.L. (1999), 'The 1996 Telecommunications Act at Three Years: An Economic Evaluation of Its Implementation by the FCC', *Information Economics and Policy*, **11** (4), 319–65.

Krattenmaker, T. and Salop, S. (1986), 'Anti-Competitive Exclusion: Raising Rivals' Costs to Achieve Power over Price', *Yale Law Journal*, **96** (2), 209–93.

Laffont, J.J., Rey, P. and Tirole, J. (1998a), 'Network Competition: I. Overview and Non-Discriminatory Pricing', *Rand Journal of Economics*, **29** (1), 1–37.

Laffont, J.J., Rey, P. and Tirole, J. (1998b), 'Network Competition: II. Price Discrimination', *Rand Journal of Economics*, **29** (1), 38–56.

Laffont, J.J. and Tirole, J. (1994), 'Access Pricing and Competition', *European Economic Review*, **38**, 1673–710.

Laffont, J.J. and Tirole, J. (1996), 'Creating Competition through Interconnection: Theory and Practice', *Journal of Regulatory Economics*, **10** (3), 227–56.

Lewis, T. and Sappington, D. (1999), 'Access Pricing with Unregulated Downstream Competition', *Information Economics and Policy*, **11** (1), 73–100.

Mandy, D. (2000), 'Killing the Goose That May Have Laid the Golden Egg: Only the Data Know [sic!] Whether Sabotage Pays', *Journal of Regulatory Economics*, **17** (2), 157–72.

Nadiri, M. and Nandi, B. (1997), 'The Changing Structure of Cost and Demand for the U.S. Telecommunications Industry', *Information Economics and Policy*, **9** (4), 319–47.

Panzar, J.C. and Sibley, D.S. (1989), 'Optimal Two-Part Tariffs for Inputs', *Journal of Public Economics*, **40**, 237–49.

Perry, M. (1989), 'Vertical Integration: Determinants and Effects', in *Handbook of Industrial Organization*, ed. R. Schmalensee and R. Willig. Amsterdam: North-Holland, pp. 183–255.

Reiffen, D. (1998), 'A Regulated Firm's Incentive to Discriminate: A Re-Evaluation and Extension of Weisman's Result', *Journal of Regulatory Economics*, **14**, 79–86.

Riordan, M.H. (1998), 'Anticompetitive Vertical Integration by a Dominant Firm', *American Economic Review*, **88** (5), 1232–48.

Salop, S., and Scheffman, D. (1983), 'Raising Rivals' Costs', *American Economic Review*, **73**, 267–71.

Sappington, D. and Weisman, D. (1996), *Designing Incentive Regulation for the Telecommunications Industry*, Cambridge, MA: MIT Press and Washington DC: AEI Press.

Sibley, D. and Weisman, D. (1998a), 'The Competitive Incentives of Vertically Integrated Local Exchange Carriers: An Economic and Policy Analysis', *Journal of Policy Analysis and Management*, **17**, 74–93.

Sibley, D. and Weisman, D. (1998b), 'Raising Rivals' Costs: The Entry of an Upstream Monopolist into Downstream Markets', *Information Economics and Policy*, **10** (4), 551–70.

Sibley, D., Doane, M. and Williams, M. (1998), 'Pricing Access to a Monopoly Input', Working Paper, University of Texas.

Sidak, J. and Spulber, D. (1997a), 'The Tragedy of the Telecommons: Government Pricing of Unbundled Network Elements under the Telecommunications Act of 1996', *Columbia Law Review*, **7** (4), 1081–161.

Sidak, J. and Spulber, D. (1997b), *Deregulatory Takings and the Regulatory Contract*, Cambridge: Cambridge University Press.

Spiller, P.T. and Cardilli, C.G. (1997), 'The Frontier of Telecommunications Deregulation: Small Countries Leading the Pack,' *The Journal of Economic Perspectives*, **11** (4), 127–38.

Vickers, J. (1995), 'Competition and Regulation in Vertically Related Markets', *Review of Economic Studies*, **62**, 1–17.

Ware, H. (1998), 'Competition and Diversification Trends in Telecommunications: Regulatory, Technological and Market Pressures', *Journal of Regulatory Economics*, **13** (1), 59–94.

Waverman, L. and Sirel. E. (1997), 'European Telecommunications Markets on the Verge of Full Liberalization', *The Journal of Economic Perspectives*, **11** (4), 113–26.

Weisman, D. (1995), 'Regulation and the Vertically Integrated Firm: The Case of RBOC Entry into InterLATA Long Distance', *Journal of Regulatory Economics*, **8**, 249–66.

Weisman, D. (1998), 'The Incentive to Discriminate by a Vertically Integrated Firm: A Reply', *Journal of Regulatory Economics*, **14**, 87–91.

Weisman, D. (1999), 'Footprints in Cyberspace: Toward A Theory of Mergers In Network Industries', *info*, **1** (4), 305–8.

Weisman, D. (2000), 'Access Pricing and Exclusionary Behavior', Kansas State University Working Paper.

Weisman, D. and Kang, J. (2000), 'Incentives for Discrimination When Upstream Monopolists Participate in Downstream Markets', Kansas State University Working Paper (May).

Weisman, D. and Williams, M. (2000), 'The Costs and Benefits of Long-Distance Entry: Regulation and Non-Price Discrimination', *Review of Industrial Organization*.

Federal Communications Commission, CC Docket 96-98 (1996), *In the Matter of Implementation of the Local Competition Provisions in the Telecommunications Act of 1996*, First Report and Order.

Federal Communications Commission, *Statistics of Communications Common Carriers* (1999), 1998/1999 edition.

Telecommunications Act of 1996. Pub. L. No. 104-104, 110 Stat. 56 (codified as amended in scattered sections of 47 U.S.C.).

Telecommunications Reports (Daily) 8 April 1996; 29 April 1996; 19 June 1997; 6 October 1997; 12 January 1998; 18 May 1998; 29 June 1998; 3 August 1998; 12 October 1998; 8 February 1999; 24 May 1999; 21 June 1999.

8. Global competition in telecommunications

Douglas A. Galbi

INTRODUCTION

In the mid-1990s a series of global alliances among established telecommunications companies was widely considered to point to the future of international telecommunications competition. Deutsche Telecom, France Telecom and Sprint formed Global One. British Telecom and MCI joined together in a venture called Concert, while AT&T put together a looser group of carriers under the names World Partners and Uniworld. Each of these ventures sought to build further links with operators around the globe in anticipation of competing to provide a one-bill bundle of services to customers around the globe.

All that has survived of the major global alliances of the mid-1990s are the brand names that they established. Competition has instead primarily developed in different directions. One direction is toward vertical industry segmentation. The Internet – meaning networks that can be connected using a particular set of globally standardized data networking protocols – is developing to provide a global platform for service competition among a wide range of independent companies. Another direction of competition is toward building wholly owned end-to-end networks among major global business centers. Such an ownership structure provides major advantages for rapidly implementing new network capabilities and rolling out services that require such capabilities.

The failure of major global telecommunications alliances highlights the rapid change and uncertainty in the telecommunications industry. Economic analysis cannot eliminate this uncertainty or provide a fail-safe means for predicting the future. However, analysis of the economics of the telecommunications industry and the responses of firms and policymakers can contribute significantly to a better understanding of the issues and possibilities. This chapter will outline key aspects of industry economics, firm organization, and public policy that are relevant to understanding the globalization of telecommunications competition.

INDUSTRY ECONOMICS

Cost Trends for Long-distance Communication and their Implications

Developments in fiber optic technology are rapidly reducing long-distance communications costs. In the 1990s the investment cost per Mbs of capacity for long-distance fiber optic communications fell by about a factor of 100 (see FCC, 1999, Table 7). This trend is likely to continue. Dense wavelength division multiplexing technology currently can provide in commercial systems 560 Gbs across a single 3600 km optical fiber without opto-electronic regeneration (for a review of technology trends see Australian Information Economy Advisory Council or AAIEAC (1999): Appendix 5). These trends have led some industry observers to speak of the 'death of distance' and 'infinite bandwidth'.

Nonetheless, it is important to recognize that the fixed costs of long-distance communications systems are significant and have not fallen. TAT-7, the last coaxial cable constructed between the United States (USA) and Europe, cost 180 million US dollars (USD) to build. In contrast, TAT-11, which went into operation in 2000 and uses advanced fiber optic technology, cost USD1500 million to construct (FCC, 1999). Laying communications wires, whether coaxial or fiber optic, is a labor-intensive operation that is not likely to experience dramatic productivity gains. Developments in fiber optic technology are reducing communications costs by greatly increasing the economies of scale in providing communications capacity.

While developments in fiber optics have tended to overshadow the role of satellites in providing long-distance point-to-point communications capacity, satellite capacity and fiber capacity differ in important ways. The fixed costs of installing long-distance satellite and fiber systems are roughly comparable and are not likely to change greatly in the future. On the other hand, fiber offers much greater economies of scale in transmission while satellites offer much more flexibility in deployment and re-deployment. An international cable needs the cooperation of authorities and operators on both ends of the cable to provide service.[1] Incumbent national operators typically are positioned between an international cable and the cable's potential end customers. In contrast, satellite systems naturally go end-to-end and can be easily redeployed for transmission between different end points. The relative value of economies of scale in capacity versus flexibility in deployment will depend on the magnitude of capacity demand and the risks associated with that demand.

Large economies of scale in international capacity have the potential to destabilize prices for international capacity. Such a development has been avoided largely in two ways. First, incumbent operators have traditionally owned international capacity through ownership shares in Intelsat and consortia created to own and operate undersea cables. Thus incumbent operators have not bought international capacity through decentralized, arms-length transactions. The incumbent operators, who naturally seek to preserve their established position, have through such organizations significantly influenced the growth of international capacity, the price of capacity and the availability of capacity to newly interested parties.

In addition, other arrangements have been established that raise transaction costs for acquiring international capacity. In particular, international capacity has traditionally been owned on a half-circuit basis. This means that in a cable connecting the United Kingdom (UK) to Japan, companies operating in the UK have been given UK half-circuits while companies operating in Japan have been given Japanese half-circuits. Establishing a connection between the UK and Japan requires matching up UK and Japanese half-circuits. As compared to full-circuit ownership shares, the convention of half-circuit ownership shares raises transaction costs and increases barriers to entry in providing international services. To acquire capacity, a new entrant must negotiate with two half-circuit owners as compared to one full-circuit owner. Such arrangements help to limit the development of a liquid market for international capacity.

A new firm can seek to establish a business limited to building and selling international communications capacity, but such a firm invariably has to confront the economies of scale in capacity and the sustainability of a business limited to selling such capacity. The trajectory of Global Crossing's business plan is instructive. Global Crossing was established as a firm that built and sold international capacity. But it soon changed its business plan and bought major local exchange carriers in the USA. It is now investing significantly in international data centers. Project Oxygen, another company that put forward ambitious plans to build a business around selling international capacity, recently folded. Until international capacity can be sufficiently differentiated to mitigate the economies of scale in its construction, a liquid market for international capacity is unlikely to develop. Instead owning international capacity will serve as an important entry ticket for providers of a range of other global telecommunications services.

Particular Economics of Wire-line and Wireless Local Links

Partisans of wire-line or wireless technology tend to put forward the view that one of these technologies will dominate. Proponents of the position that 'wires always win' argue that the capacity and quality of wire-line services, along with the fact that people spend much of their time in specific physical locations (home or office), will make wires the dominant conduit for communications. The 'only mobile' position argues that mobility is so important to users that wireless technology will trump wire-line for all but intensive communications applications such as corporate data centers and Internet service providers. These positions point to firms or organizations being divided between those that bet primarily on wire-line technology and those that bet primarily on wireless technology.

To better understand the implications for global competition, the essential differences between wire-line and wireless technologies need to be understood. Wire-line and wireless technologies essentially differ in their use of spectrum. Wire-line technology offers protected, private spectrum to a specific point. Signals are transmitted over copper wires by utilizing the electromagnetic spectrum along the wire. Signals are transmitted over fiber optic lines by utilizing their optical spectrum. In contrast, wireless technology uses public, free-space spectrum to transmit signals.

These differences in spectrum use point directly to fundamental economic differences. Since wireless signals are transmitted through free space, they give users mobility, which has considerable economic value. On the other hand, they are subject to interference from intruding users and physical phenomena, and they require public arrangements for sharing or acquiring spectrum. The trend toward spectrum auctions has made spectrum a large fixed cost associated with providing wireless service. In contrast, local wire-line spectrum can be continually installed, making local wire-line network construction nearly a constant cost technology in terms of the number of users connected. Thus, while wire-line networks are often considered to involve large sunk costs, with large wireless spectrum acquisition costs, wire-line technology provides greater potential than wireless for sustainable competition among a large number of network operators.

Even in the long term, the relative advantages of wire-line and wireless technology are likely to vary from place to place. From a static perspective, wireless spectrum acquisition costs will vary significantly from country to country based on spectrum availability and methods for allocating spectrum. Wire-line network construction costs depend significantly on local physical geography and labor costs as well as the political economics of acquiring local right-of-way and avoiding appropriation of sunk capital through a variety of public and private means. Over time, wireless technology will benefit from the 'silicon economics' of an expanding global market for standardized digital equipment. Wire-line technologies are much less subject to this dynamic, because they require much more local physical labor. But increasing consumer communication demands, which are likely to vary in nature and intensity by location, will continually re-enforce wire-line technology's advantages in capacity and quality.

Because the relative advantages of wire-line and wireless technologies differ by location, neither wireless nor wire-line technology is likely to dominate globally. This factor will constrain the scale and scope of global operators. Currently wire-line and wireless technologies are separated organizationally, and Vodaphone, one of the world's leading wireless operators, has no significant local wire-line assets. Competition will increase between wire-line and wireless networks, and service providers will increasingly look to build flexibility to operate across wire-line and wireless networks. Wire-line and wireless technology will provide an additional sustainable technological dimension of global competition in communications services.

Alternate Paths to Security and Reliability

Security and reliability of communications networks have traditionally been associated with the characteristics of physical routes. End users particularly concerned about security and reliability would demand that infrastructure providers provide physical network paths with documented provisions for redundancy and security. Security and reliability understood in terms of the physical characteristics of the underlying network tends to force service providers to integrate from the end-user service down through the physical network. This is necessary for physical network characteristics to be offered as part of service offerings to end-users.

Software-based means for providing security and reliability in communications will lessen this pressure for vertical integration. Software methods can significantly substitute for physical redundancy and security. With respect to reliability, dynamic routing using a large number of paths and fault-tolerant communications protocols can substitute for reliability in the underlying physical infrastructure. Such techniques are an important aspect of the Internet, and they are likely to continue to be important in new network protocols. With respect to security, encrypting information can substitute for the security of physical links. Software-based methods for assuring reliability and security are not perfect substitutes for physical network attributes. But in an economic sense, they do not have to be. Productivity improvements in data processing will make software the most economic way to provide the most economically relevant levels of reliability and security. This means that end-user demands for security and reliability in communications around the globe will not drive ownership of physical infrastructure.

Search for a Stable Position in the Value Distribution

The current distribution of value associated with telecommunications no longer has a solid economic foundation. Traditional voice telephony is a mature product that generates a large amount of current revenue. In the USA in 1998, consumers paid telecommunications carriers about USD200 billion for local, wireless and long-distance voice telephony (see FCC, 2000, Table 19.1). Telephony revenue, however, is highly vulnerable to industry change (see Madden and Savage, 2000, for empirical analysis of developing competition in US international services). A fundamental economic fact is that the cost of switching and transport has been plummeting.[2] Thus, for example, while US long-distance end-user revenue is about USD88 billion per annum, the capital cost of a network to provide these services probably would cost only a few billion dollars. The cost of creating mass brand awareness is much more important in providing long-distance telephone service than is network infrastructure costs (see Galbi, 1999). On-net telephony could be offered for free as part of a broader business plan and this is a likely direction for the development of Internet telephony.

Internet services offer great promise but little current revenues for network operators. The total 1998 US data communications services market[3] is about 10 per cent of telephony end-user revenues, and the cost of Internet core network services is about 1 per cent of telephony revenues.[4] Capital market valuations for Internet-focused companies are highly speculative and subject to rapid gyrations. A key challenge facing mass market Internet services is that consumers are not accustomed to paying for information and services on the Internet. Real-world economics recognize that consumer habits mediate between what consumers value and what consumers pay for. The spread and extent to which consumers will become accustomed to paying for different sorts of services received through the Internet will significantly affect the value distribution in telecommunications.

In an important and provocative paper, Odlyzko (2000) argued that communications infrastructure value will predominately be associated with business and personal communications, rather than content. Odlyzko notes that in the US telephone industry revenues are significantly greater than total advertising industry revenue, and that the US postal service generates more revenue than the entire US motion picture industry. These comparisons suggest that neither advertising nor content creation is likely to be able to provide communications infrastructure firms with revenues comparable to those today. Odlyzko's paper seems to presume that communications infrastructure revenues must remain comparable to what they are currently. But the technological trends described above suggest that infrastructure costs are likely to become less significant over time relative to creating services and marketing, ordering, and billing for them.

The distribution of value among local infrastructure, wide-area (national and global) infrastructure, communication services, content and network-based commerce is likely to be continually destabilized. Technological dynamism, unpredictable and volatile consumer preferences, and complex regulatory and political dynamics will prompt continual rearrangements in firms' business strategy. This increase in uncertainty is not merely a feature of the transition from monopolized national communications industries to global competition; it is a fundamental characteristic of global competition in communications. Galbi (2000) argues that physical coordinating points for network interconnection and content distribution will be relatively stable loci of value creation, in the same way that cities are relatively stable loci of value creation in the physical economy. While the physical locations of cities do not change rapidly, the dominant economic activities in them do: a port city becomes a financial center, a fur-trading outpost becomes a center of the software industry. Firms and policymakers in the communications industry will be continually struggling to cope with economic changes that will come increasingly rapidly.

BUSINESS STRUCTURE AND STRATEGY

Incentives to Expand Geographic Scope

Many telecommunications providers, like many other businesses, have strong incentives to expand operations across nations (Jamison, 1998). In *The Wealth of Nations* Adam Smith elaborated the logic of economic growth: the division of labor is limited by the extent of the market. In modern terms, one might say that the value of a competency is limited by the scope of the opportunities. Telecommunications providers expand internationally to exploit core competencies across a larger set of opportunities. A company that is good at managing wireless networks has all wireless networks across the globe as its scope of opportunity. In this example, a key question is whether there are important differences in the skills needed to manage wireless networks in different countries. Because of the emergence of global network equipment markets and a trend toward deregulation, technical competencies in network operations and services increasingly have global applicability. Moreover, global customers seek services that are well integrated across national borders in both technical and customer-service dimensions (provisioning, billing, maintenance and so on).

Another incentive for global expansion is risk diversification. In addition to general macroeconomic and political risks, communications industry growth rates are likely to vary significantly across countries in ways that are difficult to predict. Given low telephone penetration rates around the world and the well-established value of this service to customers from many different backgrounds and cultures, there is huge potential for telephony service growth around the globe. In the past this potential has been repressed by poorly performing government-owned telecommunications monopolies. Liberalization, industry restructuring and new regulatory approaches are rapidly removing these sorts of obstacles in many regions of the world. To lessen the risk of any particular country failing to achieve its growth potential, businesses linked to telephony service growth can position themselves to exploit growth that might occur in a number of countries.

In developed countries, future growth in the communications industry will depend on industrial and political capacity for institutional change, and consumer reaction to new services. In the development and applications of new communications technologies, the USA may increasingly lag behind other countries. Finland, Japan, Korea and Sweden are leading developments in different areas of broadband wire-line services, wireless applications and interactive TV. Companies that want to assimilate successful experience and technology, and that want to influence cutting-edge industry standards, need to be operating where leading industry developments are taking place.

In contrast to the domestic operations, participants in international commerce have to deal with important cultural differences and the absence of a common, overarching legal framework. Such differences are particularly significant in telecommunications markets because the services exchanged are complex and the regulatory framework is crucial. Internationalizing a company is a means to lessen transaction costs and to facilitate quicker reactions to new business opportunities. Organizing a new type of relationship or a new type of operation can be done more quickly and more cheaply in the context of an existing relationship. Within an existing relationship, time and capital have already been expended to establish understandings and ways of doing business. This is particularly important in the context of increasing industry uncertainty and concomitant needs for business flexibility and reaction speed.

Organizational History

The history of firms' efforts to organize to provide international service illustrates the challenges involved. While Sprint and AT&T announced domestic VPN service in 1985, it was not until 1990 that Sprint and Cable

& Wireless together attempted to set up a Global Virtual Private Network. In 1991, AT&T attempted to facilitate the implementation of advanced global services by proposing a Global Virtual Network Services Forum (GVNS). Twenty-three telecommunications companies were invited to join and ten accepted. GVNS largely failed to be a distinctive force and standardization efforts were folded back into the rather slow moving ITU-T, the International Telecommunication Unions' telecommunications standardization forum. Subsequently AT&T formed a somewhat closer group of carriers under the rubric of World Partners, although the market significance of this partnership was also questionable.

In the mid-1990s some major operators established equity-based alliances (see Galbi and Keating [1996] for further details). In 1994, BT and MCI finalized a USD4.3 billion deal whereby BT acquired 20 per cent of the equity in MCI. The deal included the establishment of a joint venture named Concert, in which BT had a 75 per cent equity stake and MCI a 25 per cent stake. In 1994 Sprint, France Telecom (FT) and Deutsche Telekom (DT) also announced an equity-based alliance. FT and DT each purchased a 10 per cent stake in Sprint. The parties established a venture named Global One that was to be 'the principal embodiment and global reference point of the International Telecommunications Services Business of the Parties'. In 1995 AT&T formed a joint venture called Uniworld, which combined Telia of Sweden, Swiss Telecom, KPN of the Netherlands and Spain's Telefonica. The exact nature of this venture was never clear, but it appeared to be an attempt to establish a tighter relationship than those previously established through large groups such as GVNS and World Partners.

None of the above equity partnerships lasted to the end of 2000. BT, after its stockholders' resistance and WorldCom's competing offer thwarted its bid to buy MCI, sold its stake in MCI, acquired full ownership of Concert, and then combined international operations with AT&T under the existing Concert brand name. Unlike agreements under the earlier alliances, the BT–AT&T venture requires each company to sell global services solely through Concert. Subsequent talks between AT&T and BT on further integration emphasize the trend toward forming a single organization for providing global services. Poor coordination and conflicts among the partners made Global One ineffective, and in early 2000 France Telecom bought out the venture. Other companies, such as MCI WorldCom, have consistently pursued a strategy of unified ownership of a global network.

Parallel to this turbulent organizational history has been the continuing importance of the structure of bilateral relations that provides for international telephone service (for more details see Volume II, Chapter 1 of this

Handbook). Since the early days of telephony international telephone calls from country A to country B have been completed based on a bilateral agreement between operators in country A and country B. Such a system of bilateral agreements has several weaknesses. One weakness is that it involves high transaction costs. The number of agreements needed rises with the square of the number of international telecommunications operators. With 233 countries and one carrier per country, about 27 000 agreements are needed for traffic exchange. The entry of new carriers in many countries has created the need for an even larger number of bilateral agreements under this organizational structure.[5]

The bilateral system has additional weaknesses. The bilateral framework does not allow for savings and innovation associated with multilateral facilities planning and routing. One study has estimated the potential savings of multilateral routing to be on the order of 10 per cent (see Nam, 1994). The most prominent weakness of the bilateral framework is that interconnection rates for international telephone calls (called settlement rates or accounting rates) have been greatly above costs. These high rates have been very significant factors in the balance sheet of carriers in developing countries (for a description of the situation in Jamaica see Myers, 1999). However, these high rates lower world welfare, generate international tension around allegations of unfair subsidies (see Melody [2000] and Stanley [2000] for discussions of some of the disputes), and attract alternative entrepreneurial ventures that might be more usefully directed elsewhere (see Scanlon, 1996; Choi et al., 1999; Malueg and Schwartz, 2000).

The Internet itself can be understood as a standards-based global partnership among a huge number of networks. Given the failure of efforts among telecommunications operators to use similar organizations to foster new international services, the Internet's growth from obscurity in the mid-1990s to huge global importance by the end of 2000 has been of major surprise to telecommunications operators and national policymakers. A noteworthy but often neglected aspect of the Internet is that domestic and international interconnection arrangements are largely undifferentiated. Interconnection protocols and institutions have the same form globally. Other organizations for international communications have distinguished domestic interconnection from international interconnection largely because of nations' incentives to try to shift rents from foreign persons to domestic persons (see Galbi, 1998). Some parties have recently expressed concern that, despite the Internet's architecture and organization, such rent shifting is also occurring on the Internet. The evaluation and treatment of such concerns may play an important role in determining whether domestic and global communications remain undifferentiated on the Internet.

ECONOMICS OF BUSINESS STRUCTURE

In considering international expansion, telecommunications firms have a range of options that include build, buy, partner and rent. Building a network in a foreign country typically requires a large amount of capital, time, and knowledge of local regulatory, political and market factors. Advantages of building include being able to install uniform, state-of-the-art network technology, tailoring the network to a specific business plan, and gaining maximum knowledge of and control over network costs. Buying an established network through an acquisition of a network owner allows faster entry and involves the acquisition of knowledgeable local employees as well as a network. Experience has shown, however, that integrating established networks and associated operating systems is a major challenge. Some companies have sought to avoid these problems by buying dark fiber in an established company's network, and then building their own switching and operating systems fabric.

While partnering or alliances in global telecommunications has a checkered history, it is likely to remain important. The computer industry illustrates how loose, rapidly changing alliances play a key role in establishing *de facto* industry standards. Standards associated with lower levels of the physical network can be established through loosely organized standards bodies. However, standards that relate more closely to service characteristics offered to end customers – products, provisioning intervals and service level agreements – are more likely to be established through closer partnerships because they are crucial to companies' competitive positions.

Equity investments across members of a partnership can be a way to redistribute returns from partnership-specific investments. Assume, for simplicity, that partnership-specific investment must be undertaken by particular partners while customers choose which partnership member will offer them service. Suppose that two carriers each require a 10 per cent return on investment and each must make a USD30 million investment in a partnership. As a result of the partnership, one carrier will earn USD4 million per annum while the other carrier will earn USD2 million per annum. So as to make the investment in the partnership worthwhile, the second carrier might acquire a 25 per cent equity share in the first carrier. More generally, the nature and scope of partnership-specific investments affects the opportunities for the partners, while transfers of common equity affect how the returns from the partnership are split between the partners.

However, as the literature on ownership rights emphasizes, the allocation of ownership rights itself can affect investment decisions (for a recent literature review, see Shleifer and Vishny, 1997). Suppose, for example, that a partnership between a French carrier and Ghanaian carrier offers a 10 per cent annual return. The partnership is structured such that the French carrier pays USD80 million and the Ghanaian carrier pays USD20 million. The French carrier in return obtains 80 per cent of the partnership earnings, while the Ghanaian carrier receives 20 per cent. Suppose that, because of asymmetric information and limitations on the scope of enforceable contracts, not all the factors that affect the earnings of the partnership are specified in the partnership agreement. Each partner thus has to make some investment decisions independently based on expected return given the partnership rules. Suppose the Ghanaian partner encounters an opportunity to invest USD10 million to increase partnership income by USD3 million annually. The return on this investment, 30 per cent, is much higher than the return that the partners require. Nonetheless, because the Ghanaian partner gets only 20 per cent of the returns from the partnership, it will not rationally choose to make the investment. Thus assignment of ownership shares in a partnership also should consider the relative opportunities of the partners to make additional non-contractible investments that benefit the partnership.

Creating opportunities to rent parts of networks has emerged as an important regulatory direction. Regulators have sought to make some incumbent network operators provide unbundled network elements (UNEs), of which unbundled network loops have attracted the most interest. In contrast to resale of service offerings, use of UNEs typically offers more flexibility in configuration and use. Where UNEs are available, using them is the least capital-intensive way to gain control over network facilities in foreign markets. However, firms have typically made UNEs available only under regulatory duress. Using UNEs may require continual regulatory battles and changes in UNE use take place in an adversarial environment. Thus while UNEs offer control over network facilities, they may not effectively offer the ability to create rapidly seamless new services that require facilities-based innovation.

STRATEGIES OF COMPETITION

Global competitive strategies will depend significantly on substitution across established network technologies, the evolution of new local competitors, and the importance of service innovation. To get a sense of the different possibilities, consider first alliances or new global carriers created from groups of incumbent same-technology (wire-line, wireless or cable) operators: territorial technological aggregation. A particular global carrier would dominate the domestic technology associated with one of its constituent incumbent carriers. Global carriers would compete across network technology through new service offerings and geographically through mergers to expand the scope of their networks. Such a competitive structure would likely involve highly contentious intra-technology interconnection agreements, low inter-technology convergence and inter-operability, and the preservation of sharp distinctions between domestic and international communications. On the other hand, such a form of competition best preserves established routines and organizations, and hence would be least disruptive to businesses and governments.

An alternative competitive structure might be an internationalization of local competition. Under such a structure, global carriers would operate across technology and global carriers would not be associated with disjoint 'home territories'. Moreover, there would not be a sharp distinction between domestic and international communications, and interconnection agreements would be negotiated primarily with regard to domestic factors. Making local competition work independently of the nationality associated with the local competitor would require considerable discipline on the part of policymakers and competitors. Domestic policymakers would need to make strenuous efforts to establish a clear, credible regulatory direction that can withstand the political clout of the domestic incumbent. New foreign entrants would need to have the vision and discipline to avoid seeking to appeal to their own national authorities as a business strategy for gaining local leverage. The internationalization of local competition is not likely to come naturally with invisible policy, but the effort to promote this form of competition is likely to bring the most rapid growth of the communications industry and the greatest benefits to consumers.

KEY GLOBAL POLICY CHALLENGES

New regulatory bodies with responsibility for the telecommunications sector are being set up in many countries in conjunction with liberalization and privatization of communications in many countries. Attracting qualified personnel and building an effective organization to meet rapidly growing policy demands is a significant challenge. This administrative constraint will itself push decision-makers toward simpler policies. Moreover, while sector-specific expertise is important for successful policy formulation and implementation, it does not necessarily have to reside only in a national regulator. Regional and multilateral organizations, universities and independent research institutes can also develop into repositories of policy experience and analysis. Such a multi-institutional approach is likely to be particularly important because of the significance of telecommunications to the economy as a whole and the need for flexible, innovative policy approaches in the face of rapid industry change.

Competition policy authorities are already playing an important role in establishing the regulatory framework in telecommunications, and they are likely to continue to do so in the future. Regulatory policy is often considered to be sector-specific while competition policy is thought to consist of generally applicable rules. As Nihoul (1998/99) points out, in practice in the European Union this is not the case: there appears to be little substantial difference between competition policy and regulatory policy. In the USA the FCC has approved mergers with substantial conditions that are similar to those established in rule-making procedures. While regulators tend to move slowly in establishing industry-wide rules, business reorganizations such as mergers and divestitures will force policymakers to make decisions promptly about how particular companies are allowed to operate. Such decisions will inevitably play an important role in shaping the overall industry regulatory framework.

MAKING MEANINGFUL POLICY DECISIONS

A largely unappreciated problem in telecommunications policy is focusing policy debate on meaningful choices. Policy debates about local loop unbundling illustrate the problem. Such policies require a huge number of detailed implementing regulations, each of which can significantly affect competitors' profitability and the incentive to make large, sunk investments in infrastructure. To assess a policy of local loop unbundling one needs to know, among other factors, the definition of rate elements, the use rights associated with them, their prices, and the regulatory regime for ensuring functionality and effectiveness in operations support systems. Knowing only that policymakers require, or will require, local loop unbundling is to know very little.

In this way a superficial homogeneity in regulatory policy around the globe may cloak large differences in interpretation and implementation. Most policymakers might advocate competition, deregulation and universal service. But what sort of competition will be promoted? What regulations will change and when? Is universal service considered to be inconsistent with competition and deregulation and if so what will policymakers do to deal with the inconsistency? Such questions will be answered. If policy is not able to address these questions directly (and experience thus far is not promising) these questions will be answered indirectly through nontransparent, unaccountable and idiosyncratic means.

Cost-based pricing for telecommunications services is an important historical example of a globally recognized policy standard. This standard has been important in structuring arguments over the level of telecommunications prices. In particular, a party makes a cost showing to support a price, and another party either makes an alternative cost showing or provides some other reason for the price level, such as universal service. The standard of cost-based pricing has provided a workable discourse for resolving pricing issues. On the other hand, cost-based pricing has tended to put into the background the consumer and industry implications of different levels of regulated prices, and cost-based pricing has led to an international distribution of telecommunications prices that is not readily rationalized in terms of underlying cost differences. With increasing attention to international comparisons and policy effects, cost-based pricing is likely to become less important as a global policy standard in the future.

Telecommunications policy has to be based on simple, significant policy levers to be transparent, accountable and amenable to homogenization globally. Statements of policy principles tend to assume that such levers are prevalent and obvious. But in fact, finding simple policy levers that can translate intelligibly policy intentions into implementations and outcomes is a major challenge. This is particularly true given the tremendous uncertainty and turmoil in telecommunications industries. Regulatory bodies that focus on analysing and assimilating experience, and learning from it, are most likely to be best able to meet this challenge. Small countries have important advantages in assimilating policy experience and pursuing innovative policies, while large countries have the advantage of economic significance in creating *de facto* global policy standards. As the thorough review in Adamska (1998) indicates, what will emerge as significant global policy standards in telecommunications remains subject to considerable uncertainty.

CONCLUSIONS

Discussions of globalization, the Internet and e-commerce typically emphasize the increasing pace of change. However, for firms, policy analysts and policymakers, a key challenge is to identify those aspects of the industry that are likely to change most slowly. A firm needs to identify its key competencies, which are exactly those skills that it builds and exploits in the midst of rapid change. Policy analysts and policymakers need to identify key industry structures that are relatively stable and will shape future industry growth. Thus the need for analysis of industry structures and the importance of industrial policy does not lessen in the turmoil of the 'new economy'. Rapid industry change, because it makes such structures less obvious, makes such analysis more important. To be intelligible, government policy, which is intrinsically slower to evolve than commercial activity, will increasingly have to focus on affecting industry structure.

ACKNOWLEDGEMENTS

The opinions and conclusions expressed in this paper are those of the author. They do not necessarily reflect the views of the Federal Communications Commission, its Commissioners or any staff other than the author.

NOTES

1. The Africa One cable system, which circles Africa, provides an example of how cable operators can mitigate some of their geographic risk. Africa One will be profitable as long as enough countries in Africa find it worthwhile to connect to the loop, while which countries connect is not important.
2. John Sidgmore, Vice Chairman of MCI WorldCom, emphasized this point in a June 1999 stock analysts' meeting. He noted that switching and transport amounted to 63 per cent of MCI's backbone investment in 1988, while in 1998 investment in switching and transport had fallen to 25 per cent. He projected that it would fall to under 10 per cent in the next few years.
3. US data communications services market size is based on data in '1999 Market Forecast', *Data Communications* (December 1988).
4. Using data released in MCI's sale of its Internet backbone to Cable & Wireless, Odlyzko (1998) notes that MCI Internet revenue in fourth quarter 1997 was USD386 million on an annual basis. Since the MCI backbone is estimated to carry 20–30 per cent of Internet backbone traffic, Odlyzko estimated that annual revenue from US domestic Internet core services is USD1.1–USD1.6 billion at end-1997. MCI's Internet revenue includes about USD60 million for dial-up access (MCI had 250000 consumer accounts, 60000 business accounts and 1300 ISP accounts (see Rickard, 1998). Odlyzko also presents other figures that support core Internet revenue (excluding revenue for dial-up access) being about USD1.5 billion at end-1997. Given an Internet growth rate of about 100 per cent per annum, mid-1998 annual revenue would be about USD2 billion.
5. The entry of new carriers has also historically involved the development of cartel-like rules such as proportional allocation among domestic carriers of incoming international traffic. For analysis of these rules see Galbi (1998) and Karikari (2000).

REFERENCES

Adamska, M. (1998), *International Telecommunications Alliances and Foreign Direct Investment as Means of Globalization: Legal and Regulatory Responses to the Emergence of Super Carriers*, Master of Law thesis, McGill University, Montreal [http://www.law.mcgill.ca/research/csri/papers/monika.html].
AAIEAC (1999), *National Bandwidth Inquiry*, Appendix 5: [http://www.noie.gov.au/projects/information_economy/bandwidth/index.htm].

Choi, H-W, Yun, K-L., Kim, I.J. and Ahn, B-H. (1999), 'On the economics of call-back services', *Journal of Regulatory Economics*, **15** (2), 165–81.
FCC (1999), *International Bureau Report: 1998*, Section 43.82: Circuit Status Data [http://www.fcc.gov/ib].
FCC (2000), Industry Analysis Division, *Trends in Telephone Service*, March [http://www.fcc.gov/Bureaus/Common_Carrier/Reports/FCC-State_Link/fcc-link.html].
Galbi, D. (1998), 'Cross-border rent shifting in international telecommunications', *Information Economics and Policy*, **10**, 515–36.
Galbi, D. (1999), 'The price of telecom competition: counting the cost of advertising and promotion', *info*, **1** (2), 133–40.
Galbi, D. (2000), 'Transforming network interconnection and transport', *CommLaw Conspectus*, **8** (2) (Summer), 203–18 [http://www.galbithink.org].
Galbi, D. and Keating, C. (1996), *Global Communications Alliances: Forms and Characteristics of Emerging Organizations*, FCC International Bureau [http://www.fcc.gov/ib, under Issues of Interest].
Jamison, M.A. (1998), 'Emerging patterns in global telecommunications alliances and mergers', Working Paper [http://www.cmcnyls.edu/Papers].
Karikari, J. (2000), 'Pricing implications of the US international settlements policy', manuscript, available from karikarij.rced@gao.gov.
Madden, G. and Savage, S. (2000), 'Market structure, competition, and pricing in United States international telephone services markets', *Review of Economics and Statistics*, **82** (2), 291–6.
Malueg, D. and Schwartz, M. (2000), 'Asymmetric telecom liberalization and gaming of international settlements via carrier alliances', manuscript, Georgetown University [available from schwarm2@georgetown.edu].
Melody, W.H. (2000), 'Telecom myths: the international revenue settlements subsidy', *Telecommunications Policy*, **24** (1), 51–62.
Myers, G. (1999), 'Squaring the circle: rebalancing tariffs whilst promoting universal service in Jamaica', paper presented at the 1999 Telecommunications Policy Research Conference [http://www.tprc.org].
Nam, K. (1994), *International Telecommunications Networks: Modeling and Analysis*, dissertation in Systems Engineering, University of Pennsylvania.
Nihoul, P. (1998/99), 'Convergence in European telecommunications: a case study on the relationship between regulation and competition (law)', *International Journal of Communications Law and Policy*, **2** [http://www.ijclp.org/].
Odlyzko, A. (1998), 'The economics of the Internet: utility, utilization, pricing, and quality of service', manuscript (7 July) [http://www.research.att.com/~amo].
Odlyzko, A. (2000), 'The history of communications and its implications for the Internet', Preliminary version (16 June) [http://www.research.att.com/~amo].
Rickard, J. (1998), 'In search of the elusive business market', *Boardwatch* (September).
Scanlon, M. (1996), 'Why is the international accounting rate system in terminal decline, and what might be the consequences?', *Telecommunications Policy*, **20** (7), 739–53.
Shleifer, A. and Vishny, R. (1997), 'A survey of corporate governance', *Journal of Finance*, **52**, 737–83.
Stanley, K. (2000), 'Toward international settlement reform: FCC benchmarks vs. ITU rates', *Telecommunications Policy*, **24** (10) [http://www.tpeditor.org].

(selected countries)

L96
F23
L14 K33

9. US settlement reform: an historic review

Michael A. Einhorn

INTRODUCTION

The issue of international settlements has suffered from more than a century of relative neglect since its inception in 1865, when twenty European nations formed the International Telegraph Union (later the International Telecommunication Union, ITU) to provide a governing framework for the settlement of international telegraph traffic. Under ITU guidelines settlement interconnected national carriers set rates, for minutes passed between nations, in bilateral negotiation. Without administrative oversight or restraint, carriers had an incentive to negotiate settlement rates well above cost so as to extract higher accounting profits. Retail prices were further inflated by domestic landline monopolies (often PTTs) that passed network infrastructure costs on to international callers.

Though settlement charges for calls between pairs of nations are generally equal regardless of call direction, revenues flow periodically from the net originator to the net receiver. The latter receives a per-minute profit equal to the difference between the settlement rates and call cost. This process is advantageous to carriers in less developed countries (LDCs), which typically receive more international message telephone service (IMTS) minutes than they originate, and disadvantage the United States (USA), which is the largest call originator. US billed international minutes grew faster than foreign IMTS in the 1980s and early-1990s (see Figure 9.1). Consequently, both billed and net carrier revenues increased (Figure 9.2) despite declines in billing and settlement rates (Figure 9.3). The revenue difference, which represents outgoing net settlement payments, grew as well. Consequently, the net revenue outflow of a billed composite of 14.1 billion United States dollars (USD) grew to USD5.6 billion in 1996. Simultaneously, investment cost per minute of high-capacity cable fell from 12.1 United States cents (USC) in 1980 to USC1.2 in 1996 (Table 9.1). Perhaps 70 per cent of the outflow represented a pure subsidy.

The winds of reform began when the ITU (1992) issued Recommendation

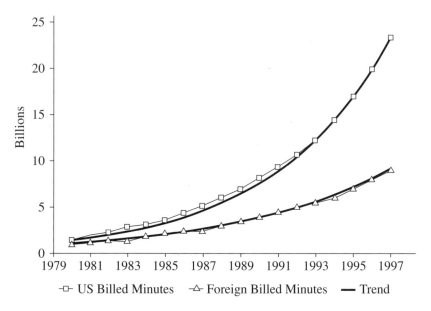

Source: *Trends in the US International Telecommunications Industry*, FCC (September 1999), Figure 4.

Figure 9.1   US and foreign IMTS minutes

D.140, which called for accounting rates to be cost-based, nondiscriminatory and transparent. The Federal Communications Commission (FCC) began its reform initiative some three to four years later. In 1997, the Commission issued a seminal order that estimated benchmark rates based on Recommendation D.140. For a review of pre-1992 US policy, see Johnson (1989), Johnson (1990) and Stanley (1991). In 1997, 69 nations also signed a World Trade Organization (WTO) treaty and thereby committed to open their domestic networks, either fully or partially, to foreign investment. There are three key dimensions to the FCC's current reform strategy. First, the Commission instituted a five-year timetable for compliance with its benchmarks of 1997. Second, the Commission offered low-cost access and liberalized termination agreements to foreign carriers that met prescribed competitive guidelines. Third, the Commission relaxed selectively its own restrictive rules that had discouraged aggressive bargaining by US carriers to terminate international calls.

The reforms of the past decade have been beneficial to US consumers. The average US settlement rate fell from USD0.59 (in 1992) to USD0.29 (in 1998) per minute. The corresponding billing rate fell from USD1.00 to

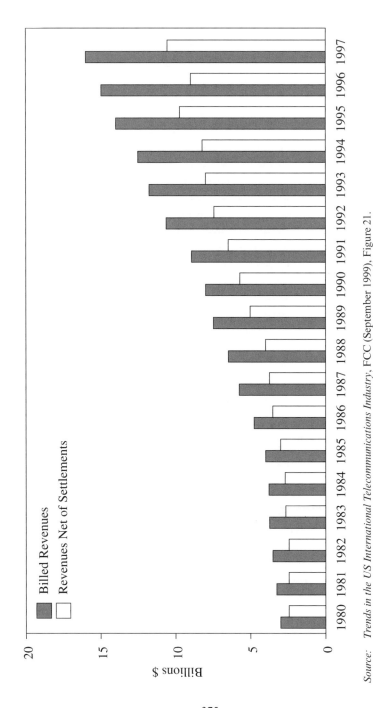

Source: *Trends in the US International Telecommunications Industry*, FCC (September 1999), Figure 21.

Figure 9.2   US carrier billed revenue before and after settlements

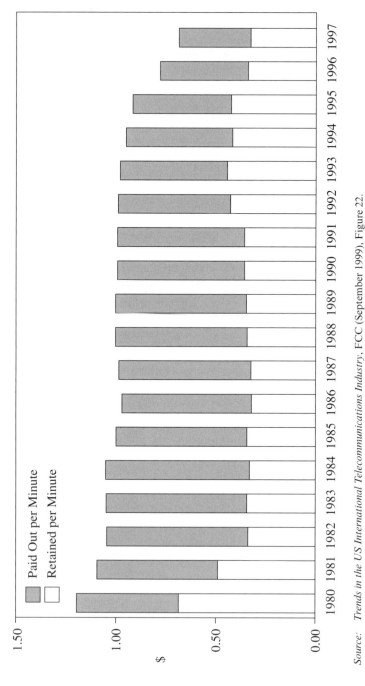

Source: *Trends in the US International Telecommunications Industry*, FCC (September 1999), Figure 22.

Figure 9.3   US retained revenue and settlement payments

Table 9.1 Transatlantic cable systems

System	Year	Technology	Estimated cost (million USD)	Total 64 kbps circuits	Usable 64 kbps circuits	Annual investment cost per usable ckt (USD)	Investment cost per minute (USD)
TAT-1	1956	Coaxial Cable	49.6	44.5	40.1	213 996	2.443
TAT-2	1959	Coaxial Cable	42.7	49.0	44.1	167 308	1.910
TAT-3	1963	Coaxial Cable	50.6	87.5	78.8	111 027	1.267
TAT-4	1965	Coaxial Cable	50.4	69.0	62.1	140 238	1.601
TAT-5	1970	Coaxial Cable	70.4	720.0	648.0	18 773	0.214
TAT-6	1976	Coaxial Cable	197.0	4000.0	3 200.0	10 638	0.121
TAT-7	1983	Coaxial Cable	180.0	4246.0	3 821.4	8 139	0.093
TAT-8	1988	Fiber Optic	360.0	7 560.0	6 048.0	10 285	0.117
TAT-9	1992	Fiber Optic	406.0	15 120.0	10 584.0	6 628	0.076
TAT-10	1992	Fiber Optic	300.0	22 680.0	18 144.0	2 857	0.033
TAT-11	1993	Fiber Optic	280.0	22 680.0	18 144.0	2 667	0.030
TAT-12	1996	Fiber Optic	378.0	60 480.0	60 480.0	1080	0.012
TAT-13	1996	Fiber Optic	378.0	60 480.0	60 480.0	1080	0.012
Gemini	1998	Fiber Optic	520.0	241 920.0	241 920.0	371	0.004
AC-1	1998	Fibre Optic	850.0	483 840.0	483 840.0	304	0.003

Source: *Trends in the US International Telecommunications Industry*, FCC (September 1999), Table 12.

USD0.59. Reductions were largest for calls to industrialized nations that liberalized more rapidly. When made available, data beyond 1998 will confirm the trend continued to the present and expanded to include LDCs. This chapter reviews contemporary US policy concerning international settlements of voice and data telephone traffic. Accordingly, the Commission's previous settlements policies and the problems that it posed to US carriers and callers are examined. The FCC's strategies for reform are then detailed. The track record of the market under more competitive regimes is also considered. An overview of future concerns concludes the chapter.

ACCOUNTING RULES

The key technical relationship between a pair of interconnected international telephone networks is that calls originate at both ends and must be transferred between them. Correspondent networks, that pass traffic between their respective nations, must agree on rates for call interconnection. Under common international practice, interconnected carriers bilaterally negotiate a per-minute accounting rate, which is specified in USD, gold francs or SDRs. The originating carrier credits its interconnection partner for an amount equal to minutes transferred times a settlement rate, that is, usually 50 per cent of the accounting rate. Rates are not cost-based and do not generally vary with usage or time of day.

Originating carriers bill called minutes with collection rates. Each outgoing toll minute presents a marginal cost to its carrier that is equal to the sum of the settlement rate, domestic transport costs and any access charge paid to the local network. Including the access charge, which is not really a component of economic marginal cost, the FCC estimates that the average cost of an outgoing international minute is between USD0.06 and USD0.09. The apportioned per-minute investment cost of the highest capacity fiber optic cable (AC-1), which is the best measure of actual marginal cost, is now USC0.3. Traffic minutes and revenue credits between two carriers may move in both directions, however, these do not necessarily relate directly to monetary flows. Rather, a carrier only compensates a correspondent on the basis of net traffic outflow (or revenue credit). Total revenues due from an outgoing carrier are the product of the relevant international settlement rate and the net outflow of minutes. The direction of compensation between nations depends on their relative economic development and collection charges. Collection charges can also be affected by domestic competitiveness. As a wealthy country with domestic competition, the USA is the largest exporter of both IMTS minutes and settlement revenues. Consequently, it has the greatest financial incentive to lower settlement rates. Network partners with traffic surpluses, particularly LDCs, have an incentive to maintain high rates. Incoming payments to less affluent nations can be particularly important means of financing their infrastructure. Table 9.2 displays net settlement payments (in million USD) from the USA and related accounting charges to individual countries. The charts confirm that US revenue paid to all regions increased substantially in the period 1985 to 1992. As explained above the increase resulted from higher levels of outgoing traffic flow, US settlement rates to developing countries exceeded rates to industrial nations, outgoing revenue paid from

Table 9.2   US net settlement revenues (USD millions)

Region	1985	1992	1997
Western Europe	171	617	388
Eastern Europe	41	112	210
Africa	32	130	317
Bermuda / West Indies	92	305	477
Latin America	176	584	933
Asia / Oceania	388	933	2439
Canada / Mexico	230	800	785
Total	1129	3492	5552

US accounting rates (in current USD)

Industrial countries			
Canada (Stentor)	0.42	0.28/0.24	0.20/0.12
France	1.70	0.96	0.20
Germany	1.32	1.10	0.20
Italy	1.79	1.65	0.21
Japan (KDD)	2.35	1.31	0.28
United Kingdom (BT)	1.06/0.76	0.74/0.52	0.21/0.13
Less developed countries			
Brazil	2.50	1.50	0.60
China	3.23	3.37	0.99
India	2.70	2.00	1.28
Indonesia	3.25	1.80	0.60
Nigeria	1.50	1.50	1.15
Saudi Arabia	2.36	2.20	1.33

Note: Accounting rates for Canada and the UK differ by time of day.

the USA to Western Europe (and Japan) declined in the period 1992 to 1997, and US revenues paid to remaining regions, which include more LDCs, increased steadily in the period 1992 to 1997. For econometric investigations of the key determinants of US accounting rates, see Acton and Vogelsang (1992) and Madden and Savage (1998, 2000).

INTERNATIONAL SETTLEMENTS POLICY

Another key aspect of the FCC reform strategy entails its international settlements policy. In 1936, the FCC refused to allow Norway's PTT to route all telegraph traffic to a single US carrier. The concern was that competing US carriers could be aggressively 'whipsawed' into less generous deals in 'take-it-or-leave-it' bargaining. With a reiterated fear of whipsawing in 1986, the Commission established its uniform-settlement policy, now international settlements policy (FCC, 1986). This international settlements policy has three key dimensions: equal rates – accounting rates to any nation must be equal for all US carriers; equal shares – the settlement rate to each carrier must be 50 per cent of the accounting rate; and proportional return – all US carriers must be credited with a share of incoming revenue from any nation in proportion to its respective share of outgoing revenue to that nation. The Commission also disallowed US carriers from entering into special service arrangements with foreign carriers.

The equal rate and share rules are evidently anticompetitive. They disallow US carriers from competing with one another to lower accounting rates. Consequently, carriers may actually have an incentive to raise accounting rates to offset profit losses that arise when domestic competition forces collection rates down (Yun et al., 1997). The proclivity to maintain higher accounting rates hurts both foreign and US callers. The rule of proportionate return is problematic as well. Under the rule, a competing US provider that settles an additional outgoing minute increases its 'settlement share' and therefore earns additional revenue from incoming traffic. The resulting gain in return revenue then decreases the carrier's perceived marginal payments that result when it settles outgoing calls. However, the perceived reduction in payment does not reflect any reduction in actual costs. This rule has three unwanted consequences (Galbi, 1997). All US carriers have an incentive to settle some calls that would be better routed over bypass lines at lower cost. Further, the perceived revenue gain from incoming traffic would be greatest for US domestic carriers with the smallest outgoing market share and therefore distorts competition. Finally, foreign carriers have an incentive to argue for higher accounting rates as incoming settled US traffic increases.

REFORM AND ACCOUNTING RATES

As regards the accounting rate benchmark reforms, Recommendation D.140 identified three components for international cost-based charges: international transmission, international switching and national extension (domestic transport and termination). A 1993 follow-up, Recommendation D.300R (ITU, 1993), set forth a procedure for estimating transit prices based on distances between national borders and on the degree of digitization in each country. Another report (ITU, 1997) advocated financial aid for developing countries faced with revenue decline resulting from settlement reform. Facing a 1996 imbalance in US settlement revenue that exceeded USD5 billion, the FCC began to consider establishing benchmarks of accounting rates using the ITU's guidelines from Recommendation D.300R. Using service cost and price data from 65 countries (FCC, 1996b, Appendix), the International Bureau measured three sets of numbers. Transmission charges were based on per-minute converted prices of international T1 or E1 circuits that were sold for private line service. National extension charges were based on domestic tariffs that were charged for direct dialed telephone service, aggregated appropriately over time of day, day of week, mileage band, and so on. Switching charges were estimated from cost information that participant countries provided in D.300R. The specified costs differed by the proportion of switch digitization: USC4.8 (0–30 per cent), USC3.4 (31–60 per cent), and USC1.9 (61–100 per cent), which were then assigned to individual countries based on their per capita income.

Based on International Bureau estimates, a 1997 FCC Order (FCC, 1997b) set forth settlement rate benchmarks for high-, middle-, and low-income countries of USC15, USC19, and USC23 per calling minute, respectively. The order took effect on 1 January 1998. It required compliance by high-income countries by 1999 and established four subsequent annual deadlines for less affluent nations. Before the first deadline became effective, the FCC allowed foreign carriers to route US-bound calls over incoming private lines (thereby avoiding settlement charges) only when 50 per cent or more of its reverse traffic was settled at or below the FCC's benchmark for the particular country. Furthermore, the Commission required that foreign affiliates of US carriers must charge settlement rates at or below the applicable benchmark if they route US-bound traffic over facilities-based switched or private line service.

DOMESTIC ENTRY AND RECIPROCAL COMPENSATION

The FCC Foreign Entry Order (FCC, 1995) allowed foreign providers to compete directly in US domestic markets. However, a particular foreign provider could enter US markets only when its country offered effective competitive opportunities to US carriers. Nations that provided effective competitive opportunities needed to meet four criteria (FCC, 1995). They must allow US carriers to offer international facilities-based services to their domestic customers; competing US carriers faced nondiscriminatory rates and terms for interconnection to the domestic landline network; competitive safeguards existed to protect against anticompetitive practices with regard to the second criterion; and an independent regulatory body would administer the above rules.

The scope of the Foreign Entry Order was pared back considerably in 1997. In February of that year, the WTO completed its *Basic Telecommunications Services Agreement*, which affected 95 per cent of world telecommunications revenue. In the international agreement, 52 fully compliant nations consented to allow foreign investment in their domestic markets by other signatories. The fully compliant nations, and 17 more, also committed to allow landline interconnection at all technically feasible points with nondiscriminatory and transparent rates and independent regulatory oversight. Table 9.3 lists the WTO signatories and their respective calling volume to and from the USA.

By signing the WTO agreement, the USA was committed to open its domestic markets to foreign entrants. In compliance, the Commission instituted its *Foreign Participation Order* (FCC, 1997b) in November 1997. The Order waived the effective competitive opportunities requirement entirely for WTO signatories that sought to provide international facilities-based competition or to resell services in the USA. However, the requirements remained effective for non-WTO members. Further, the Order also relaxed the 'No Special Concessions' rule and allowed US carriers to enter into special arrangements, including affiliations, with non-dominant foreign carriers, as now determined by having 50 per cent or less of the market share in international transport, intercity transport and local access facilities or services. Any foreign carrier with less than a 50 per cent market share in its respective domestic market was deemed non-dominant and granted a presumptive allowance that contending parties may attempt to rebut. The Order also granted an allowance to dominant carriers that lacked market power to affect US competition. Finally, the Order avoided prescribing a rule that would have disallowed foreign affiliates of US carriers from reselling switched services to US markets unless their respective settlement rates were at or below the applicable benchmark

Table 9.3   US billed IMTS traffic with WTO signatories (minutes)

WTO signatory	Calls to the USA		Calls from the USA	
	1997	1998	1997	1998
Antigua and Barbuda	7115819	7107775	54506678	666719925
Argentina	38596380	46086064	226585250	233981041
Australia	131569530	261265578	389186489	445122839
Austria	24951748	26127656	58888931	75985975
Bangladesh	4469717	4718364	60076079	52729727
Barbados	12667958	12575598	38785436	47696085
Belgium	50477727	39674103	122795210	116719060
Belize	4170792	4972453	14686178	15061084
Boliva	5385806	5456986	34427178	52838539
Brazil	159248397	180184691	495323734	591195594
Brunei Darussalam	685163	881120	2362320	8229089
Bulgaria	1344957	1584034	12538600	14556691
Canada	3145941566	3358956938	3922086328	3881035995
Chile	61028462	81136627	112992559	128227790
Colombia	67640760	51227342	263350037	224266040
Cote d'lvoire	2772729	3385163	15120099	16716097
Cyprus	4102808	4564433	9827450	11093247
Czech Republic	8815481	9934223	27014751	30295227
Denmark	27994754	26519824	76302051	75841746
Dominica	2212662	2108923	17124731	29295426
Dominican Republic	107817532	111412775	379150470	439057321
Ecuador	17413209	13563630	184854451	113526248
El Salvador	12743151	16048993	155245851	198892640
Estonia	1311042	1523455	2930421	5624290
Finland	17787971	14064555	29464238	32070084
France	216746922	270817706	502561833	575472724
Germany	325240908	471169441	994884808	1408763612
Ghana	5240219	7762385	50269789	44832311
Greece	35148254	42788898	97898305	116252178
Grenada	2375226	2325612	17054700	63825297
Guatemala	15375065	17913027	126547737	145642331
Chinese Hong Kong	78379209	214669927	671797709	598108183
Hungary	14863885	12682758	40909624	39380507
Iceland	6918811	8890797	10562442	14060636
India	49707870	58470938	574380674	755297617
Indonesia	28175746	28637808	119051535	103364245
Ireland	48663111	68971405	134329538	161006122
Israel	121555694	171390354	214880358	221874627
Italy	117814252	133973684	476352714	529505067
Jamaica	49902389	51903485	254091318	291487463

Table 9.3   (cont.)

WTO signatory	Calls to the USA		Calls from the USA	
	1997	1998	1997	1998
Japan	342893941	336428346	849358999	808781369
Kenya	3480276	4122039	25698678	26485742
Korea	199932652	219729719	426476913	384149732
Kyrgyz Republic	8110	4892	1746448	825642
Latvia	1209108	974333	3186037	4644611
Luxembourg	6154177	6894882	11075994	16018922
Malaysia	23602566	25158025	86630077	103651471
Mauritius	487927	598379	6762451	14018985
Mexico	942158920	1086310773	2766488494	3020570877
Morocco	5805115	6372647	14401988	74918,664
Netherlands	98661319	416747866	222710538	280795926
New Zealand	37596723	54672922	110777088	75621277
Norway	52504621	51832479	66451714	90625278
Pakistan	10518252	15610182	161013491	213480603
Papua New Guinea	662939	873510	2376641	1803485
Peru	24500128	31425645	164810714	185465
Philippines	33363428	50688174	448866511	574574981
Poland	27311276	27880730	170558158	189479771
Portugal	14972676	23130262	55159003	71957330
Romania	5248333	21967892	26884438	76112046
Senegal	1717016	2016631	19342020	19925012
Singapore	61419122	65635330	202681488	175699888
Slovak Republic	3335617	3335101	10612303	11218818
South Africa	31119827	36484345	111618488	130778758
Spain	85309869	69495669	191010953	254022029
Sri Lanka	1987980	3054378	16938724	21174993
Suriname	868592	644589	9923684	3431142
Sweden	185679847	113844782	143818519	150968300
Switzerland	98017798	102139822	246280971	199877106
Thailand	26280859	32215132	117177476	130337078
Trinidad and Tobago	29319430	31508545	94345185	105862920
Tunisia	1513284	1135704	4470221	3786847
Turkey	24986937	28379378	62531049	113736357
Uganda	703254	886897	4812428	10645215
UK	946020819	806808330	1553813287	1169563822
Venezuela	59069810	60244323	213508910	233340641

Source: Report on International Telecommunications Markets 1999 Update, FCC,
Attachment 5.

INTERNATIONAL SETTLEMENTS POLICY

The Commission began to liberalize its international settlements policy in 1996. Under the Flexibility Order (FCC, 1996a) US long-distance carriers were permitted to bid competitively to terminate inbound traffic from a foreign country provided that the correspondent nation offered effective competitive opportunities to US carriers.

Alternatively, competitive bidding was also permitted when the resulting rates could be expected to enhance competition. Consistent with statutory obligations imposed in the Telecommunications Act of 1996, the Commission in 1998 identified its international settlements policy (ISP) as a rule that was no longer necessary in the public interest.

A subsequent Notice of Proposed Rulemaking (FCC, 1998a) faulted the policy because: by mandating equal settlement rates and shares for all US carriers, the policy reduced the incentive for any carrier to negotiate lower charges; the law of proportionate return awarded carrier credits based on their respective share of outgoing settled traffic, encouraging oversettling and establishing unequal cost structures; and collection rates increased when competition in settlements declined.

The Commission substantially modified its Flexibility Order in April 1999 (FCC, 1999a). The revised order eliminated altogether the international settlements policy and contract filing requirements for non-dominant foreign carriers. It also eliminated ISP restrictions for dominant foreign carriers that terminate 50 per cent or more of incoming US traffic at settlement rates equal to or below 75 per cent of the FCC benchmark. Further, the Commission relaxed a restriction in the Notice of Proposed Rulemaking that initially would have limited deregulation to WTO member countries. In June 1999, the Commission identified 190 nations where at least one carrier was presumed to be dominant and not eligible for relief from the policy. In July, the Commission issued a list of eleven international routes that satisfied criteria for relief. The list is encouraging because of the economic importance and calling volume of each partner: Canada, Chinese Hong Kong, Denmark, France, Germany, Holland, Ireland, Italy, Norway, Sweden and the UK.

THE STATE OF COMPETITION

To mark the first anniversary of the WTO pact, the International Bureau issued a December 1998 report that summarized important developments in international markets (FCC, 1998b). It followed up with a subsequent report in January 2000 (FCC, 2000a). At the time of the report, benchmark rates had been accepted in 47 upper-income countries and were actually achieved in 30 of them. The affected shares of total settled minutes were 62 per cent and 38 per cent, respectively. The US average accounting rate had fallen from USC63 to USC55 per minute and the average collection rate had fallen from USC70 to USC60 per minute. Until December 1998, the Commission had granted over 700 applications to new providers of international service to US customers. Of the applicants, 175 had foreign ownership of 10 per cent or more, 48 were foreign carriers, and 18 were foreign dominant carriers. The Commission allowed incoming private line service on eleven new routes, including France, Germany, Italy, and Japan, plus the original five (Australia, Canada, New Zealand, Sweden and the UK). The number of European countries permitting basic voice competition, including international access, grew from three to 13. For the world the total had expanded to 23.

By the time of the second report, 99 per cent of net settled minutes for upper-income countries, scheduled for 1 January 1999, were in compliance with FCC benchmarks. The compliance rate for the next tier was 86 per cent. The average settlement rate for WTO signatories fell from USC53.5 in 1996 to USC30.4 in 1999, a decline of 43.2 per cent. The average settlement rate in WTO signatory countries with (without) full access commitments declined from USC31.9 (USC68.5) in 1996 to USC16.8 (USC39.9) in 1999, a decline of 47.4 per cent (41.8 per cent). Table 9.4 illustrates individual rates for the largest correspondents with the USA. The price decline benefited US callers, as the average US collection rate fell further to USC55 per minute. The UK–US rate reached USC10 and collection rates on other very large routes also fell significantly (Table 9.5).

Some additional data confirm the success of the WTO agreement to open networks to foreign investment. There are now 14 non-incumbent multinational carriers in the world (see Table 9.6). Thirteen had a positive growth rate in 1999 and five grew faster than in 1998. The WTO agreement is now fully in force in 25 countries, with Ireland, Peru and Portugal being the most recent signatories. Each country now allows foreign entities to own a majority interest in facilities used to provide international service, including voice and data. Other aspects of domestic deregulation were also successful. The total number of provider companies increased in the five largest foreign markets (1998, 1999): Japan (9, 42), Germany (32, 44), the UK (122, 158), France (33, 48) and Italy (14, 20). Based on EU agreements, interconnection rates for intra-continental traffic in that continent fell during the period.

Table 9.4   US settlement rates for WTO signatories with full market access (USD)

WTO Signatory	1996	1997	1998	1999	Benchmark	Effective
Australia	0.225	0.210	0.150	0.150	0.15	1/1/99
Austria	0.215	0.205	0.135	0.135	0.15	1/1/99
Belgium	0.280	0.185	0.135	0.135	0.15	1/1/99
Canada	0.110	0.100	0.100	0.100	0.15	1/1/99
Chile	0.500	0.500	0.350	0.350	0.19	1/1/00
Denmark	0.145	0.135	0.110	0.110	0.15	1/1/99
Dominican Republic	0.450	0.350	0.300	0.190	0.19	1/1/01
El Salvador	0.550	0.440	0.385	0.385	0.19	1/1/01
Finland	0.255	0.205	0.160	0.135	0.15	1/1/99
France	0.175	0.130	0.105	0.105	0.15	1/1/99
Germany	0.115	0.100	0.105	0.105	0.15	1/1/99
Guatemala	0.500	0.450	0.385	0.320	0.19	1/1/01
Iceland	0.470	0.375	0.240	0.135	0.15	1/1/99
Italy	0.260	0.165	0.110	0.110	0.15	1/1/99
Ireland	0.175	0.165	0.105	0.105	0.15	1/1/99
Japan	0.455	0.430	0.145	0.145	0.15	1/1/99
Korea	0.615	0.490	0.425	0.355	0.19	1/1/00
Luxembourg	0.290	0.135	0.135	0.135	0.15	1/1/99
Malaysia	0.445	0.395	0.395	0.190	0.19	1/1/00
Mexico	0.485	0.395	0.370	0.190	0.19	1/1/00
Netherlands	0.180	0.135	0.095	0.070	0.15	1/1/99
New Zealand	0.215	0.135	0.135	0.135	0.15	1/1/99
Norway	0.145	0.110	0.080	0.080	0.15	1/1/99
Peru	0.615	0.500	0.350	0.330	0.19	1/1/01
Philippines	0.500	0.500	0.360	0.285	0.19	1/1/01
Portugal	0.415	0.300	0.197	0.150	0.15	1/1/99
Spain	0.320	0.240	0.135	0.135	0.15	1/1/99
Sweden	0.085	0.060	0.060	0.060	0.15	1/1/99
Switzerland	0.255	0.170	0.135	0.135	0.15	1/1/99
UK	0.110	0.070	0.060	0.060	0.15	1/1/99
Ave. settlement rate	0.319	0.259	0.198	0.168		
Percentage change		−18.6%	−23.5%	−15.6%		

Source: Report on International Telecommunications Markets 1999 Update, FCC, Attachment 3.

*Table 9.5   US rates for low-cost calling plan (USD per outgoing billed
              minute)*

AT&T

	November 1997	November 1998	December 1999
Canada	0.05–0.12	0.05–0.10	0.05–0.10
Mexico	0.49	0.44	0.35
Germany	0.35	0.25–0.29	0.17
Japan	0.47	0.30	0.16
India	0.80	0.75	0.55
Monthly fee	3.00	3.00–5.95	5.95

WorldCom

	November 1997	November 1998	December 1999
Canada	0.12	0.05–0.12	0.05–0.07
Mexico	0.61	0.44	0.42
Germany	0.35	0.29	0.09–0.17
Japan	0.48	0.35	0.35
India	0.80	0.80	1.22
Monthly fee	3.00	3.00	4.95

Sprint

	November 1997	November 1998	December 1999
Canada	0.10–0.25	0.10	0.10
Mexico	0.55–0.75	0.47	0.63
Germany	0.30–0.70	0.27	0.25
Japan	0.43–0.87	0.46	0.39
India	1.05–1.40	0.78	0.72
Monthly fee	3.00	3.00	5.95

Note: The range represents the gap between peak and off-peak rates.

Source: *Report on International Telecommunications Markets 1999 Update*, FCC, p. 5.

CONCLUSION

The Commission has set its sights on reducing settlement rates paid to the
remaining nations outside the first benchmark tier. In 1999, the
Commission directly ordered US carriers to Cyprus (FCC, 1999b) and
Kuwait (FCC, 1999c) to reduce settlement rates to the benchmark of
USC15 per minute (down from USC37 and USC78, respectively). It also

Table 9.6   Non-incumbent multinational US and foreign carrier revenue

Carrier	Quarter 1, 1998 (thousand USD)	Annual change (%)	Quarter 1, 1999 (thousand USD)	Annual change (%)
Teleglobe	860	26	747	–13
Equant	>200/320	200	444	39
RSL Communications	132	212	340	159
Star Telecom	129	52	228	67
Global Telesystems	78	54	171	270
Energis	66	57	168	73
IDT Corporation	87	153	161	127
Pacific Gateway	105	104	141	34
COLT Telecom	59	134	137	130
Primus Telecom	118	24	131	64
Viatel	21.2	46	62	190
WorldCom	2350	39	n.a.	n.a.
Telegroup	86	16	n.a.	n.a.
Espirit Telecom	24	36	n.a.	n.a.

Source: Report on International Telecommunications Markets 1999 Update, FCC, Attachment 2.

rejected a petition from AT&T and some Philippine parties to reconsider the Benchmarks Order (FCC, 1999d). In September 1999, the Commission supplemented its ratemaking activity by authorizing direct carrier access to INTELSAT, an organization of 143 national signatories that owns and operates a fleet of satellites (FCC, 1999g). Most voice and data traffic routed to satellites is destined for developing countries. Previously, US carriers had to purchase access to INTELSAT through the Comsat Corporation, which was the US signatory to the INTELSAT treaty that represented a bottleneck and extra administrative expense. The Commission simultaneously began the privatization of Comsat with a 49 per cent share. Anticipated rate reductions from the INTELSAT reform range from 10 per cent to 71 per cent, depending on route.

Shortfalls in international settlements revenues can reduce funds needed for infrastructure development. The Commission has focused its attention upon competitive reform and technical assistance. On 2 June 1999, FCC Chairman Kennard unveiled the Kennard Development Initiative with the intention of providing telecommunications policy and regulatory assistance to a few key less-affluent nations. The purpose was to achieve universal access by implementing the stated goals of the WTO Basic

Telecommunications Services Agreement by promoting competition, liberalizing markets, encouraging investment, and adopting regulatory policies that are transparent and pro-competitive. Selected countries for FCC assistance were chosen on the basis of general regulatory environment, need for development and regulatory assistance, potential as a regional leader, and compatibility with US objectives. The Commission published an on-line manual (FCC, 1999e) that stressed the importance of independent regulation, open markets, market competition, spectrum management, licensing, and cost-based interconnection. To initiate the program, the Chairman visited southern Africa in August 1999 and met with telecommunications representatives from 12 countries in the region. Kennard challenged them to sign the WTO Agreement and stressed the importance of satellite and wireless systems to their networking needs. Annual bilateral work plans were signed subsequently with Ghana, South Africa and Uganda, all of whom are WTO members that meet the FCC's stated criteria. A recent report (FCC, 2000b) details the Commission's recent activity in Africa. Plans are under way for Asia, Central Europe and Latin America in the remainder of 2000.

To summarize, the Commission's policymaking in international telephony (in non-ideological fashion) has encompassed four distinct aspects since it began reform efforts in 1996. In 1997, the Commission was the classic regulator, establishing benchmark prices for international routes and timetables for meeting them. In 1996–99, the Commission served as chief US negotiator, offering liberalized pricing, interconnection, and settlements policy to foreign carriers that complied with its competitive guidelines. In 1999, the Commission became a unilateral deregulator, authorizing direct carrier access to INTELSAT and selling off part of Comsat. In 2000, the Commission became an international consultancy, offering technical assistance to developing nations in an effort to hasten their network build-out and liberalize their telecommunications sector. Price rules will become less meaningful as carriers continue to deploy IP telephony that defies per minute categorizations. The Commission can be expected to expedite the process as a negotiator/deregulator in calling partnerships with industrial nations.

ACKNOWLEDGEMENTS

The expressed views are personal and do not necessarily reflect the positions of other experts at LECG.

REFERENCES

Acton, J.P. and Vogelsang, I. (1992), 'Telephone demand over the Atlantic: evidence from country-pair data', *Journal of Industrial Economics*, **40**, 1–19.

Cave, M., and Donnelly, M.P. (1996), 'The pricing of international telecommunications services by monopoly operators', *Information Economics and Policy*, **8**, 107–23.

Einhorn, M.A. (2002), 'International telephony: a review of literature', *Information Economics and Policy*, **14** (1), 51–74.

FCC (1986), 'Implementation and scope of the international settlements policy for parallel routes', CC Docket No. 85-204, Report and Order, Washington, DC.

FCC (1995), 'In the matter of market entry and regulation of foreign-affiliated entities', IB Docket No. 95-22, Report and Order, Washington, DC.

FCC (1996a), 'In the matter of international settlement rates', IB Docket No. 96-261, Notice of Proposed Rulemaking, Washington, DC.

FCC (1996b), 'In the matter of regulation of international accounting rates', Docket No. CC 90-337, Phase II, Fourth Report and Order, Washington, DC.

FCC (1997a), 'In the matter of international settlement rates', IB Docket No. 96-261, Order, Washington, DC.

FCC (1997b), 'In the matter of rules and policies on foreign participation in the US telecommunications markets', Order and Notice of Proposed Rulemaking, IB Docket No. 97-142, Washington, DC.

FCC (1998a), 'In the matter of 1998 Biennial Regulatory Review – Reform of the international settlements policy and associated filing requirements', Notice of Proposed Rulemaking, IB Docket No. 98-148, Washington, DC.

FCC (1998b), 'Developments in international telecommunications markets', International Bureau, Report No. IN 98-58, Washington, DC.

FCC (1999a), 'In the matter of regulation of international accounting rates', CC Docket No. 90-337, Report and Order, Washington, DC.

FCC (1999b), 'In the matter of petition for enforcement of international settlements benchmark rates for service with Cyprus', IB Docket No. 96-261, Washington, DC.

FCC (1999c), 'In the matter of petition for enforcement of international settlements benchmark rates for service with Kuwait, IB Docket No. 96-261, Washington, DC.

FCC (1999d), 'In the matter of international settlement rates', IB Docket No. 96-261, Washington, DC.

FCC (1999e), 'Connecting the globe: a regulator's guide to building a global infrastructure community', Washington, DC.

FCC (1999f), 'Trends in the US international telecommunications industry', Washington, DC.

FCC (1999g), 'In the matter of direct access to the INTELSAT system', IB Docket No. 98-192, Washington, DC.

FCC (2000a), 'Report on international telecommunications markets: 1999 update', International Bureau, Report No. DA 00-87, Washington, DC.

FCC (2000b), 'Connecting the globe: the Africa initiative', Washington, DC.

Galbi, D. (1997), 'The implications of bypass for traditional international interconnection', presented at Telecommunications Policy Research Conference, Alexandria, Virginia (25 September).

Hundt, R. (1997), 'Separate statement on International Settlement Rates Report and Order' (7 August), FCC, Washington, DC.

ITU (1992), 'Accounting rate principles for international telephone services', Recommendation D.140, Geneva.

ITU (1993), 'Recommendations for regional application: determination of accounting rate shares in telephone relations between countries in Europe and the Mediterranean Basin', Recommendation D.300R, Geneva.

ITU (1997), 'Report of the Informal Expert Group on International Telecommunications Settlements', unpublished document, Geneva and London.

Johnson, L.L. (1989), 'Competition, pricing, and regulatory policy in the international telephone industry', Rand Corporation, R-3790-NSF/MF, Santa Monica.

Johnson, L.L. (1990), 'Dealing with monopoly in international telephone service: a US perspective', *Information Economics and Policy*, **4**, 225–47.

Madden, G. and Savage, S.J. (1998), 'Interconnection in international telecommunications: an empirical investigation of United States settlement rates', *Telecommunications Policy*, **22**, 1953–61.

Madden, G. and Savage, S.J. (2000), 'Market structure, competition, and pricing in United States international telephone services markets', *Review of Economics and Statistics*, **82**, 291–6.

Stanley, K.B. (1991), 'Balance of payments, deficits, and subsidies in international communications services: a new challenge to regulation', *Administrative Law Review*, **43**, 411–38.

Yun, K.L., Choi, H.W. and Ahn, B.H. (1997), 'The accounting revenue division in international telecommunications: conflicts and inefficiencies', *Information Economics and Policy*, **9**, 71–92.

L96
H54

10. Telecommunications infrastructure and economic development

M. Ishaq Nadiri and Banani Nandi

INTRODUCTION

The term 'infrastructure' generally describes large social overhead capital such as roads, bridges, sewer facilities, electricity generation and distribution, and communications networks. These infrastructures provide the basic framework for a nation to support essential public services in order to achieve higher economic growth and a better quality of life. The general characteristics of such infrastructural capital are such that the development and upgrading of infrastructure systems require large amounts of long-term investment and that they generate network externality benefits as the number of users connected to the infrastructure network increases. These infrastructure systems can be owned either by the public or private sector. Since investments in such systems are highly risky, public financing is the more common. Therefore, the existing econometric studies in this area focus primarily on publicly owned infrastructure systems. Numerous studies (Holz-Eakin, 1988; Aschauer, 1989a, 1989b, 1990; Munnell, 1990a, 1990b, 1992; Tatom, 1991; Berndt and Hansson, 1992; Morrison and Schwartz, 1994, 1996; Nadiri and Mamuneas, 1994, 1996), have measured the contribution of publicly funded infrastructure capital to economic growth. In general the results show a positive contribution but its magnitude and degree of significance vary among the studies.

Telecommunications infrastructure capital is highly dynamic, embodying new technologies, and has pervasive effects on society. It not only provides facilities for communications but saves time, energy, labor and capital by condensing the time and space required for production, consumption, market activities, governmental operation, educational and health services.[1] The quality and extent of this type of infrastructure system is critical for the functioning of a modern society and is an important contributor to economic growth in developing nations. Also, information and particularly telecommunications infrastructure capital stands in some contrast with other types of infrastructure capital. Its effect is very pervasive, it has experienced very high productivity and technical change, it has attracted a sizeable amount of investment capital, and it has linked world economies and societies, thereby making a significant contribution to the globalization of world economies and politics.

In recent years the world has experienced an explosive growth in information technology and its applications, particularly in the telecommunications industry, and it is extremely important for public policy to evaluate the contribution of the telecommunications infrastructure capital to economic and productivity growth.[2] There are, however, very few studies available that explore the role of telecommunications infrastructure services in various industries and the economy. The telecommunications network is composed of intricate and interdependent components, exhibits positive consumption and production externalities, and is technologically very advanced. In this study we assemble and evaluate the available literature on the contribution of telecommunications infrastructure capital to the growth of output and productivity. We specifically focus on the following three sets of issues: the contribution of telecommunications capital to growth of various economies; the effects of telecommunications infrastructure capital services on the production structure of other industries (on the input mix and output decisions in various industries); and the social rate of return to investment in telecommunications infrastructure capital.

The remaining discussion is organized as follows. The following section describes the nature of the telecommunications infrastructure capital and the services it provides. Studies on the contribution of telecommunications as one of the sources of economic growth is briefly summarized thereafter. The next section is devoted to studies based on an Input–Output (I–O) framework, and econometric cost models that describe the telecommunications capital effects on the cost structure of various industries and productivity growth of the national economy. The social rate of return to telecommunications investment is also discussed in this section. The conclusions are contained in the last section.

SERVICES PROVIDED BY THE TELECOMMUNICATIONS INFRASTRUCTURE

The telecommunications infrastructure network permits the point-to-point two-way transmission of information in accordance with user's choice. This network encompasses the transmission media, which include wired and wireless networks, satellites and antennas, together with routers, and other devices that control the transmission path of information. It also includes the software that is used to send, receive and manage signals that are transmitted and the various types of end user equipment that enable users to originate and terminate communications. The telephone network is often divided into two broad categories based on geographic coverage of service provisioning: national and international networks. National networks include local, regional and long-distance networks: long-distance networks connect different local and regional networks, and international networks connect networks of different countries and facilitate communications among countries.

Telecommunications networks like other infrastructure networks exhibit positive consumption and production externalities. A positive (or network) externality means that the benefits of telecommunications networks increase with the number of units sold or used. In the case of telephone networks, this kind of externality effect could appear due to an increase in penetration rate or an increase in telephone mainlines per population, so that more individuals access and use the network. The former kind of expansion takes place mostly in developing countries, whereas the later case of expansion is more common in developed countries where universal telephone service has existed for many years.

Different components of the telecommunications infrastructure have evolved dramatically over the last few decades. Rapid technological progress has occurred in wired networks, from copper wires to coaxial cables to optical fiber cable. Message conversation and signaling systems have changed from analog to digital systems through the introduction of advanced switching and transmission systems in many countries. Digital technology has brought great improvements into the telecommunications network and allowed the introduction of packet switching. The advantage of packet switching is that, unlike analog systems, it does not require unique wired paths between callers and recipients. Internet networks are the results of successful implementation of the packet transmission technology. The digital switching system also processes the traffic much faster and improves the quality of sound reproduction.

Simultaneously rapid technological progress is occurring in the area of radio-based telecommunications systems due to an increased demand for mobile telecommunications. Personal network systems are still new and have a shorter range than do cellular systems but are cost effective. The development of Global Mobile Personal Satellite is in progress, with the introduction of satellites, and will provide coverage of wireless telecommunications all over the world.

Recently, in most countries with advanced telecommunications systems, attention has been given to building broadband networks. In order to use these broadband facilities efficiently, technologists are designing methods to transport voice, data and video in an integrated fashion through the same switch and transmission machinery. The new broadband technology will integrate all three networks into one unified telecommunications network. In addition, the recent improvements in computer and telecommunications technology, together with the revolutionary improvement in information technology are leading to the convergence of computer and telecommunications infrastructure into one unified 'information infrastructure'. The current Internet infrastructure is the first step towards the future 'information superhighway'. Internet overlay augments the telecommunications network, which facilitates Internet telephony. Although different segments of the modern telecommunications network experience different rates of technological advance, the network as a whole remains highly interdependent and integrated.

With this rapid evolution of the telecommunications infrastructure, communications are becoming easier and more efficient and cost effective in facilitating complicated activities. With great speed and accuracy some of the important tasks handled by the modern telecommunications infrastructure are:

1. it improves management communication and implementation of decisions within firms;
2. it is a primary and indispensable contributing factor towards globalization of different corporations and the interconnection of world markets;
3. at the household level, the availability of better information helps in obtaining better education, better health services and making better decisions towards improving economic well being;
4. it facilitates the acquisition and transfer of information in the most cost-effective and time-efficient manner and at the same time minimizes the obstacles or barriers due to distance. This is particularly important in the diffusion of new ideas and knowledge and successful development strategies. Also it plays a significant role in narrowing regional development gaps in income and productivity;
5. telecommunications services as production inputs lead to economizing on costs, such as labor and other types of capital, and facilitates expansion of various types of information capital involved in the processing of information (Antonelli, 1990). A telecommunications network has additional advantages: it increases the efficiency of market operations, it lowers equilibrium idle resources in the economy (Norton, 1992), decreases transaction costs and increases aggregate output (Leff, 1984) and decreases cost of capital by increasing the arbitrage opportunities that make financial markets more efficient; and
6. telecommunications infrastructure generates significant spillover effects and externalities. The externalities involve lower search costs,

increased arbitrage abilities, more information on the distribution of prices and services and diffusion of key technological innovations to other sectors or economies. The impact of the communications infrastructure on the economy is similar to the impact of increased innovation (McNamara, 1991; Antonelli, 1993; Röller and Waverman, 1996).

AGGREGATE CONTRIBUTION OF TELECOMMUNICATIONS CAPITAL

There are several studies that use aggregate time-series, cross-section data to investigate the relationship between telecommunications investment and economic development. The analyses are based on statistical correlation or single regression equations linking telephone penetration ratio, that is, telephone lines per capita in a country, and some index of economic well being, such as GDP per capita. Jipp (1963) studied the relationship between income and telephone density using data from a number of countries and found a positive correlation. Bebee and Gilling (1976), using data from 29 countries at different stages of development for the year 1970, found the empirical relation:

$$Y = 5.928 - 9.078(1/X_1) - 7.093(1/X_2)$$

where Y is an index of economic performance, X_1 is an index of development support factor and X_2 is an index of telephone use and availability. Their statistical analysis shows a strong positive relationship between the telephone index and the economic performance index.

These results have been confirmed in other studies (Moss, 1981). Hardy (1980) estimated an econometric model and regressed the GDP per head for the current period on one period lagged values of GDP per head, number of telephones per million population and radio per thousand population. The coefficient of lagged telephone variable appeared to be statistically significant. Saunders, Warford and Wellenius (1994) summarized the benchmark work carried out by the CCITT of the International Telecommunication Union (ITU). The CCITT group (1968) estimated a log-linear relationship between the telephone line density (d) and GDP per capita (g):

$$\log(d) = \log(a) + b\log(g)$$

using cross-section data from the ITU for 1955, 1960 and 1965. The estimated value of b was about 1.4 for each of the three-year cross-section estimates. They also estimated the same equation for Sweden using annual time-series data for 1900 through 1965. In this estimate they obtained the value of b coefficient as 3.2 for the period 1900 through 1915, and 1.5 for the period 1920 through 1965. The former time-period better reflects the early stage of telephone service introduction. Although no detailed analysis is available, the studies of CCITT (1968) and Bebee and Gilling (1976) report that the telephone utilization rate increases more quickly as a country becomes more developed. The lowest telephone utilization rate is associated with the poorest of the developing countries. Though more research is needed the results of these two studies suggest that the rate of cost savings from telecommunications use is positively correlated with the stages of economic development.

In another study (Dhokalia and Harlam, 1994), using cross-section data for 50 states in the USA, found a positive correlation between telecommunications infrastructure investment and average annual pay per employee. Availability of telephones measured by the number of business access lines per employee explained 67 per cent of the variance in 1990 average annual pay of each state. Several other studies have shown that economic growth and development often involve multiple development inputs, such as education, telecommunications, and other physical infrastructure investment, and that telecommunications infrastructure capital works cooperatively with other inputs to promote economic development.

Madden and Savage (1998) estimated the relationship between gross fixed investment, telecommunications infrastructure investment and economic growth for a sample of transitional economies of Central and Eastern Europe. They estimated an economic growth equation, proposed by Kormedi and Meguire (1985), using a sample of eleven countries for annual data from 1991 to 1994. Their results showed that the share of telecommunications investment in GDP was a statistically significant variable explaining the rate of growth of GDP.

It is important to recognize that the relationship between investment in the telephone infrastructure network and economic growth and development is not unidirectional. Some recent studies suggest two-way causal relationships where telecommunications investment enhances economic activity and economic growth in turn stimulates the demand for telecommunications. Lee (1994), using the Granger causality test on Korean data, supported the hypothesis of two-way causality. Similar results are reported, for post-war US data, by Cronin et al. (1991) and DRI/McGraw-Hill (1991).

These aggregate level analyses indicate a positive relationship between telecommunications investment and economic development. The CCITT study shows that the teledensity ratio increases faster than per capita GDP growth as the elasticity of teledensity with respect to per capita GDP is greater than unity, or about 1.4, which suggests a very high rate of return on telecommunications investment. These high rates of return are likely to be overestimates and, like the early high rates of return estimated by Aschauer (1989a, 1990) for public capital infrastructure investment, are likely to be revised downward using more disaggregated analyses. None the less, the social rate of return to telecommunications investment, based on I–O framework and disaggregated industry studies, still show very impressive rates of return. The empirical results of the aggregate studies indicate that the contribution of telecommunications investment varies among countries. Investment in telecommunications infrastructure and the benefits it provides depend on the stage of development of each country, the regulatory environment and the degree of integration of the telecommunications services with other sectors of the economy.

The aggregate level studies, though very useful, do not distinguish between the direct and indirect contribution of telecommunications investment to growth of output and productivity and do not fully capture the network characteristics of telecommunications capital. There are few relevant studies available at the disaggregate level, particularly for developing countries and even for the majority of Western economies. The studies reported in the next section are based on US data that illustrate the role played by the telecommunications network as a basic infrastructure in a modern economy.

DIRECT AND INDIRECT EFFECTS OF INVESTMENT ON TELECOMMUNICATIONS CAPITAL

The contribution of telecommunications infrastructure capital and the distinction between its direct and indirect effects on various industries and the aggregate economy can best be analysed in a disaggregated framework. The direct contribution of telecommunications investment to economic growth arises from the productivity and growth of the telecommunications sector itself. As is well documented, investment in the telecommunications sector as a percentage of total investment, and the share of this industry in total output, has been rising in the past few years in both developed and developing countries (ITU 1999). Also, the productivity growth of this sector has been one of the highest in most economies. For example, Cronin et al. (1993a) report a productivity growth rate of 3 per cent per annum for the US telecoms industry for the period 1965 to 1991. Fuss and Waverman (1981), Jorgenson et al. (1987), Christensen et al. (1991), Crandall (1991), and Nadiri and Nandi (1999) report similar results. The direct contribution of the telecommunications sector to the total economy is due to its superior productivity growth, often the highest, among all industries and its increasing share in total national output.

The high investment rate and the superior productivity performance of the telecommunications industry have resulted in a substantial fall in relative prices of telecommunications services in most economies. Using the US experience as an example, the rate of output price growth of telecommunications was by far the lowest, averaging only 60 per cent of the next lowest sector for the period 1963 to 1991 (Cronin et al., 1997). The fall in the price of telecommunications services is even more dramatic when account is taken of the impressive advances in quality of telecommunications services. Given the substantial decline in relative prices of telecommunications services over time all industries, including telecommunications, have substituted telecommunications input for relatively higher priced traditional inputs such as labor and capital. A number of studies have examined the degree of substitution and complementarity of telecommunications input in the production function of user industries (Cronin et al., 1993b; Nadiri and Nandi, 1998). The general pattern is that telecommunications input services are often a substitute for labor input but highly complementary with capital input. As a result, the intensity of telecommunications has increased in production. For example, Cronin et al. (1993b) reported that in 1987 the average industry used approximately USD0.62 worth of telecommunications services to produce USD100 of output compared to USD0.27 to produce the same output in 1965, an almost 250 per cent increase. Some industries (industry codes) are more intense users of telecommunications services, such as finance, insurance, business services, wholesale and retail trade; these industries are some of the main contributors to the growth of total national output.

Improvement in the telecommunications sector has indirect effects that generate further savings throughout the economy. Improvement in telecommunications infrastructure investment reduces the cost and increases the quality of production in an industry that in turn reduces costs in other industries. In other words, improvement in one industry using telecommunications services begets improvements in others, thereby multiplying the indirect effects. One way to capture the indirect effect is to use an I–O framework. Cronin and his associates in several studies (1991, 1993b, 1997) have used this framework to assess the direct and indirect effects of investment in telecommunications infrastructure. Another approach to capture the contribution of telecommunications infrastructure is based on an econometric model such as that developed by Nadiri and Nandi. These approaches are briefly discussed below.

INPUT–OUTPUT APPROACH

The savings from investment in modernization and expansion in the telecommunications sector can be analysed by using I–O analysis based on Leontief (1953) and Peterson's (1979) sectoral I–O model. As noted earlier, Cronin et al. (1993b, 1997), in several studies have used an I–O framework to calculate a total benefit in resource savings from advances in the telecommunications sector. Cronin et al. (1997) followed a two-step estimation procedure in their recent study using US telecommunications data for the period 1963–91. In this study, first the I–O system is solved to obtain the relevant input requirements corresponding to actual 1991 relative prices and telecommunications technology. In the second step the system is solved under the assumption that the relative price and the technology of telecommunications services had remained at its initial 1963 level.

To calculate the resource saving the I–O coefficient matrix, A, a (36×36) matrix of material inputs required to produce a unit of output in 1991 is defined. The primary input coefficients matrix, B, is a (2×36) matrix of labor and capital inputs required to produce the 1991 output. Finally, the 1991 final demand vector C, is a (36×1) vector of actual 1991 final deliveries by each of 36 industries. The I–O model is based on the relationship:

$$V = B(I - A)^{-1}C \qquad (10.1)$$

where V is a (2×1) vector of actual labor and capital used in 1991 to produce the 1991 final demand. The above system (10.1) is solved again under the restriction that the technology of telecommunications services has not changed since 1963. Matrix A and vector B are redefined as $A\#$ and $B\#$ which are the hypothetical matrices that reflect the situation if the relative prices and telecommunications technology had stayed at their 1963 levels and the solution vector is redefined as $V\#$. Based on V and $V\#$ vectors, they calculated the total benefits of the advancement of telecommunications infrastructure since 1963. Total benefits are decomposed into production and consumption effects. Production effects are resource savings from telecommunications industry productivity gains due to modernization. The consumption effect, on the other hand, captures efficiency gains to end-users. Gains arise from the introduction of new and enhanced services, and lower prices. However, the production and consumption effects are not independent. To capture this complex relationship, Cronin et al. (1997) calculate the consumption effect and the total (production and consumption) effect separately by redefining matrix $A\#$ and matrix $\#B$. The consumption effect is inferred by subtracting the consumption from the total effect.

Cronin et al. (1997) report a total benefit of USD134.4 billion in resource savings in 1991 due to advances in telecommunications since 1963. The estimated benefit to the 1991 economy of the consumption effect was USD61.1 billion. The difference between the total effect and the consumption effect is USD73.3 billion, which represents the benefits to the 1991 economy due to the production effect. The distribution of these benefits as shown in Table 10.1 is highly variable across industries and between the production and consumption effects. The main benefits of telecommunications investment occur in industries (codes) such as finance and insurance, wholesale and retail trade, other services such as construction, agriculture, transportation and motor vehicles. However, the effect of productivity improvement in the telecommunications industry accounts for a large proportion of the aggregate saving.

Using this methodology, the authors generate a time-series of resource saving attributable to telecommunications modernization and expansion since 1963. The total benefit due to telecommunications capital expansion and improvement reveals a steady increase over the period 1963 to 1991. The social rate of return on telecommunications investment is computed as a function of the benefits and costs of such investment. The cost is the economy-wide investment in telecommunications, and the benefits, as noted above, are the stream of benefits stemming from increased usage of telecommunications since 1963. The rate of return has been rising rapidly over the period 1964–91 and is in the range of 31 to 40 per cent. Comparing this return to yield on long-term utility bonds, the telecommunications network investment yielded, on average, 27 per cent more per annum than the bond market.

AN ECONOMETRIC COST MODEL

Nadiri and Nandi developed a model that incorporates the effects of telecommunications infrastructure and publicly financed infrastructure on the cost structure of 34 sectors and industries of the US economy. The model is estimated using the time-series data for 1950 through 1990. The benefits of the telecommunications infrastructure captured in this model is in addition to the fees paid directly by the non-telecommunications industries to the telecommunications service providers. They are due to the externality gains from the telecommunications network not paid for directly by users. The cost function considered for each industry is:

$$C = C(q,\ Y,\ T,\ S_1,\ S_2) \tag{10.2}$$

Table 10.1 *1991 direct and indirect resources saved nationwide due to telecommunications advances since 1963 (billions of 1991 US dollars)*

Industry	Total savings	Savings due to increased tele-communications consumption	Savings due to improved tele-communications productivity
Finance and insurance	20.3	17.8	2.4
Wholesale and retail trade	19.3	14.3	5.0
Real estate	9.0	8.1	1.0
Personal and miscellaneous services	7.4	4.5	3.0
Construction	3.6	2.0	1.7
Agriculture, food and tobacco	3.1	1.9	1.2
Transportation and warehousing	2.9	2.1	0.8
Motor vehicles and miscellaneous	2.8	1.8	1.0
Comp. office and non-electrical machinery	2.5	1.1	1.4
Electric and electronic equipment	2.0	1.2	0.8
Business services	1.7	0.5	1.2
Utilities	1.6	1.3	0.3
Amusements	1.2	1.0	0.2
Chemicals and products	0.9	0.5	0.4
Printing and publishing	0.8	0.5	0.3
Textiles	0.6	0.3	0.3
Instruments	0.6	0.3	0.3
Lumber and wood products	0.5	0.3	0.2
Fabricated metals	0.5	0.3	0.2
Crude petroleum, mining and refining	0.5	0.4	0.1
Furniture	0.4	0.2	0.2
Automotive repair	0.3	0.0	0.3
Miscellaneous manufacturing	0.3	0.2	0.1
Rubber and plastics	0.2	0.2	0.1
Paper and paperboard	0.2	0.1	0.1
Stone, clay and glass	0.1	0.1	0.0
Leather	0.1	0.0	0.0
Radio and TV	0.1	0.0	0.0
Mining	0.0	0.0	0.0
Primary metals	−0.1	−0.1	0.0
Non-telecommunications subtotal	83.4	60.9	22.5
Telecommunications	51.0	0.2	50.8
US Total	134.3	61.1	73.3

Source: Cronin et al. (1997).

where q denotes the vector of input prices for inputs labor L, private capital K and materials M, Y denotes output quantity, T is an index of time representing disembodied technical change, S_1, the flow of services of the telecommunications infrastructure capital and S_2 the flow of services from public infrastructure capital. The fees paid for telecommunications services by each industry are included as part of the payment for materials. It is assumed that the demand for private inputs fully adjusts to their cost-minimizing levels within one period. There are no market prices for the externality effects of the infrastructure capital, S_1 and S_2. However, one can determine the shadow price or willingness-to-pay for S_1 and S_2, that is, the savings in private production costs associated with them. The shadow values are the measure of the potential cost savings from a decline in variable cost.

The marginal benefits of the two types of infrastructure capital for each industry are given by:

$$U_{S_1} = -\partial C / \partial S_i \qquad i = 1, 2 \tag{10.3}$$

which shows that an increase in any of the infrastructure capitals results in savings of U_{S_i} monetary unit of total production cost. These cost reductions are defined as the marginal benefit of each infrastructure capital. The aggregate marginal benefit derived from the telecommunications infrastructure is obtained by summing the marginal benefits of all industries and is defined as $SMBS_1$. Similar estimates of the aggregate marginal benefits derived from the public infrastructure capital are defined as $SMBS_2$.[3]

The average cost elasticity (η_{CS1}) with respect to telecommunications infrastructure capital S_1 is shown in Column 1 of Table 10.2. This shows that an increase in telecommunications capital stock reduces costs in all manufacturing as well as non-manufacturing industries. The magnitudes of the cost elasticities (η_{CS1}) vary considerably among industries. Using these estimates and equation (10.3), we calculate the marginal benefits from the telecommunications infrastructure capital in each industry. They show the magnitude of cost reduction experienced by an individual industry as a result of an increase in this type of infrastructure capital service. The average marginal benefit (MB) of the telecommunications capital in real terms (that is, in terms of materials prices) for each industry over the sample period is shown in Column 5 of Table 10.2. They indicate each industry's willingness-to-pay for an additional unit of the telecommunications infrastructure capital services. As noted earlier this benefit is exclusive of the direct payments each industry makes for telecommunications services to the providers of these services. These estimates measure the network externality benefits of the telecommunications infrastructure capital in each industry.

Table 10.2 Elasticities of cost and conditional input demand, and marginal benefit (mean values), 1950–91

Industry Code	Industry Title	η_{CS1}	η_{LS1}	η_{KS1}	η_{MS1}	MBS_1
1	Agriculture, forestry and fisheries	−0.0104	−0.0062	0.0005	−0.0152	0.0072
2	Metal mining	−0.0116	−0.0074	−0.0047	−0.0171	0.0003
3	Coal mining	−0.0108	−0.0080	0.0001	−0.0170	0.0007
4	Crude petroleum and natural gas	−0.0093	−0.0031	−0.0056	−0.0165	0.0028
5	Nonmetallic mineral mining	−0.0119	−0.0081	−0.0044	−0.0175	0.0004
6	Construction	−0.0091	−0.0061	0.0128	−0.0141	0.0120
7	Food and kindred products	−0.0100	−0.0027	0.0149	−0.0135	0.0088
8	Tobacco manufactures	−0.0119	−0.0032	−0.0038	−0.0162	0.0006
9	Textile mill products	−0.0111	−0.0045	0.0135	−0.0147	0.0023
10	Apparel and other textile products	−0.0108	−0.0072	0.0241	−0.0150	0.0022
11	Lumber and wood products	−0.0122	−0.0084	0.0166	−0.0165	0.0021
12	Furniture and fixtures	−0.0115	−0.0082	0.0105	−0.0162	0.0010
13	Paper and allied products	−0.0103	−0.0059	0.0022	−0.0148	0.0027
14	Printing and publishing	−0.0102	−0.0073	0.0050	−0.0155	0.0029
15	Chemicals and allied products	−0.0100	−0.0045	−0.0016	−0.0145	0.0045
16	Petroleum refining	−0.0090	0.0110	0.0155	−0.0122	0.0036
17	Rubber and plastic products	−0.0102	−0.0067	0.0217	−0.0146	0.0028
18	Leather and leather products	−0.0115	−0.0080	−0.0007	−0.0160	0.0005
19	Stone, clay and glass products	−0.0102	−0.0070	0.0030	−0.0155	0.0016
20	Primary metals	−0.0081	−0.0022	0.0132	−0.0119	0.0042
21	Fabricated metal products	−0.0101	−0.0066	0.0058	−0.0149	0.0030
22	Machinery, except electrical	−0.0085	−0.0052	0.0078	−0.0135	0.0047
23	Electrical machinery	−0.0097	−0.0065	0.0067	−0.0147	0.0038

305

Table 10.2 (cont.)

Industry Code	Industry Title	η_{CS1}	η_{LS1}	η_{KS1}	η_{MS1}	MBS_1
24	Motor vehicles	−0.0095	−0.0038	0.0103	−0.0134	0.0046
25	Other transportation equipment	−0.0102	−0.0073	0.1605	−0.0150	0.0034
26	Instruments	−0.0088	−0.0060	0.0049	−0.0146	0.0019
27	Miscellaneous manufacturing	−0.0114	−0.0079	0.0069	−0.0160	0.0010
28	Transportation and warehousing	−0.0099	−0.0073	0.0012	−0.0165	0.0067
30	Electric utilities	−0.0104	−0.0049	−0.0063	−0.0177	0.0033
31	Gas utilities	−0.0098	−0.0006	−0.0019	−0.0141	0.0019
32	Trade	−0.0078	−0.0057	0.0033	−0.0167	0.0181
33	Finance, insurance, and real estate	−0.0087	−0.0057	−0.0029	−0.0169	0.0136
34	Other services	−0.0085	−0.0061	−0.0009	−0.0168	0.0196
35	Government enterprises	−0.0107	−0.0078	−0.0030	−0.0177	0.0023

306

All the marginal benefits estimates are positive, indicating that the shadow value of the telecommunications infrastructure capital service is positive in all industries. However, the magnitudes of these benefits vary considerably among industries. Low rates of marginal benefits are observed in metal mining (industry code 2), coal mining (industry code 3), non-metallic mining (industry code 5) and some industries like furniture and fixtures (industry code 12), leather products (industry code 18) and miscellaneous manufacturing (industry code 27). The magnitudes of marginal benefits are high for the service sectors. The top two sectors are trade (industry code 32), and finance, insurance and real estate (industry code 33). The relatively high values of MB in the service industries reflect the high information intensities of these industries. These estimates are consistent with findings based on I–O models.

As noted earlier, increased demand for telecommunications services in response to its fairly steep price decline leads to changes in the input mix of each industry. The effect on the demand for other inputs such as labor and capital will depend on the substitutions or complementarities embedded in the production structure of each industry. Nadiri and Nandi calculate the degree of substitution and complementarity of capital, labor and materials with an increase in telecommunications infrastructure capital, S_1. The relevant elasticities are shown in Column 2 to Column 4 of Table 10.2. The magnitudes and signs of the average estimated elasticities with respect to telecommunications capital differ for labor, private capital stock and material inputs. The signs of the labor and material input elasticities are negative suggesting a substitution relationship for telecommunications infrastructure capital. The elasticities with respect to capital are generally positive suggesting complementarity between S_1 and private capital. In general, telecommunications investment is labor and materials saving and capital using in almost all US industry.

The increase in the telecommunications infrastructure capital raises the production efficiency of all industry and thereby increases productivity at national level. The contributions of the telecommunications infrastructure capital at the economy level are estimated by aggregating the industry level estimates. The average cost elasticity, social marginal benefits and the net rate of return to telecommunications infrastructure capital is shown in Table 10.3. The average aggregate elasticity of cost with respect to the telecommunications infrastructure (η_{CSI}) is – 0.0091. This suggests that a 1 per cent increase in telecommunications capital will reduce costs by about 0.01 per cent.

The aggregate elasticities of labor, capital and materials with respect to the telecommunications infrastructure capital are shown in Panel B of Table 10.3. The average elasticity of labor with respect to telecommunications

Table 10.3 *Average elasticity and social and net rate of return of telecommunications capital*

Panel A

η_{CS1}	$SMBS_1$	$GRRS_1$	$NRRS_1$
−0.0091	0.1514	0.2571	0.1713

Panel B

η_{LS1}	η_{KS1}	η_{MS1}
−0.0057	0.0020	−0.0147

Panel C

$\bar{\eta}_{LS1}$	$\bar{\eta}_{KS1}$	$\bar{\eta}_{MS1}$
0.0019	0.0062	−0.0031

---

infrastructure capital η_{LS1} is about − 0.0057, while with respect to materials η_{MS1} is also negative, but larger, that is − 0.0147. The size of the elasticity with respect to capital η_{KS1} is positive but smaller, about 0.002. These elasticities suggest that telecommunications infrastructure is mainly materials saving. On average an increase of 1 per cent in S_1 generates a saving of about 0.6 per cent in labor but increases demand for capital by about 0.2 per cent per annum. When the level of output is allowed to vary in response to the reduction of costs due to the increase in telecommunications infrastructure capital, S_1, the demand for traditional factors of production would be affected. The increased level of output would require an increase in demand for the inputs. The total input elasticities including the output expansion effects and substitution (complementarity) effects are shown in Panel C of Table 10.3. These elasticities suggest that telecommunications infrastructure capital increases demand for labor and particularly capital while it is still materials saving. That is, the effect of this type of capital on the production structure of the economy is not neutral, but biased against the demand for materials and increases the demand for labor and capital.

The net rate of return to the telecommunications infrastructure capital is calculated using the expression

$$NRRS_1 = \sum_f \left(\frac{\partial C_f}{\partial S_1} \right) / P_{S1} - \delta \qquad (10.4)$$

where $\sum_f \partial C_f / \partial S_1$ is the sum of marginal benefits of all industries, P_{S1} is the acquisition price and δ is the depreciation rate of the telecommunications infrastructure capital. Applying (10.4), we estimate the net rate of return to this capital. Table 10.3 also shows the values of the sum of marginal benefits of all industries ($SMBS_1$). The magnitude of $SMBS_1$ suggests an average marginal benefit from telecommunications infrastructure to the national economy of about 15 per cent per annum. The value of $SMBS_1$ has been rising since the 1950s, and in 1991 was about 35 per cent. The average gross rate of return to S_1, $GRRS_1$, is about 26 per cent while the net rate of return, $NRRS_1$, is approximately 17 per cent. Both gross and net rates of return $GRRS_1$ and $NRRS_1$, respectively, have consistently increased over time. In 1991 they were 33 per cent and 24 per cent, respectively. This suggests an impressive social rate of return to the telecommunications infrastructure investment. The total rate of return, consisting of direct payments received by the telecommunications firms from other industries and the social rate of return on the telecommunications infrastructure network, as noted, are even more impressive.

CONCLUSIONS AND UNRESOLVED ISSUES

The evidence presented in this study suggests a strong positive relationship between the growth of telecommunications infrastructure capital, the growth of output and productivity. This relationship is evident in the estimated results at the industry and aggregate economy levels. The study also measures the externality effects of telecommunications infrastructure capital on production. The evidence clearly shows that increases in size and modernization of telecommunications networks are cost reducing in all industries of the national economy. The marginal benefits of this capital in each industry are positive and fairly large. These benefits vary considerably across industries. At the aggregate level the total marginal benefit from this infrastructure capital is also rather sizeable, about 30 to 40 per cent. Moreover, telecommunications infrastructure capital influences production structure at both industry and aggregate economy level. Its effect on an input such as labor, materials and capital is not neutral. It is mainly labor and materials saving but it increases demand for capital. Finally, the gross and net rates of return to this infrastructure capital are very high and rising over time.

Our results have important policy implications. The evidence of a significant contribution of telecommunications infrastructure to growth of output and productivity suggests that through regulation, tax and subsidy programs, governments should encourage efficient use of existing infrastructure. The fact that marginal benefits derived from the telecommunications infrastructure vary across industries and since the effects of this type of capital on input is shared among various industries, has important policy implications for the growth of different industries.

In this chapter we provide a broad overview of the relationship between telecommunications infrastructure capital and economic development based on the existing empirical evidence. However, several important issues require further research and closer examination. Some of the more important issues are:

1. Telecommunications infrastructure capital also affects consumer decisions and leads to greater demand for various products of different industries, and demand for new services. This can generate an externality effect on the demand-side and can further enhance the output expansion effect due to telecommunications infrastructure capital. There is clearly a need for a general equilibrium model to account for effects of telecommunications infrastructure on industries' costs and demands. A more complete analysis would incorporate the link between investments in telecommunications infrastructure capital and demand for telecommunications services from business and residential users. More detailed analysis of the short- and long-run impacts of this capital on employment, information capital and other factors of production will be worthwhile.

2. Rapid advancement in telecommunications technology not only lowers costs but also affects the quality dimension of the services it provides. The existing studies do not focus on this issue. Serious measurement obstacles need to be overcome in terms of data and modeling to capture the full benefits of telecommunications (see McNamara, 1991). Also, the long-term dynamic efficiency of the improvement and expansion of telecommunications infrastructure cannot be captured fully by the static model. Some aspects may be addressed by using growth models based on endogenous technological progress that incorporate the benefits of knowledge and cost of acquiring it (Innis, 1950, 1951; Lucas, 1988; Romer, 1990; Dudley, 1999).

3. So far, empirical research has focused on the impact of investment in telecommunications infrastructure on domestic economic development. However, there is an important global feature of national telecommunications infrastructure. The question is, what are the linkages among various countries' telecommunications infrastructure systems and how do they affect the global economy? There is broad consensus that telecommunications is a powerful tool for the integration of developed and developing economies. This integration will increase the efficiency of the global economy by increasing the international division of labor and the efficiency of the international

capital market and through the rapid transfer of new technologies and knowledge from one country to another (Engelbrecht, 1997). Successful implementation of a global infrastructure and the proper distribution of economic benefits derived from it among different countries need more thought and coordination among nations.

4. A related issue is to identify a reasonable regulatory framework which enables each country to enjoy the full economic and social benefits of modern telecommunications technology. There is no unique model of financing telecommunications infrastructure and no single set of regulatory policies regarding the ownership structure of networks and pricing structures which will be appropriate for all countries. Optimum policy regarding any infrastructure is country specific and to a great extent depends on the availability of investment funds, stage of development and the institutional structure of the telecommunications sector. The host of issues of coordination and regulation of telecommunications on a world basis is an important policy design challenge that needs close attention.

ACKNOWLEDGEMENTS

Support from the C.V. Starr Center for Applied Economics of New York University is gratefully acknowledged. The authors also wish to thank Hyunbae Chun and Ayda Erbal for their valuable help on this paper.

NOTES

1. See NTIA Infrastructure Report (1991), Maddock (1997), ITU (1999).
2. A number of recent studies have examined the contribution of the general information infrastructure capital to output and productivity growth of various industries (Wildman, 1992; Khan, 1993; Loveman, 1994; Gera et al., 1998). A number of studies specifically measure the contribution of information technology equipment, such as computers, to the performance of the overall economy (Bresnahan, 1986; Jorgenson and Stiroh, 1995, 1999; Morrison and Schwartz, 1997). These studies suggest some interesting findings but they do not cover the potential benefits of the entire telecommunications infrastructure network to all the US sectors and industries.
3. For econometric specification of a similar model and its estimation see Nadiri and Nandi (1998).

REFERENCES

Antonelli, C. (1990), 'Information Technology and the Demand for Telecommunications Service in Manufacturing Industry', *Information Economics and Policy*, **4**, 45–55.

Antonelli, C. (1993), 'Investment, Productivity Growth and Key-Technologies: The Case of Advanced Telecommunications', *The Manchester School*, **LXI** (4), 386–97.

Aschauer, D.A. (1989a), 'Is Public Expenditure Productive?', *Journal of Monetary Economics*, **23**, 177–200.

Aschauer, D.A. (1989b), 'Public Investment and Productivity Growth in the Group of Seven', *Journal of Economic Perspectives*, **13** (5), 17–25.

Aschauer, D.A. (1990), 'Why Is Infrastructure Important?' in A. Munnell' (ed.), *Is There a Shortfall in Public Capital Investment?*, Boston, MA: Federal Reserve Bank of Boston, 21–50.

Bebee, F.L. and Gilling, E.J.W. (1976), 'Telecommunications and Economic Development: A Model for Planning and Policy Making', *Telecommunications Journal*, **43** (7), 537–43.

Berndt, E.R. and Hansson, B. (1992), 'Measuring the Contribution of Public Infrastructure in Sweden', *Scandinavian Journal of Economics*, **94**, S152–S172.

Bresnahan, T. (1986), 'Measuring the Spillovers from Technical Advance', *American Economic Review*, **76** (4), 742–55.

CCITT (International Consulting Committee on Telephone and Telegraph) (1968), *GAS-5 Handbook: Economic Studies at the National Level in the Field of Telecommunications*, Geneva: International Telecommunication Union.

Christensen, L.D., Christensen, D.C. and Schoech, P.E. (1991), *Total Factor Productivity in the Bell System, 1947–1979*, Wisconsin: Christensen Associates Inc.

Crandall, R.W. (1991), 'Efficiency and Productivity', in Berry G. Cole (ed.), *After the Breakup: Assessing the New Post AT&T Divestiture Era*, New York: Columbia University Press, 409–18.

Cronin, F.J., Parker, E.B., Colleran, E.K. and Gold, M.A. (1991), 'Telecommunications Infrastructure and Economic Growth: An Analysis of Causality', *Telecommunications Policy*, **15** (6), 529–35.

Cronin, F.J., Colleran, E.K., Herbert, P.L. and Lewitzky, S. (1993a), 'Telecommunications and Growth', *Telecommunications Policy*, **17** (9), 677–90.

Cronin F. J., Colleran, E.K., Gold, M.A., Herbert, P.L. and Lewitzky, S. (1993b), 'Factor Prices, Factor Substitution, and the Relative Demand for Telecommunications Across US Industries', *Information Economics and Policy*, **5**, 73–85.

Cronin F. J., Colleran, E.K., Gold, M.A., Herbert, P.L. and Lewitzky, S. (1997), 'The Social Rate of Return from Telecommunications Infrastructure Investment', mimeo.

Dholakia, R.R. and Harlam, B. (1994), 'Telecommunications and Economic Development', *Telecommunications Policy*, **18** (6), 470–77.

DRI/McGraw-Hill (1991), 'The Contribution of Telecommunications Infrastructure to Aggregate and Sectoral Efficiency', mimeo.

Dudley, M.L. (1999), 'Communications and Economic Growth', *European Economic Review*, **43**, 595–619

Engelbrecht, H.J. (1997), 'The International Economy, Knowledge Flows and Information Activities', in D. Lamberton, (ed.), *The New Research Frontier of Communications Policy*, Amsterdam: Elsevier, 19–42.

Fuss, M.A., and Waverman, L. (1981), 'Regulation and Multi-Product Firms: The Case of Telecommunications in Canada', in Gary Fromn (ed.), *Studies in Public Regulation*, Cambridge, MA: MIT Press.

Gera, S., Su, W. and Lee, F.C. (1998), 'Information Technology and Productivity Growth: An Empirical Analysis for Canada and the United States', Microeconomic Analysis Working Paper No. 20, Industry Canada.

Hardy, A.P. (1980), 'The Role of the Telephone in Economic Development', *Telecommunications Policy*, **4** (4), 278–86.

Holz-Eakin, D. (1988), 'Private Output, Government Capital and the Infrastructure Crisis', Discussion Paper Series No. 394, New York: Columbia University.

Innis, H.A. (1950), *Empire and Communications*, Oxford: Clarendon Press.

Innis, H.A. (1951), *The Bias of Communication*, Toronto: University of Toronto.

International Telecommunication Union (ITU) (1999), *Telecommunications and Economic Growth*, Reports from the Seminars jointly organized by Telecommunications Development Bureau, International Telecommunication Union (ITU) in 1998 and International Relations Program, Webster University Geneva, in 1997, Published by Department of Telecommunications Bureau, International Telecommunication Union (electronic book).

Jipp, A. (1963), 'Wealth of Nations and Telephone Density', *Telecommunications Journal*, **30** (1), 199–201.

Jorgenson, D.W., Gallop, F.M. and Fraumeni, B.M. (1987), *Total Factor Productivity and US Economic Growth*, Cambridge, MA: Harvard University Press.

Jorgenson, D. and Stiroh, K.J. (1995), 'Computers and Growth', *Economics of Innovation and New Technology*, **3** (3–4), 295–316.

Jorgenson, D. and Stiroh, K.J. (1999), 'Information Technology and Growth', *American Economic Review*, **88** (2), 109–15.

Kahn, B. (1993), 'The Ebbs and Flows of Infrastructure Modernization', Paper presented at the Advanced Workshop in Regulation and Public Utility Economics' Twelfth Annual Conference, Massachusetts.

Kormedi, R. and Meguire, P. (1985), 'Macroeconomic Determinants of Growth', *Journal of Monetary Economics*, **6**, 41–63.

Lee, C. (1994), 'The Causal Relationship between Telecommunications Investment and Economic Development in Korea', paper presented to the International Telecommunications Society Meeting, Sydney.

Leff, N.H. (1984), 'Externalities, Information Costs, and Social Benefit-Cost Analysis for Economic Development: An Example from Telecommunications', *Economic Development and Cultural Change*, **32**, 255–76.

Leontief, W.W. (1953), *Studies in the Structure of the US Economy*, Oxford: Oxford University Press.

Loveman, G. (1994), 'An Assessment of the Productivity Impact of Information Technologies', in Allen, T.J. and Morton, M.S. (eds), *Information Technology and the Corporation of the 1990s: Research Studies*, Oxford: Oxford University Press.

Lucas, R.E. (1988), 'On the Mechanism of Economic Development', *Journal of Monetary Economics*, **22**, 3–42.

Madden, G. and Savage, S.J. (1998). 'CEE Telecommunications Investment and Economic Growth', *Information Economics and Policy*, **10** (2), 173–95.

Maddock, R. (1997). 'Telecommunications and Economic Development', in D.M. Lamberton (ed.), *The New Research Frontier of Communications Policy*, Amsterdam: Elsevier, 159–74.

McNamara, J.R. (1991), *The Economics of Innovation in the Telecommunications Industry*, New York: Quorum Books.

Morrison C.J. and Schwartz, A.E. (1994), 'Distinguishing External from Internal

Scale Effect: The Case of Public Infrastructure', *Journal of Productivity Analysis*, **5** (3), 249–70.

Morrison C.J. and Schwartz, A.E. (1996), 'State Infrastructure and Productivity Performance', *American Economic Review*, **8** (5), 1095–111.

Morrison C.J. and Schwartz, A.E. (1997), 'Assessing the Productivity of Information Technology Equipment in US Manufacturing Industries', *Review of Economics and Statistics*, **79** 3), 471–81.

Moss, M.L. (ed.) (1981), *Telecommunications and Productivity*, Reading: Addison-Wesley.

Munnell, A.H. (1990a), 'Why has Productivity Declined? Productivity and Public Investment', *New England Economic Review*, Jan./Feb., 3–22.

Munnell, A.H. (1990b) 'How Does Public Infrastructure Affect Regional Preference?', *New England Economic Review*, Sept./Oct., 11–32.

Munnell, A.H. (1992), 'Policy Watch: Infrastructure Investment and Economic Growth', *Journal of Economic Perspectives*, **6** (4), 189–98.

Nadiri, M.I. and Mamuneas, T. (1994), 'The Effects of Public Infrastructure and R&D Capital on the Cost Structure and Performance of the US Manufacturing Industries', *Review of Economics and Statistics*, **76** (1), 22–37.

Nadiri, M.I. and Mamuneas, T. (1996), 'Contribution of Highway Capital to Industry and National Productivity Growth', mimeo.

Nadiri, M.I. and Nandi, B. (1998), 'The Communications Infrastructure and Economic Growth in the Context of the US Economy', mimeo, paper presented at the International Telecommunications Society's annual conference, Stockholm, Sweden.

Nadiri, M.I. and Nandi, B. (1999), 'Technical Change, Markup, Divestiture and Productivity Growth in the US Telecommunications Industry', *Review of Economics and Statistics*, **81** (3), 488–98.

NITA Infrastructure Report: Telecommunications in the Age of Information (1991) Washington, DC: US Department of Commerce.

Norton, S.W. (1992), 'Transaction Costs, Telecommunications and the Micro-economics of Macroeconomic Growth', *Economic Development and Cultural Change*, **41** (1), 175–96

Peterson, W. (1979), 'Total Factor Productivity in the UK: A Disaggregated Analysis', in *Measurement of Capital*, Basingstoke: Macmillan, 212–23.

Röller, L.H. and Waverman, L. (1996), 'The Impact of Telecommunications Infrastructure on Economic Development', in P. Howitt (ed.), *The Implications of Knowledge Based Growth for Micro-Economic Policies*, Calgary: University of Calgary Press.

Romer, P.M. (1990), 'Endogenous Technological Change', *Journal of Political Economy*, **98**, 71–102.

Saunders, R.J., Warford, J.J. and Wellenius, B. (1994), *Telecommunications and Economic Development*, Second Edition, A World Bank Publication, Baltimore: Johns Hopkins University Press.

Tatom, J. (1991), 'Public Capital and Private Sector Performance', *St. Louis Federal Reserve Bank Review*, **73** (3), 3–15.

Wildman, S. (1992), 'Information Technology, Productivity, and Trade Implications', Working Paper No. 521, Columbia Institute for Tele-Information.

11. Universal service (US) L96 L98

James H. Alleman and Paul N. Rappoport

INTRODUCTION

Digital divide, digital inclusion, universal service, universal service obligation (USO), and national information infrastructure (NII) initiative. These
expressions all have the sound of virtue. Who could be against closing the
digital divide or expanding universal service? But in fact when one explores
the meaning of these terms in greater detail and, more importantly, the
manner in which they are implemented and funded the concepts become
much less virtuous. These are basically political clichés that have clouded
the economic goals that underlie the terms. In this chapter it is argued that
what is addressed by these phrases is really a resource allocation issue. The
programs designed to implement these allocation goals have for the most
part been unsuccessful. For clarity and expository purposes the notion of
universal service and the digital divide are separated into: the desired goals;
the available implementation methods – the instrument used and those proposed; and the funding mechanisms utilized.

The definition and the rationales offered for universal service (including:
increasing telephone penetration, network externalities, income redistribution and infrastructure development) are explored. The rationales are
judged against economic theory, empirical evidence, the instruments used
to obtain these goals and the efficacy of the mechanism itself. All are found
deficient. Original and received empirical analyses support this conclusion.
Although a global policy this chapter concentrates on its treatment in the
United States (USA) because of the availability of data and the changing
nature of the policy with the passage of the Telecommunications Act of
1996. The chapter is organized as follows. An overview of the concept of
universal service is given. This discussion considers why this policy concern
remains an issue. Then the changing nature of the definition, focusing on a
brief history of the concept, is examined. A theoretical model is explained
and empirical research reported. Finally, an alternative method for funding
universal service programs is briefly explored. The chapter concludes with
policy recommendations.

DEFINITION

Universal service is a phrase that many people in the telecommunications industry recognize but may have different ideas about. In the USA universal service is the social obligation imposed on the telecommunications industry to ensure that residential exchange rates are low and that rural telephone rates are not higher than urban rates. It is the principle that exchange service will be subsidized in order, it is alleged, to increase telephone penetration. The definition of universal service has expanded to include access to Internet service.[1] In the current international context of technology change, privatization and deregulation are forcing a reconsideration of universal service policy. The goal is being challenged by alternative technology and the instrument is inconsistent with the promotion of competition while the funding mechanism lacks sustainability in the face of competition. Nevertheless, the policy is expanding to include the Internet. However, the definition has been shifting ever since it was first introduced in 1907 when it encompassed the universality of the nationwide AT&T system to the exclusion of competitors (Mueller, 1997). In this discussion the definition of universal service used is the provision of exchange telephone access at prices that are below cost. However, this definition embodies the notions of the goal, the instrument and the funding mechanism. The notions are addressed separately. The confounding of definition with these notions has in large part led to misunderstanding of the underlying policy issues.

Goals

Even though several rationales are offered for universal service policy and universal service has become part of national telecommunications policy in many countries, the objective of the policy has never been entirely clear. Several objectives have been offered (Johnson, 1988; Crandall and Waverman, 2000; Laffont and Tirole, 2000). Justifications include increased subscription, network externality benefits, income redistribution, regional planning and infrastructure development (Crandall, 1998; Laffont and Tirole, 2000). All have the objective, at least in principle, of improving economic outcomes. Here the focus is on the overarching goals of increasing telephone penetration, infrastructure expansion and income redistribution. Until recently, the regulator did not need to recognize the distinction between those aspects of traditional regulation intended to control the incumbent's exercise of market power, and those designed to achieve socially desirable policy objectives, such as universal service. Indeed, to some extent it may have been useful for regulatory authorities to obscure the costs of certain social policies by embedding them within the pervasive regulation of an incumbent firm.

Instruments

Although many methods are available to implement universal service, two generic methods are used here. The first is pricing the service below cost, thus bringing the marginal subscriber on to the network (a service subsidy). The second method is equivalent to giving consumers funds to purchase the telephone connection. An analogy is food stamps in the USA (a targeted subsidy). The major distinction between these is who receives the subsidy. In the former case, where consumers benefit only indirectly, it is the telephone company. In this latter case, consumers benefit directly. The USA has practiced both methods. The means chosen by most countries to reach the goal of universal service has been to price exchange access below cost, or provide the service at affordable rates. The implicit assumption is that lower rates meet the goal. To be an effective instrument, below cost pricing of the service should be efficient, it should be cost effective and competitively neutral. However, it is not the most effective method as it misses the target, is expensive to fund and, coupled with the funding mechanism, may be counterproductive.

Funding Mechanism

1. *Implicit tax:* Pricing other telecommunications services significantly above cost has been the predominant method of funding the below-cost pricing of universal service. This method acts exactly like a tax on the over-priced telecommunications services. A policy question that arises with this method is whether this is an efficient method to raise the funds? The answer is no. This chapter considers whether there are more effective methods to raise the funds for universal service.
2. *Cross subsidization:* In the absence of government subsidies or intervention to promote universal service, residential basic monthly service charges can be under-priced only if other telephone services are over-priced, assuming the telephone companies are allowed to and do earn a fair rate of return on their invested capital. Thus, business rates for exchange service have been much higher than residential rates for the same service offering and arguably well above the direct cost of providing that service. Trunk or long-distance rates, both domestic and international, have traditionally exceeded their direct cost by enough to be major generators of revenue to support exchange service or other enterprises such as the postal service. In short, an elaborate system of

cross subsidies has, in general, been put in place over time to shelter residential exchange rates.

A Moving Target

The above situation changes with the decision to promote competition in
telecommunications markets. For the transition to competition to succeed,
asymmetric measures to control market power should be phased out as the
incumbent's market power diminishes. However, when the regulator wishes
to maintain some market interventions in the newly competitive market, in
order to meet social policy goals, then a new method has to be devised –
one that does not rely on the market power of the incumbent, that will be
sustainable in an environment with more than one firm and that will be
minimally distorting to the market outcome.[2] Whatever the definition, the
rationale for universal service is generally laudable – improve economic
outcomes, increase telephone or Internet penetration. But in the cases
examined here the instrument is not effective in obtaining the goal either
because the funding mechanism is in conflict with the goal, the instrument
is too broad and/or the action does not achieve the desired result. The conclusion is that the policy is ill advised and, not only does it not achieve the
desired ends, it is counterproductive.

HISTORY OF THE USO CONCEPT

Early Years

The history of the concept of universal service goes back to the time of
Theodore Vail; however, it has changed over time to suit industry needs.
Vail, the head of the Bell System, coined the term 'universal service' in the
early years of the twentieth century when he was attempting to rescue
the Bell System from its financial troubles and consolidate the network. The
term that he used was 'one system, one policy, universal service'. In this
context he was promoting a marketing technique to ensure that potential
subscribers would choose the Bell system (Mueller, 1997, pp. 93–103). Vail
offered to end his competitive wars with independent telephone companies,
to interconnect with them and to accept a framework of exclusive fran-
chises and government regulation (Mueller, 1997, p. 108). By this motto
Vail meant that service would be universal only in the sense that any subscriber could place a call to any other subscriber because networks would
be interconnected (Mueller, 1997, p. 96). The cross subsidies that maintain
universal service policies today could not exist then for the simple reason

that telephone companies offered few services beyond local service and there was thus no source of revenue to fund such cross subsidies (Mueller, 1997, p. 37–42).

When Congress passed the Communications Act of 1934 establishing the Federal Communications Commission (FCC), the term 'universal service' did not appear anywhere in the Act, although the principle that service should be widely available was affirmed.[3] At that time Congressional records contained no mention of telephone penetration levels (Mueller, 1997, p. 157). During the 1940s and 1950s, as long-distance service developed, the revenue from long-distance provided a source of funds that regulators could use to keep local rates low.[4] Since the Bell System provided both services until 1984 the cross subsidy from long-distance service to local rates was accomplished as a matter of bookkeeping within the system.[5]

In the early-1970s as competition was entering the Bell System market, the term was revised in order to argue that the Bell System was engaged in desirable cross subsidies, which supported this universal service concept. The argument then was that the competitors for long-distance were cream skimming, that is, only targeting the high-margin long-distance service. This practice would eliminate the ability of the System to raise these desirable cross subsidies. This was the defense that the company used throughout the 1970s in order to slow down competitive inroads into its markets. When AT&T was divested in 1984 it lost the local service market and kept the long-distance market. Thus this flow of funds between the services had to be handled on an arms-length basis. The old subsidy flow was replaced by the access charges that local companies charged long-distance carriers to originate or terminate long-distance calls.[6] Although the campaign of the 1970s was not altogether successful it imbued the legislatures and certain members of the public with the virtues of universal service. So much so that when the Telecommunications Act of 1996 was passed as an Amendment to the 1934 Act, it was finally embodied in this legislation (Mueller, 1997, pp. 167–70).[7] The revised Telecommunications Act of 1996 enshrined universal service as a national policy goal; however, the term had assumed a very different meaning from the one Vail had used earlier (Mueller, 1997, pp. 167–70).[8]

Current Environment

Competitive forces, technology, and the convergence of industries such as telephony, broadcast media, publishing and computers are transforming the economies of the world. The regulatory structure in each industry has been distinct, with different methods of social control, goals and objectives.

The convergence of these industries has created problems and issues for policymakers and analysts.[9] The traditional telephone monopolies are disappearing, although vestiges of their market power may continue for some time.[10] New regulatory tools of incentive regulation and competitive entry are replacing the traditional rate-based, rate-of-return regulation and rate structure-setting methods (Laffont and Tirole, 2000, pp. 16–17). Many issues arise because of this transition. Are the competing regulatory structures at odds with one another? What market structure will emerge? What market structure is desired? One element of traditional regulation, however, remains and still impedes the development of an effective competitive transition – the USO.

GOALS

Increasing Penetration and Promoting Subscription

The objective most often cited for universal service is that of increased penetration or promoting subscription.[11] This justification takes several forms. Some suggest that subsidization of local service corrects for externalities associated with the network, in that the value of the telecommunications system is increased for all subscribers when more people participate, an effect that may not be fully internalized in each individual's decision with respect to subscription (Johnson, 1988; Laffont and Tirole, 2000, p. 219; Mueller and Schement, 1996, pp. 275–6) Furthermore, widespread subscription may be seen as a public good, valued for its own sake, or as a matter of perceived fairness. Finally, some argue that wider subscription facilitates the delivery of public services and participation in community and political affairs.

In the USA, as in most western European countries, the vast majority of households now subscribe to telephone service. It is difficult to argue that the external benefit to existing subscribers is high when new subscribers are added.[12] Furthermore, if such effects were significant, telecommunications firms would partially internalize them, since they would increase demand for services by infra-marginal subscribers (Laffont and Tirole, 2000, pp. 229–30). The story is less clear in developing countries. This issue has been addressed in several fora, such as the International Telecommunication Union and the World Bank, with mixed results. Generally, a strong relationship exists between increased telephone penetration and economic growth but it is not necessarily causal. While a mixture of evidence points to telecommunications having an impact on economic development it is also clear that promoting universal service as traditionally practiced is an inefficient method to achieve that goal.

Externalities

A rationale for increasing penetration has been that of externalities. There are three externalities associated with telephone service: network or access, call and calls-beget-calls.

1. *Network externality:* A network becomes more valuable as additional subscribers are added to it. This is known as network externality (Littlechild, 1975). The network externality arises when an additional subscriber joins the network because there is another party the previously existing network can reach. It is thought that this benefit continues to accrue as the network increases in size.[13] This externality may be more critical in developing countries where the systems have very low penetration. However, when a country has obtained a high level of penetration this argument loses much of its force. There are large proportions of subscribers whose access, insofar as it enables one to reach and be reached by them, is irrelevant except to the extent that the costs to all are reduced due to economies of scale. However, someone having access can still give rise to benefits for others, whether subscribers or not. Where service is priced at its marginal cost it would not have the socially optimal number of subscribers because of network externalities. Hence, it is alleged that the service should be price subsidized.[14] When subscribers as a group can internalize this externality, that is, subsidize potential subscribers who would otherwise not join the network in order that they would join the network, all subscribers benefit.
2. *Call externality:* A call externality arises from the impact, positive or negative, upon the called party which is in addition to the benefit received by the caller who must regard the benefit to be at least as great as the cost. Obviously, not all calls received are regarded as beneficial by those called, and one manifestation of this is unlisted numbers. But on balance the call externality is likely to be positive.
3. *Calls-beget-calls:* Taylor (1994) identifies and estimates the calls-beget-calls phenomenon. It arises when a call creates another call. For example, if I call someone but only reach an answering machine she will hopefully call back. This is in contrast with the traditional notion of the call externality in which a person receives a positive externality (or negative in the case of a telemarketer) when she receives a call but does not have to pay for it. While these externalities may provide a rationale for pricing exchange service below cost, Coasian alternatives exist to deal with the issue (Crandall and Waverman, 2000, p. 25).
4. *Infrastructure development:* Closely related to both the externality and

the penetration arguments is the infrastructure development rationale. The idea is that economic growth and development can be promoted by the expansion of the telecommunications network. Some see subsidies to telecommunications infrastructure in rural areas as a policy to promote business development and employment. For example, in the USA universal service policy strives to maintain equity between rural and urban consumers.[15]

5. *Income redistribution:* An objective of universal service policy could be the redistribution of income to households perceived to be needy because of their low income (Laffont and Tirole, 2000, pp. 219–29). Subsidizing the price of local service could provide a mechanism for redistribution to the extent that the subsidized service is a larger proportion of the expenditures of the customers the government wishes to favor.

CRITIQUE OF USO GOALS AND INSTRUMENTS

Increase Penetration

The goals of universal service and the instruments used to obtain the goals cannot be easily parsed out.[16] If promoting subscription were the goal of universal service policy, then subsidizing rates for local service generally is an inefficient means to do so. Infra-marginal subscribers do not need a subsidy to remain on the service. Only the marginal subscribers have to be subsidized to remain on the system at prices that cover the cost of access. For example, if at full subscriber access cost 10 per cent of current subscribers would no longer subscribe these customers could be given a direct subsidy for one-tenth of the cost of the current subsidy to all subscribers. In Figure 11.1, the current subsidy required is the sum of the shaded areas (A + B) to support the service subsidy required to add the incremental subscribers (Q – Q'). The same increment could be added by subsidizing only these marginal subscribers by directly giving the dark shaded area (B) to the incremental subscribers (Q – Q').

The preferred method of addressing the problem is through targeting subsidies directly to the individuals rather than subsidizing the service for all customers.[17] In the USA state commissions have taken steps in this direction. Most states have programs funded jointly by them and by a federal subsidy scheme that defray part of the cost of local service for qualifying low-income subscribers. There are inherent difficulties in targeting subsidies to marginal subscribers but a qualification test with any power must be more efficient than simply subsidizing everyone

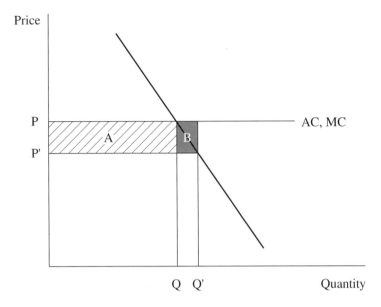

Figure 11.1   Service subsidy

(Hausman, 1998, pp. 15–16; Laffont and Tirole, 2000, pp. 231–5). The policy is particularly egregious since the service is among the most inelastic of all telecommunications services at between –0.01 and –0.02 (Taylor, 1980, pp. 200–204). This implies that a decrease in price of 10 per cent will only expand the system by 0.1 to 0.2 per cent. Furthermore, the current policy implicitly assumes that the decision to subscribe depends entirely on the price of local service and not the prices of other telecommunications services. In the USA empirical analyses indicate a significant negative cross elasticity between the long-distance and subscriber access, that is, higher toll service prices lessen demand for subscriber access.[18] When the only reason to subscribe to the telephone system is to make or receive a call this makes sense. Access is a derived demand from the demand for local, toll and international use. As these prices increase the demand for subscriber access is less. Received studies suggest that the most important factor prompting households to disconnect phone service is the accumulation of high toll bills (Mueller and Schement, 1996, pp. 186–290). Thus, it is counterproductive to raise the price of long-distance service in order to finance subsidies for local service. Targeted programs such as toll blocking, which help households control their toll bills, may be more effective in promoting subscription.

Other studies concur with this observation, for example, Crandall and

Waverman (2000) find universal service pricing policy has a minimal effect on subscriber penetration, does not address the causes of non-subscription, installation and disconnection due to unpaid long-distance bills, and the subsidy is raised via surcharges on local and long-distance calls, which feeds back to non-subscription. Moreover, an adverse impact on economic welfare occurs because the services to which the surcharges are applied are more elastic than exchange access. Crandall and Waverman estimate the effects of rebalancing the rates to costs on telephone penetration and the redistribution of income from rich to poor, and urban to rural subscribers. They use the various cost models developed in the USA. Welfare gains accrue when the services are re-priced because of the highly inelastic nature of the subscriber price elasticity and the relatively higher long-distance prices. Further states with a large rural population may not gain from the current practice due to the higher long-distance charge the rural resident incurs to support the exchange subsidy.

Income Redistribution

An objection in using universal service to promote income redistribution is that when the objective is to transfer income it would be more efficient to do this directly by giving money to the targeted households, rather than indirectly by subsidizing a service they may or may not wish to consume.[19] Another objection is that a subsidy for the price of a given service poorly targets benefits to needy groups. Different customers consume differently – for example, long-distance service – but the consumption is strongly related to income differences. Segmenting income shows a mix of high- and low-volume customers within income groups. For the lowest income segment Crandall examines, those with an annual household income below USD10000, 45 per cent of the average monthly bill represents charges for long-distance service (Crandall, 1998, p. 403). Thus a system that relies on high long-distance prices to fund a local service price could result in poor households that make many long-distance calls subsidizing wealthy households that do not. Data presented below amplify this conclusion. Crandall and Waverman explore the demographics of telephone use based on data from the USA, Canada and the UK (Crandall and Waverman, 2000, Chapter 3). Here again, it is noted that low-income subscribers purchase a significant amount of long-distance service. Moreover, within strata there is a wide dispersion of calling activity that is hidden by only examining the means. This within-strata skewness implies that generating a subsidy through a surcharge on long-distance rates results in an income transfer from poor to rich, and from heavy to light users, respectively. Using US data, a rebalancing of the rates produces a consumer surplus gain of 1.5

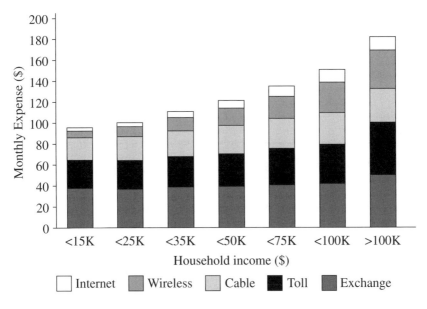

Source: PNR & Associates ReQuest™ (2000).

Figure 11.2 Expenditures by income strata

billion US dollars (USD). All income strata gain, other than the lowest three. These lower strata lose approximately USD10 per annum (p.a.) per household. Crandall and Waverman conclude that to hold exchange rates below cost through high long-distance prices is inefficient. Further, examination of detailed US telecommunications use and billing data for wireless, cable, Internet, local and long-distance calling, as well as traditional wireline telephony access (stratified by income) confirms that consumers, even those with low incomes, choose to purchase packages of wireless, cable and other services with prices at least as high as local telephone prices would be in the absence of the current subsidy.

Expenditure

Examination of discretionary spending by US low-income subscribers in Figure 11.2 shows the level of spending is greater than that for subscriber access. Hence it remains unclear whether raising the price of discretionary services to fund subsidies for local service is an effective means of transferring wealth to those customers in most need.[20]

Table 11.1   Telecommunications expenditures by household

Household income	Less than USD15000	Greater than USD15000
Exchange	38.20	39.74
Toll	25.95	32.33
Cable	22.23	26.51
Wireless	5.89	17.90
Internet	3.02	7.69

Source: PNR & Associates ReQuest™ (1999).

Similarly, Crandall and Waverman explore the burden of household telecommunications expenditures by country. Their empirical evidence shows spending on durables in most cases has higher penetration than telephone service but that these durable goods are not subsidized.[21] They also note that installation and calling charges may be more important than subsidized access as a driver of penetration.

Remaining Issues

Universal service policy in the USA is not practiced in other regulated industries (Crandall and Waverman, 2000, Chapter 3 and Chapter 4). Except for electricity no support for other regulated industries exists. Electricity service is supported via a government entity, the Rural Utilities Service (RUS) of the Department of Agriculture, and a few publicly operated companies and government sponsored power generators. None of the support is through the distortion of the rate structure. However, only the telephone industry is burdened by the enormous cross subsidies paid by businesses, some urban residents, and heavy long-distance users, designed to keep rates for rural high-cost areas artificially low (Crandall and Waverman, 2000, p. 88). In this sense telephone service is unique in the political demands it places on regulators for cross subsidization (Crandall and Waverman, 2000, p. 79).

EVALUATION OF USO INSTRUMENTS

Distorting the prices of telecommunications services is a particularly costly method of financing universal service subsidies. The services with elevated prices are generally those for which demand is more elastic than for local services. Current universal service policy thus represents reverse Ramsey pricing. Because the burden of this funding is concentrated on certain telecommunications services, rather than drawn from general revenues, the base of the tax is relatively narrow and the markups on the prices of the services generating the subsidy are quite high. Finally, the telecommunications sector is undergoing rapid change, as new technology appears and as competition is introduced. Hausman estimates that for every USD1 raised by increasing the price of long-distance service the welfare cost to consumers is USD1.65 (Hausman, 1998, p. 13–14). This far exceeds estimates of the comparable deadweight loss associated with USD1 of general revenue, which is approximately USD0.4. Crandall finds that if rates were to be rebalanced, without any universal service mechanisms to cushion the effects, the average welfare loss among the lowest income group would be only about USD6 p.a. compared with an average expenditure for telecommunications among that group of about USD490 p.a. (Crandall, 1998, pp. 405–6).

EVALUATION OF USO FUNDING MECHANISMS

Pricing above Costs

In many countries, as part of the pervasive regulation, incumbent authorities have held rates for basic local service at levels below those that would have been set by firms in a competitive market (Laffont and Tirole, 2000, pp. 217–20). To ensure revenue sufficiency regulators have allowed incumbent firms to set relatively high prices for other services, such as inter-exchange access, termination of international calls, long-distance service and vertical services such as calling features.[22] The effects of this policy in the USA are striking. Figure 11.3 provides an overview of market intervention on prices, by major service category, for the areas in 28 states where GTE (now part of Verison) provides local telephone service as the incumbent. The bars show contribution, calculated as the difference between revenue at current rates and direct or TSLRIC cost, in USD p.a. by category.[23] Figure 11.3 shows large positive contributions for interstate-switched access, intrastate-switched access, toll calls within GTE's serving areas and vertical services.[24] The right-hand side of the figure shows a large negative contribution for residential local service. It is unlikely that this pattern of prices represents profit-maximizing behavior by the firm, since local service is generally found the least elastic of the service categories shown (Taylor, 1980, p. 162). For the purpose of comparison Figure 11.4 shows what the contributions would look like were rates rebalanced to yield the same total revenue but with a constant percentage markup over TSLRIC by service category.[25]

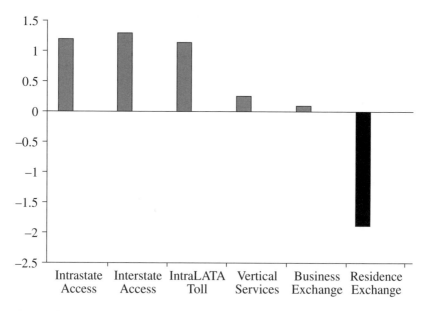

Figure 11.3   Contribution by service category

It is difficult to know what the market equilibrium rates would be with certainty, if there were effectively competitive markets for each of the service categories shown in Figure 11.4. If it is assumed, however, that all firms in the market have cost levels similar to GTE's, and thus consistent with the revenue level in Figure 11.4, and no assumptions are made as to how Ramsey prices might differ from the constant markup prices, then Figure 11.4 can serve as an approximate guess as to the contribution of service categories at market rates.[26] The differences between the figures provide an indication of the degree to which regulation has intervened to displace the current rates from these market or cost-based levels. For interstate-switched access the difference in revenue between the price levels is about USD1.2 billion p.a. for GTE, and for all US local carriers it is about USD5.9 billion (Weller, 1999a, p. 650).[27]

While Figure 11.3 and Figure 11.4 reveal a large flow of funds across services, they obscure reality in important ways. First, the figures aggregate across geographic areas and so they fail to show differences in cost across regions. Recent cost models suggest costs can vary substantially even within the service area of a rural central office. Figure 11.4 suggests that on average residential local rates are almost USD23 per month below market levels. In some low-density rural areas, however, this difference might be as much as

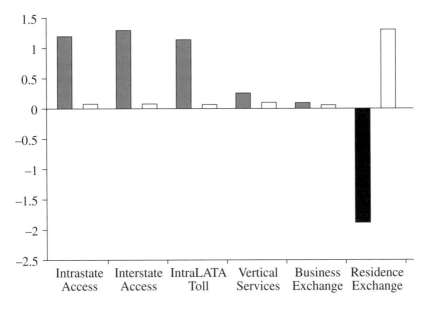

Figure 11.4   Contribution by service category, rebalanced

several hundred USD per month. In some urban areas residential rates may be much closer to and even above costs. Second, the figures average across customers and, while the price vectors yield the same revenue, the distribution of use for access, long-distance and vertical services across customers is highly skewed, for example, only 6 per cent of the end-user locations served by GTE generate almost half the demand for inter-exchange access. As a result the price vectors produce very different revenues when the demand of an individual customer, or market segment, is evaluated. When a new local carrier enters GTE's serving area in Texas and attempts to serve all of GTE's local residence customers it would find that the revenue from all services would fail to cover costs for 78 per cent of those customers.[28] This system of price manipulation is sustainable, if inefficient, in a sole-provider environment. With the introduction of competition new concerns arise. First, the high margins for services will induce firms to focus their entry strategies on the minority of customers who have high demand for those services. Second, the current low price for local service will largely preclude entry into local service markets.[29] This will especially be true for rural areas where costs are high relative to the average rates.

Other Mechanisms

Crandall and Waverman (2000) examine the effect of a tax in the US context to accomplish universal service goals but find it does not do as well as rebalancing the rates. They review other FCC programs and note that residential competition will likely not come about without a rebalance of residential rates to cost. Providing a competitive environment by raising exchange rates cannot be easily accomplished in the political and regulatory environment in which the FCC operates. They suggest the simplest proposal is to ratchet up local rates, given the ineffectiveness of the other alternatives they examined (Crandall and Waverman, 2000, p. 140).[30] In the USA the Telecommunications Act of 1996 requires that rates in rural areas be reasonably close to those in urban areas. However, since rural customers generally rely more heavily on long-distance service raising long-distance rates to subsidize rural subscribers is counterproductive. Further, it is not clear that all rural subscribers are needy, for example, in Colorado rural areas contain some of the most affluent regions in the country. Communities such as Aspen and Vail receive preferential rates and the serving companies receive funds from the telephone industry's Universal Service Fund.[31]

AN ALTERNATIVE

Regulators continue to intervene in local service markets. This means that the generally available rate for basic service in many areas is held below market levels. There may also be requirements with respect to quality, tariffs and other non-price attributes of service. An alternative is the use of auctions to determine which carriers should undertake a USO and what compensation they should receive for performing this function. The auction would reveal carriers' valuations of the USO, determine the number of USO providers endogenously and provide an alternative to traditional cost-of-service regulation. Weller (1999a, pp. 645–8) suggests a process of competitive bidding could serve this purpose.[32] The regulator would define the market intervention it wished to impose in the form of a USO. An auction would then determine which carriers should undertake this obligation and the compensation those carriers should receive in return. The auction would be a single round, sealed bid; the form of the bid would be the per customer support amount the carrier would require. The low bidder would win the rights to serve that market and receive the support. The auction would allow more than one carrier to win, with the number of winners in a given area determined endogenously. A limited form of conditional bidding is used to take account of possible economies of density. Repeated auctions over time allow this framework to adapt to changes in technology, costs, or policy objectives.[33]

Most of the discussion about universal service, in the USA and elsewhere, has focused on estimating the cost of the basic service and deriving support levels by comparing this cost to some estimate or assumption regarding revenue. Compared to this alternative, an auction offers a number of advantages over traditional cost-of-service approaches as a means to select universal service providers, and to determine the level of support payments. Auctions have the advantage of speed of sale, revealing information about buyers' valuations (in this case, sellers' valuations), preventing dishonest dealing between the seller's agent and the buyer (here, the buyer's agent and the seller).[34] The application of auctions to universal service has these advantages.

CONCLUSIONS

The major distortion in the telephone industry is universal service or the subsidization of subscribers' access to the network. This chapter shows that universal service is inefficient as a means of reaching its intended goal as it is not directed to the marginal subscribers and it is costly to support because it is not targeted. Fundraising through cross subsidies from other services is counterproductive as higher prices for services providing the subsidies reduce the demand for subscriber access from the group that it is intended to aid. The subsidies inhibit effective competition because of artificially low prices for subscriber access and high prices for other services, thus preventing the market from testing the efficiency of the provider. This outcome can lead to inefficient entry in the high-priced markets and preclude efficient, low-cost entry in the subsidized markets. This is incompatible with competition policy. When a democratic process determines that subsidies are desirable they should be targeted to end-users and funded directly through government. While the myth of universal service, as currently embedded in regulatory policies, is without economic foundations universal service arguments nevertheless continue to plague the telecommunications industry to the detriment of business, the public and potential competitors.

Clearly, universal service is poor public policy based on subsidizing telephone carriers rather than the subscribers directly affected. The subsidy does not distinguish between rich and poor subscribers and, to the extent that the subsidy comes from services with strong cross elasticity, the policy is self-defeating. Moreover, because the policy distorts the incumbent's rates the current subsidies distort the development of competition. High rates for services that generate the subsidy, such as inter-exchange access, long-distance, and calling features create an artificial incentive for entry into markets for customers such as large businesses with high concentrations of these services. At the same time, low rates for local service make it unattractive for new firms to compete for customers with lower usage levels, especially for residential customers. Furthermore, the rate distortion makes it difficult for the market to reveal whether the incumbent is an efficient provider of service. When an entrant can offer lower rates to a business customer in an urban area, that may indicate the new firm is more efficient than the incumbent, but it may simply mean that the incumbent has charged high rates for that customer as part of the pervasive scheme of cross subsidy. Similarly, when no entrant can match the incumbent's rate for local service to a residential subscriber in a rural area, this may show the incumbent is the most efficient provider, but it probably means only that the local rate to that customer has been below the incumbent's costs. This anticompetitive effect of universal service can be reduced by minimizing the subsidy itself, and also by making the necessary subsidies explicit and portable among the local carriers chosen as universal service providers

A more focused policy is suggested. Importantly, the instrument for obtaining the universal service goals should be examined. As the policy currently stands, 'universal service policies are a very imprecise system of income redistribution and have a large cost in terms of economic welfare' (Crandall and Waverman, 2000, p. 167). The source of the subsidy should be examined. The current situation based on implicit taxes on long-distance service is inefficient and possibly inequitable. Liberalization will change the focus and subsidies will not be sustainable in this competitive environment. It is recommend that the universal service programs be reduced and targeted. To summarize, the current universal service policy is inefficient as a means of attaining its intended goal as it is not directed to marginal subscribers, it is not directed to needy subscribers, it may not be desired or necessary, the pricing practice does not reach the desired goal and the means of raising the funds to support the subsidy may be counter-productive. Should regulators insist on subsidizing the services the auctions approach offers a more efficient solution.

ACKNOWLEDGEMENTS

The authors wish to thank Dennis Weller for reviewing the chapter and providing invaluable suggestions and Soontaraporn Techapalokul for research assistance.

NOTES

1. This type of effort has begun to be called the digital divide. While perhaps more catchy and market oriented than universal service it suffers from the same infirmities this paper addresses (Laffont and Tirole, 2000, pp. 16–17).
2. In terms of the taxonomy developed by Cherry and Wildman (1996), the traditional pervasive regulation imposes a USO on the incumbent, which is asymmetric and unilateral. The challenge for the regulator is to develop a new approach in which the obligation is symmetric (in that it can be applied to firms other than the incumbent) and multilateral (it involves a transaction entered into voluntarily, in which each carrier takes on the obligation in return for compensation).
3. The wording in the preamble is, 'to make available, so far as possible, to all people a rapid, efficient, nation-wide, and world-wide wire and radio communications service at reasonable charges' 47 USC § 151 (Mueller, 1997, p. 157).
4. State regulators also relied on revenues from private line services, business lines, touchtone service and later calling features such as call-forwarding and call-waiting to subsidize local service (Laffont and Tirole, 2000, p. 3: Mueller, 1997, pp. 156–157).
5. See Hausman (1998, p. 16) and Gabel (1967, pp. 129–32). Gabel provides a discussion of the process known as separations.
6. See 47 C.F.R. pt. 69 (1999); Access Charge Order, 93 FCC 2d 241 (Feb. 28, 1983).
7. The telephone companies implied that the 1934 Telecommunications Act had required them to provide universal service. They relied on the term 'affordable service', which appeared in the 1934 Act. However, the Act contains no mention of universal service nor of pricing exchange service below cost. Indeed, the confusion remains today, for example, Crandall and Waverman (2000, p. 165) stated that universal service was embedded in the Telecommunications Act of 1934. Affordable rates were mentioned, not below-cost rates. Affordable rates could simply mean rates with no monopoly rents. It certainly does not indicate an embrace of the universal service subsidy.
8. By 1996 it was clear that the Internet was emerging as a significant force. Thus, the US Congress added access to the Internet for schools and libraries to the 1996 Act. That is, in addition to including residential voice telephony as part of the universal service requirement the Act included the provision of Internet access for schools and libraries at reduced rates.
9. See Kahn (1988) for a discussion of pre-incentive regulation. For a review of the more recent incentive regulation see Laffont and Tirole (2000, p. 37–96).
10. See Laffont and Tirole (2000, pp. 265–72) for most recent documentation of this convergence.
11. See Laffont and Tirole (2000, pp. 217–18).
12. Households who choose not to subscribe are those for whom the value of subscription is low. It is likely that the external benefits to other households of having those households subscribe is correspondingly low as well.
13. In the USA this could be the possible rationale for charging subscribers to larger exchanges more for access than subscribers to smaller exchanges.
14. Littlechild (1975) addresses the issue. Willig and Klein (1977) consider the proper pricing of this externality. This work was ultimately incorporated into Willig (1979).
15. See 47 USC § 254 (Supp. III 1997).
16. While the arguments are illustrated for the telephone industry they apply equally to the digital divide or NII. A similar concept is applied to the NII as noted in the following quotation from the Telecommunications Act: '(E)xtend the "universal service" concept to ensure that information resources are available to all at affordable prices. Because information means empowerment – and employment – the government has a duty to ensure that all Americans have access to the resources and job creation potential of the Information Age.' 47 USC § 254 (Supp. III 1997).
17. Universal service is the acknowledged goal of the subsidy; however, as pointed out two decades ago and repeatedly since then, regulators have not addressed the incidence of

the subsidy. Targeted subsidies are preferable, if subsidies are supported at all, to the service subsidies currently applied in the industry. See Johnson (1988, p. 74), Laffont and Tirole (2000, p. 219) and Hausman (1998, p. 17).

18. See Taylor (1994, pp. 200–204) and Hausman et al. (1993, p. 178).

19. Laffont and Tirole (2000, pp. 228–32) discuss *caveats* to Atkinson's and Stiglitz's (1976) results where some assumptions underlying the result may not be met, for example, if the inequality the government wishes to correct cannot be observed directly.

20. ReQuest™ is a national residential survey that provides market information concerning consumer behaviors, attitudes, switching probabilities and price sensitivity. PNR annually surveys over 45 000 US households to collect information on household expenditure, penetration rates, and attitudes on telecommunication products and services. Households are randomly selected from a national panel of households and are weighted to correspond to census distributions for age, income, household size and marital status. ReQuest™ covers local telephony, short-distance toll, long-distance, wireless (cellular, PCS and paging), cable, Internet, calling card, coin and international long-distance. The data displayed in the accompanying figures and table are based on surveys conducted during the first quarter 1999 (PNR, 2000).

21. Low-income consumers without the need for complex subsidies purchase other discretionary services as well.

22. These cross subsidies are well known in the industry and have been addressed recently by Laffont and Tirole (2000, pp. 144–7).

23. Telecommunications service long-run incremental cost (TSLRIC) is the additional cost to the firm attributable to offering a given increment of service (Laffont and Tirole, 2000, pp. 24–25). Here, the increment is a broad category of service. As a practical matter it is difficult to estimate TSLRIC costs with any accuracy (Alleman, 1999, pp. 159–79). The estimates used here were developed internally by GTE (now part of Verison) and reported in Weller (1999b). If the figures were constructed using cost estimates from a different model the absolute values of the bars might change somewhat but the general pattern would remain much the same.

24. This is access provided through the switched telephone network for long-distance calls that originate in one state and terminate in another. The FCC regulates these rates (47 USC § 15 1994). Note that, if the dark and light bars were summed, the result would be a positive number. This is because the sum of the contributions from GTE's different service offerings must cover GTE's common costs of about USD2 billion.

25. Since the aggregate revenue in Figure 11.3 is the same as that in Figure 11.2, and all of the costs are also the same, the rebalanced rates in Figure 11.3 also generate about USD2 billion in contribution toward the common costs of the firm (Weller, 1999a, p. 649, n. 11).

26. Ramsey pricing is a means to maximize consumer welfare while ensuring firm breakeven. In its simplest form Ramsey pricing prices services above cost in proportion to the inverse of the services' demand elasticities. In effect, one can think of the constant-margin rates as the result of mindless Ramsey pricing by a firm unable to take account of differences in demand, and the current rates as reverse Ramsey prices since the pattern of markups is exactly the opposite of what would be expected from a firm setting Ramsey prices (Laffont and Tirole, 2000, pp. 60–69). For this reason, the constant-markup rates provide, if anything, a conservative reference point. If firms did take demand into account in setting prices it might be expected that the markup for local service rates would be higher than the uniform level shown here.

27. This figure includes only rates for the larger service areas designated non-rural under the Telecommunications Act of 1996 (Telecommunications Act of 1996, Pub. L. No. 104-104, 110 Stat. 56 (1996) (codified at 47 USC § 254 (Supp. III 1997))). The national total would thus be somewhat higher. On the other hand, the current access rates include recovery of a portion of the costs of a new fund to subsidize telecommunications services for schools and libraries (47 USC § 254(c)(1)). The Fifth Circuit has recently reversed a requirement established by the FCC that contributions to these funds be recovered through access rates. See Texas Office of Public Util. Counsel v. FCC, 183 F.3d 393 (5th Cir. 1999).

28. For this analysis, it is assumed that the competitor costs are the prices for unbundled network elements established on an interim basis by state regulators in Texas. This is a conservative basis for calculating the entrant's costs; if GTE were to sell its current output vector in Texas at those prices, GTE's revenue would be about 36 per cent lower than it is currently.
29. The observed pattern of entry into local telecommunications markets in the USA appears to be consistent with these incentives. Significant entry has already occurred in markets for local business services. At the same time, there has been little competitive activity in local markets for residential service (Baumol and Sidak, 1994, p. 7).
30. Since the publication of Crandall and Waverman (2000), the FCC has adapted the CALLS proposal, which is a step in the prescribed direction. Federal Communications Commission Sixth Report and Order in CC Docket Nos. 96-262 and 94-1 Report And Order in CC Docket No. 99-249 Eleventh Report and Order in CC Docket No. 96-45, May 31, 2000.
31. Vail is in Eagle County, Colorado and has an average per capita assessed property value of about USD51500. See Eagle County Government – Eagle County, Colorado at http://www.eagle-county.com/frames/gov.htm (lists the total taxable assessed property value USD1647562700); Eagle County Government – Eagle County, Colorado http://www.eagle-county.com/frames/vis.htm (listed population of Eagle County is about 32000).
32. Laffont and Tirole (2000, pp. 244–50) provide a discussion of this work, referring to it as the GTE proposal. See also Klemperer (1999, pp. 227–31).
33. For a more detailed description see Weller (1999a, pp. 655–73).
34. See Wolfstetter (1996, p. 369).

REFERENCES

Alleman, J.H. (1999), 'The Poverty of Cost Models, the Wealth of Real Options', in Alleman, J.H. and Noam, E. (eds), *The New Investment Theory of Options and Its Implications for Telecommunications Economics*, Boston: Kluwer Academic, 159–79.

Atkinson, A.B. and Stiglitz, J. (1976), 'The design of tax structure: direct and indirect taxation', *Journal of Public Economics*, **6** (1,2), 55–75.

Baumol, W.J. and Sidak, G. (1994), *Toward Competition in Local Telephony*, Cambridge, MA: MIT Press.

Cherry, B.A. and Wildman, S.S. (1996), 'Executive summary of need for a typology of economic regulation', in *A Framework for Managing Telecommunications Deregulation While Meeting Universal Service Goals*, available at http://www.benton.org/Policy/Uniserv/Conference/Frame/frame-exec.html.

Crandall, R.W. (1998), 'Telephone subsidies, income redistribution, and consumer welfare', in Noll, R.G. and Price, M.E. (eds), *A Communications Cornucopia: Markle Foundation Essays on Information Policy*, Washington, DC: Brookings Institution Press.

Crandall, R.W. and Waverman, L. (2000), *Who Pays for Universal Service: When Subsidies Become Transparent*, Washington, DC: Brookings Institution Press.

Gabel, R. (1967), 'Development of separations principles in the telephone industry', East Lansing: Institute of Public Utilities, Michigan State University.

Hausman, J. (1998), *Taxation by Telecommunications Regulation: The Economics of the E-Rate*, Washington, DC: AEI Press.

Hausman, J., Tardiff, T. and Belinfante, A. (1993), 'The effects of the breakup of

AT&T on telephone penetration in the United States', *American Economic Review*, **83** (2), 178–98.

Johnson, L.L. (1988), *Telephone Assistance Programs for Low-Income Households: A Preliminary Assessment*, Santa Monica, Rand.

Kahn, A.E. (1988), *The Economics of Regulation: Principles and Institutions*, Cambridge, MA: MIT Press.

Klemperer, P. (1999), 'Auction theory: a guide to the literature', *Journal of Economic Surveys*, **13** (3), 227–86.

Laffont, J.-J. and Tirole, J. (2000), *Competition in Telecommunications*, Cambridge, MA: MIT Press.

Littlechild, S.C. (1975), 'Two-part tariffs and consumption externalities', *Bell Journal of Economics and Management Science*, **6** (2), 661–70.

Mueller, M.L. (1997), *Universal Service: Competition, Interconnection, and Monopoly in the Making of the American Telephone System*, Cambridge, MA: MIT Press.

Mueller, M.L. and Schement, J.R. (1996), 'Universal service from the bottom up: a study of telephone penetration in Camden, New Jersey', *Information Society*, **12** (3), 273–92.

PNR & Associates (2000), 'Bill Harvesting ReQuest', available at http://www.pnr.com/ProductsServices/Market_Information/Consumer_Market_Info/Bill_Harvesting/bill_harvesting.htm.

Posner, R.A. (1971), "Taxation by regulation", *Bell Journal of Economics and Management Science*, **2** (1), 22–50.

Taylor, L.D. (1980), *Telecommunications Demand: A Survey and Critique*, Cambridge, MA: Balinger Publishing Company.

Taylor, L.D. (1994), *Telecommunications Demand in Theory and Practice*, Boston: Kluwer Academic Publishers.

Weller, D. (1999a), 'Auctions for universal service obligations', *Telecommunications Policy*, **23** (9), 645–74.

Weller, D. (1999b), 'Universal service – the policy that was', conference presentation at the London Business School (30 April).

Willig, R.D. (1979), 'The theory of network access pricing', in Trebing, H.M. (ed.), *Issues in Public Utility Regulation*, MSU Public Utilities Papers Series, East Lansing, Michigan State University.

Willig, R. and Klein, R. (1977), 'Network externalities and optimal telecommunications pricing: a preliminary sketch', in *Proceedings of Fifth Annual Telecommunications Policy Research Conference*, NTIS, vol. 2, 475–505.

Wolfstetter, E. (1996), 'Auctions: an introduction', *Journal of Economic Surveys*, **10** (4), 367–420.

Index